SBIR at the National Science Foundation

Committee on Capitalizing on Science, Technology, and Innovation:
An Assessment of the Small Business Innovation Research Program—
Phase II

Board on Science, Technology, and Economic Policy

Policy and Global Affairs

The National Academies of
SCIENCES · ENGINEERING · MEDICINE

THE NATIONAL ACADEMIES PRESS
Washington, DC
www.nap.edu

THE NATIONAL ACADEMIES PRESS 500 Fifth Street, NW Washington, DC 20001

This activity was supported by grant number DMI-0221736 from the National Science Foundation. Any opinions, findings, conclusions, or recommendations expressed in this publication do not necessarily reflect the views of any organization or agency that provided support for the project.

International Standard Book Number-13: 978-0-309-31196-0
International Standard Book Number-10: 0-309-31196-9

Additional copies of this report are available for sale from the National Academies Press, 500 Fifth Street, NW, Keck 360, Washington, DC 20001; (800) 624-6242 or (202) 334-3313; http://www.nap.edu.

Printed in the United States of America

Suggested citation: National Academies of Sciences, Engineering, and Medicine. 2015. *SBIR at the National Science Foundation*. Washington, DC: The National Academies Press.

The National Academies of
SCIENCES • ENGINEERING • MEDICINE

The **National Academy of Sciences** was established in 1863 by an Act of Congress, signed by President Lincoln, as a private, nongovernmental institution to advise the nation on issues related to science and technology. Members are elected by their peers for outstanding contributions to research. Dr. Ralph J. Cicerone is president.

The **National Academy of Engineering** was established in 1964 under the charter of the National Academy of Sciences to bring the practices of engineering to advising the nation. Members are elected by their peers for extraordinary contributions to engineering. Dr. C. D. Mote, Jr., is president.

The **National Academy of Medicine** (formerly the Institute of Medicine) was established in 1970 under the charter of the National Academy of Sciences to advise the nation on medical and health issues. Members are elected by their peers for distinguished contributions to medicine and health. Dr. Victor J. Dzau is president.

The three Academies work together as the **National Academies of Sciences, Engineering, and Medicine** to provide independent, objective analysis and advice to the nation and conduct other activities to solve complex problems and inform public policy decisions. The Academies also encourage education and research, recognize outstanding contributions to knowledge, and increase public understanding in matters of science, engineering, and medicine.

Learn more about the National Academies of Sciences, Engineering, and Medicine at **www.national-academies.org**.

Committee on Capitalizing on Science, Technology, and Innovation: An Assessment of the Small Business Innovation Research Program— Phase II

Jacques S. Gansler (NAE), *Chair*
Roger C. Lipitz Chair in Public Policy and Private Enterprise
Director of the Center for Public Policy and Private Enterprise
School of Public Policy
University of Maryland

David Audretsch
Distinguished Professor
Ameritech Chair of Economic Development
Director of the Institute for Development Strategies
Indiana University

Gene Banucci*
Executive Chairman, ret.
ATMI, Inc.
(Member: 6/26/2009-4/23/2014)

Thomas J. Bond*
Grant and Proposal Director
Association for Manufacturing Technology
(Member: 6/26/2009-5/21/2014)

Michael Borrus
Founding General Partner
XSeed Capital

J. Michael Brick
Vice President and Co-Director of Survey Methods
Westat

Gail H. Cassell (NAM)
Senior Lecturer
Department of Global Health and Social Medicine
Harvard Medical School

M. Christina Gabriel
President
University Energy Partnership

Charles E. Kolb (NAE)
President and Chief Executive Officer
Aerodyne Research, Inc.

Virginia Lesser
Professor of Statistics
Department of Statistics
Director, Survey Research Center
Oregon State University

Henry Linsert, Jr.
Chairman and CEO
Columbia Biosciences Corporation

W. Clark McFadden II
Senior Counsel
Orrick, Herrington & Sutcliffe, LLP

*Not involved in writing of the report

Duncan T. Moore (NAE)
Vice Provost for Entrepreneurship
Rudolf and Hilda Kingslake
Professor of Optical Engineering
The Institute of Optics
University of Rochester

Linda Powers*
Managing Director
Toucan Capital Corporation
(Member: 6/26/2009-10/13/2011)

Donald Siegel
Dean and Professor
School of Business
University at Albany, SUNY

Jeffrey E. Sohl
Professor and Director of the Center for Venture Research
Peter T. Paul College of Business and Economics
University of New Hampshire

Tyrone C. Taylor
President
Capitol Advisors on Technology, LLC

John P. Walsh
Professor of Public Policy
School of Public Policy
Georgia Institute of Technology

Patrick H. Windham
Principal
Technology Policy International

Project Staff

Sujai J. Shivakumar
Study Director

David E. Dierksheide
Program Officer

Karolina E. Konarzewska
Program Coordinator
(through September 2015)

Gail E. Cohen
Board Director

Frederic A. Lestina
Senior Program Assistant

Natacha R. Montgomery
Senior Program Assistant
(through October 2015)

For the National Academies of Sciences, Engineering, and Medicine, this project was overseen by the Board on Science, Technology, and Economic Policy (STEP), a standing board established in 1991. The mandate of the Board on Science, Technology, and Economic Policy is to advise federal, state, and local governments and inform the public about economic and related public policies to promote the creation, diffusion, and application of new scientific and technical knowledge to enhance the productivity and competitiveness of the U.S. economy and foster economic prosperity for all Americans. The STEP Board and its committees marshal research and the expertise of scholars, industrial managers, investors, and former public officials in a wide range of policy areas that affect the speed and direction of scientific and technological change and their contributions to the growth of the U.S. and global economies. Results are communicated through reports, conferences, workshops, briefings, and electronic media subject to the procedures of the Academies to ensure their authoritativeness, independence, and objectivity. The members and staff of the STEP Board* are listed below:

* As of October 2015.

Preface

Today's knowledge economy is driven in large part by the nation's capacity to innovate. One of the defining features of the U.S. economy is a high level of entrepreneurial activity. Entrepreneurs in the United States see opportunities and are willing and able to assume risk to bring new welfare-enhancing, wealth-generating technologies to the market. Yet, although discoveries in areas such as genomics, bioinformatics, and nanotechnology present new opportunities, converting these discoveries into innovations for the market involves substantial challenges.[1] The American capacity for innovation can be strengthened by addressing the challenges faced by entrepreneurs. Public-private partnerships are one means to help entrepreneurs bring new ideas to market.

The Small Business Innovation Research (SBIR) program is one of the largest examples of U.S. public-private partnerships. An underlying tenet of the program is that small businesses are a strong source of new ideas, and therefore economic growth, but that it is difficult to find financial support for these ideas in the early stages of their development. The SBIR program was established in 1982 to encourage small businesses to develop new processes and products and to provide quality research in support of the U.S. government's many missions. By involving qualified small businesses in the nation's research and development (R&D) effort, SBIR grants stimulate innovative technologies to help federal agencies meet their specific R&D needs in many areas, including health, the environment, and national defense.

[1]See Lewis M. Branscomb, Kenneth P. Morse, Michael J. Roberts, Darin Boville, *Managing Technical Risk: Understanding Private Sector Decision Making on Early Stage Technology Based Projects* (Gaithersburg, MD: National Institute of Standards and Technology, 2000).

The U.S. Congress tasked the National Research Council (NRC)[2] with undertaking a "comprehensive study of how the SBIR program has stimulated technological innovation and used small businesses to meet federal research and development needs" and with recommending further improvements to the program.[3] In the first round of this study, an ad hoc committee prepared a series of reports from 2004 to 2009 on the Small Business Innovation Research program at the Department of Defense (DoD), the National Institutes of Health (NIH), the National Aeronautics and Space Administration (NASA), the Department of Energy (DoE), and the National Science Foundation (NSF)—the five agencies responsible for 96 percent of the program's operations.[4]

Building on the outcomes from the first round, this second round examines topics of general policy interest that emerged during the first round as well as topics of specific interest to individual agencies. The results will be published in reports of agency-specific and program-wide findings on the SBIR and Small Business Technology Transfer (STTR) programs to be submitted to the contracting agencies and Congress. In partial fulfillment of these objectives, this volume presents the committee's second review of the NSF SBIR program's operations.[5]

PROJECT ANTECEDENTS

The current two-phase assessment of the SBIR program follows directly from an earlier analysis of public-private partnerships by the National Research Council's Board on Science, Technology, and Economic Policy (STEP). From 1990 to 2005, the Committee on Government-Industry Partnerships prepared 11 volumes reviewing the drivers of cooperation among industry, universities, and government; operational assessments of current programs; emerging needs at the intersection of biotechnology and information technology; the current experience of foreign government partnerships and opportunities for international cooperation; and the changing roles of government laboratories, universities, and other research organizations in the national innovation system.[6]

---

[2]Effective July 1, 2015, the institution is called the National Academies of Sciences, Engineering, and Medicine. References in this report to the National Research Council, or NRC, are used in a historic context identifying programs prior to July 1.

[3]See the SBIR Reauthorization Act of 2000 (H.R. 5667, Section 108).

[4]For the overview report, see National Research Council, *An Assessment of the SBIR Program* (Washington, DC: The National Academies Press, 2008). See also National Research Council, *An Assessment of the SBIR Program at the National Science Foundation* (Washington, DC: The National Academies Press), 2008. The committee also prepared reports on the SBIR programs at DoD, DoE, NIH, and NASA.

[5]The formal Statement of Task is presented in Chapter 1 of this report.

[6]For a summary of the topics covered and main lessons learned, see National Research Council, *Government-Industry Partnerships for the Development of New Technologies: Summary Report* (Washington, DC: National Academy Press, 2002).

This analysis of public-private partnerships includes two published studies of the SBIR program. Drawing from a 1998 workshop, the first report, *The Small Business Innovation Research Program: Challenges and Opportunities,* examined the origins of the program and identified operational challenges to its future effectiveness.[7] The report also highlighted the relative paucity of research on the SBIR program.

After the release of this initial report, the DoD asked the Academies to compare the operations of its Fast Track Initiative to those of its regular SBIR program. The resulting report, *The Small Business Innovation Research Program: An Assessment of the Department of Defense Fast Track Initiative,* relying on case study and survey research, found that the DoD SBIR program was achieving its legislated goals. The report also found that the Fast Track Initiative was achieving its objective of greater commercialization and recommended that it be continued and expanded where appropriate.[8] The report recommended that the SBIR program overall would benefit from further research and analysis, a recommendation subsequently adopted by Congress.

ACKNOWLEDGMENTS

On behalf of the Academies, the committee expresses its appreciation and recognition for the valuable insights and close cooperation extended by NSF staff, the survey respondents, and case study interviewees among others. The committee gives particular thanks to its lead researcher, Robin Gaster of Innovation Competitions LLC, as well as to Rosalie Ruegg of TIA Consulting, and to Peter Grunwald of Grunwald Associates LLC, which conducted the surveys and described the results presented in this volume. David Dierksheide of the STEP staff is specially recognized for his important contributions to operation of this study and the preparation of this report.

Acknowledgment of Reviewers

This report has been reviewed in draft form by individuals chosen for their diverse perspectives and technical expertise, in accordance with procedures approved by the National Academies of Sciences, Engineering, and Medicine's Report Review Committee. The purpose of this independent review is to provide candid and critical comments that will assist the institution in making its published report as sound as possible and to ensure that the report meets institutional standards for objectivity, evidence, and responsiveness to the study charge. The

---

[7]See National Research Council, *The Small Business Innovation Research Program: Challenges and Opportunities* (Washington, DC: National Academy Press, 1999).

[8]See National Research Council, *The Small Business Innovation Research Program: An Assessment of the Department of Defense Fast Track Initiative* (Washington, DC: National Academy Press, 2000).

review comments and draft manuscript remain confidential to protect the integrity of the process.

We wish to thank the following individuals for their review of this report: Robert Barnhill, Arizona State University; Frank Douglas, BioInnovation Institute; Ronald Fecso, Ernst & Young LLP; Glenn Firebaugh, Pennsylvania State University; Alastair Glass, Tyndall National Institute, Cork, Ireland; Philip Neches, California Institute of Technology; Richard Nelson, Columbia University; Colm O'Muircheartaigh, University of Chicago; Kathie Olsen, Science Works DC; Diane Palmintera, Innovation Associates; Juan Rogers, Georgia Institute of Technology; David Spencer, wTe Corporation; and Glendowlyn Thames, Connecticut Innovations.

Although the reviewers listed above have provided many constructive comments and suggestions, they were not asked to endorse the conclusions or recommendations, nor did they see the final draft of the report before its release. The review of this report was overseen by Ed Przybylowicz, Eastman Kodak (Retired) and Irwin Feller, Pennsylvania State University. Appointed by the Academies, they were responsible for making certain that an independent examination of this report was carried out in accordance with institutional procedures and that all review comments were carefully considered. Responsibility for the final content of this report rests entirely with the authoring committee and the institution.

Jacques S. Gansler Sujai J. Shivakumar

Contents

Summary

Created in 1982 through the Small Business Innovation Development Act, the Small Business Innovation Research (SBIR) program remains the nation's single largest innovation program for small business. The SBIR program offers competitive awards to support the development and commercialization of innovative technologies by small private-sector businesses. At the same time, the program provides government agencies with technical and scientific solutions that address their different missions.

Adopting several recommendations from the 2008 Academies study of the SBIR program, Congress reauthorized the program in December 2011 for an additional 6 years. In addition, Congress called for further studies by the Academies. In turn, the National Science Foundation (NSF) requested the Academies to provide a subsequent round of analysis, focused on operational questions with a view to identifying further improvements to the program.

This study therefore seeks to understand how the NSF SBIR program—which pioneered the SBIR program—is currently working and how it could work even better in the future. Drawing on the methodology developed in its previous study, an ad hoc committee issued a revised survey of SBIR companies, revisited some case studies and developed new ones, and interviewed agency managers and other stakeholders to provide a second snapshot of the program's progress toward achieving its legislative goals. Survey instruments and case studies are found in the report appendixes. Case studies provide a rich description of the program from the user's perspective.

This study recognizes that the NSF SBIR program is relatively unique in terms of scale, integrity, and mission focus. Therefore, it focuses on the SBIR program at NSF, and it does not purport to benchmark the NSF SBIR against

SBIR programs at other agencies or non-SBIR programs in the United States or abroad. Further, the study does not consider if the NSF SBIR should exist or not; rather, it assesses the extent to which the SBIR program at NSF has met the Congressional objectives set for the program, examining the extent to which recent initiatives have improved program outcomes, and providing recommendations for further improving the program to meet its objectives.

FOCUS ON LEGISLATIVE OBJECTIVES

The SBIR programs are unique efforts designed by Congress to provide funding via government agencies in pursuit of four objectives[1]:

(1) Stimulate technological innovation;
(2) Use small business to meet Federal research and development needs;
(3) Foster and encourage participation by minority and disadvantaged persons (including woman owners of small business) in technological innovation; and
(4) Increase private-sector commercialization innovations derived from federal research and development.

It is important to note at the outset that this volume—and this study—does not seek to provide a comprehensive review of the value of the SBIR program, in particular measured against other possible alternative uses of federal funding. This is beyond the study scope. Our work is focused on assessing the extent to which the SBIR program at NSF has met the Congressional objectives set for the program, to determine in particular whether recent initiatives have improved program outcomes, and to provide recommendations for improving the program further.

Thus, this study does "not to consider if SBIR should exist or not"—Congress has already decided affirmatively on this question, most recently in the 2011 reauthorization of the program. Rather, the committee is charged with "providing assessment-based findings of the benefits and costs of SBIR . . . to improve public understanding of the program, as well as recommendations to improve the program's effectiveness.

KEY FINDINGS

Based on this research, the Committee finds that with one exception the NSF SBIR program is meeting its overall legislative and mission-related goals. The exception is the important legislative goal to foster and encourage participation by women and minorities, which has not been met. Key findings with regard to the legislative goals of the SBIR program are highlighted and cross referenced below. Chapter 7 of this report lists the Committee's findings in full.

[1]Pub. L. 97–219, § 2, July 22, 1982, 96 Stat. 217.

- **Commercialization**
  - o *SBIR projects at NSF commercialize at a substantial rate.* According to the responses to the 2011 Survey, about 70 percent of Phase II projects reported some sales and an additional 19 percent anticipated future sales. (Finding I-A)
  - o *NSF's Phase IIB program supports the accelerated commercialization of SBIR-funded research.* Information from surveyed recipients and from case studies strongly suggests that the program serves as a catalyst, attracts additional funding, and has a positive effect on company activities and outcomes. (Finding V)
- **Participation by Women and Minorities**
  - o *Levels of participation by underserved groups are low and not rising.* (Finding II-A) Data from NSF indicate that the success rates of minority owned small businesses (MOSB) applicants are strikingly lower than those of non-MOSB applicants. Woman-owned small businesses (WOSBs) and minority-owned small businesses (MOSBs) not only submit fewer proposals but also have lower success rates than non-WOSB and non-MOSB groups. WOSBs and MOSBs received approximately 6 percent and 4 percent of awards, respectively. These percentages have not increased during the past decade.
  - o In the 2011 Survey, companies reported that 11 percent of Principal Investigators (PIs) were minority (the same as the 2005 survey). However, further analysis indicates that only 1 percent of PIs were Hispanic, and less than 0.5 percent were African American, and none were Native American. (Finding II-A)
  - o *NSF's efforts to "foster and encourage" the participation of woman-owned and minority-owned small businesses have not been effective.* (Finding II-B)
- **Stimulating Technological Innovation and Meeting Agency Mission Needs**
  - o *The NSF SBIR program supports the development and adoption of technological innovations.* Selection of topics and individual projects for funding maintains a strong focus on technological innovation. (Finding III-A)
  - o *The NSF SBIR program continues to connect companies and universities in a variety of ways.* Nearly 60 percent of survey respondents reported a link to a university. About 35 percent of projects involved university faculty (not as principal investigator [PI]), 30 percent of projects employed graduate students, and almost 25 percent of projects had universities serving as subcontractors. (Finding III-B)
  - o *NSF SBIR projects generate substantial knowledge-based outputs.* More than 70 percent of survey respondents reported filing at least one patent related to the surveyed project. About 80 percent of survey

respondents reported at least one resulting peer-reviewed publication. (Finding III-C)

- **Fostering Innovative Companies**
    - Forty-five percent of survey respondents indicated that the company was founded entirely or in part because of the SBIR program. (Finding IV-A)
    - Thirty-five percent of survey respondents indicated that the NSF SBIR program had a "transformative" effect on the company. Another 54 percent reported a "substantial positive long-term effect." (Finding IV-C)

KEY RECOMMENDATIONS

The Committee's key recommendations are highlighted below:

- **NSF should immediately enhance efforts to address the clear Congressional mandate to foster the participation of underserved populations in the SBIR program (one of the four major goals of the program as stated in Chapter 1).**
    - *NSF should not develop quotas for the inclusion of selected populations in the SBIR program.* Such an approach is not necessary to meet Congressional intent and is likely to reduce program effectiveness. (Recommendation I-A)
    - *NSF should develop new benchmarks and metrics.* Measures of the participation of socially disadvantaged groups should be disaggregated by race/ethnicity, and attention should be focused on the clear Congressional intent to support "minority" participation. (Recommendation I-B)
    - *NSF should undertake an investigation to understand better why its efforts to date to expand the participation of underserved populations have been largely unsuccessful.* (Recommendation I-C)
- **NSF should continue to operate the Phase IIB program.** (Recommendation III-A)
    - *NSF should consider expanding the size of the Phase IIB program.* The strongly positive impact suggests that further projects would benefit from Phase IIB funding. (Recommendation III-B)
- **NSF should improve its efforts in data collection, assessment, and reporting.**
    - *NSF should improve current data collection methods and standardize key questions to improve program evaluation.* This data collection effort should address the entire range of congressionally mandated outcomes, not just commercialization. (Recommendation IV-A)
    - *NSF should provide a comprehensive annual report to Congress and the public on its SBIR/STTR program operations.* (Recommendation IV-D)

1

Introduction

Small businesses continue to be an important driver of innovation and economic growth,[1] despite the challenges of changing global environments and the impacts of the 2009 financial crisis and subsequent recession.[2] In the face of these challenges, supporting small businesses in their development and commercialization of new products is essential for U.S. competitiveness and national security.

The SBIR was started as a pilot program at the National Science Foundation (NSF) in the late 1970s. It received legislative authorization in 1982, through the Small Business Innovation Development Act, and subsequently

---

[1]See Zoltan Acs and David Audretsch, "Innovation in Large and Small Firms: An Empirical Analysis," *The American Economic Review* 78(4):678-690. See also Zoltan Acs and David Audretsch, *Innovation and Small Firms* (Cambridge, MA: The MIT Press, 1991); Erik Stam and Karl Wennberg, "The Roles of R&D in New Firm Growth," *Small Business Economics* 33(2009):77-89; Eileen Fischer and A. Rebecca Reuber, "Support for Rapid-Growth Firms: A Comparison of the Views of Founders, Government Policymakers, and Private Sector Resource Providers," *Journal of Small Business Management* 41(4):346-365; Magnus Henrekson and Dan Johansson, "Competencies and Institutions Fostering High-Growth Firms," *Foundations and Trends in Entrepreneurship* 5, no. 1 (2009): 1-80.

[2]Citing recent data, some analysts have recently questioned the strength of small businesses as a driver of economic growth. See, for example, Newmark et al. "Do Small Businesses Create More Jobs? ..." in *The Review of Economics and Statistics* 93, issue 1, pp 16-23. Others have pointed out the negative but temporary impact of the 2008 financial crisis on credit and investment on small business led growth. See Daniele Archibugi, Andrea Filippetti, and Marion Frenz, "Economic Crisis and Innovation: Is Destruction Prevailing Over Accumulation?" *Research Policy* 42, no. 2 (March 2013): 303-314. The authors show that "the 2008 economic crisis has severely reduced the short-term willingness of firms to invest in innovation" and that it "led to a concentration of innovative activities within a small group of fast growing new firms and those firms already highly innovative before the crisis." They conclude that "the companies in pursuit of more explorative strategies towards new product and market developments are those to cope better with the crisis."

the Small Business Research and Development Enhancement Act of 1992, and the Small Business Innovation Research Program Reauthorization Act of 2000. The Small Business Innovation Research (SBIR) program is the nation's largest innovation program for small business. Through FY2014, it had made more than 150 thousand awards totaling nearly $40 billion.[3]

The SBIR program offers competitive awards to support the development and commercialization of innovative technologies by small private-sector businesses. At the same time, the program provides government agencies with technical and scientific solutions that address their different missions.

Currently, the SBIR program provides funding in three phases:

- Phase I provides limited funding (up to $100,000 prior to the 2011 reauthorization and up to $150,000 thereafter) for feasibility studies.
- Phase II provides more substantial funding for further research and development (typically up to $750,000 prior to 2012 and $1 million after the 2011 reauthorization).[4]
- Phase III reflects commercialization without providing access to any additional SBIR funding, although funding from other federal government accounts is permitted.

Congress mandated four goals for the program: "(1) to stimulate technological innovation; (2) to use small business to meet federal research and development needs; (3) to foster and encourage participation by minority and disadvantaged persons in technological innovation; and (4) to increase private sector commercialization derived from federal research and development."

The research agencies have pursued these goals through the development of SBIR programs that differ from each other in many respects, utilizing the administrative flexibility built into the general program to address their unique mission needs.

SBIR awards are highly competitive. Between 2003 and 2012,[5] about 18 percent of Phase I applications and 44 percent of Phase II applications to the National Science Foundation (NSF) resulted in an award.

Over time, through a series of reauthorizations, SBIR legislation has required federal agencies with extramural research and development (R&D) budgets in excess of $100 million to set aside a growing share of their budgets for the SBIR program. The set-aside reached 2.5 percent following the 2000

[3]Small Business Administration, *<http://www.sbir.gov/past-awards>*.

[4]All resource and time constraints imposed by the program are somewhat flexible and are addressed by different agencies in different ways. For example, the National Institutes of Health (NIH), and to a much lesser degree the Department of Defense (DoD), have provided awards that are much larger than the standard amounts, and NIH has a tradition of offering no-cost extensions to see work completed on an extended timeline.

[5]National Science Foundation (NSF) data provided to the Academies.

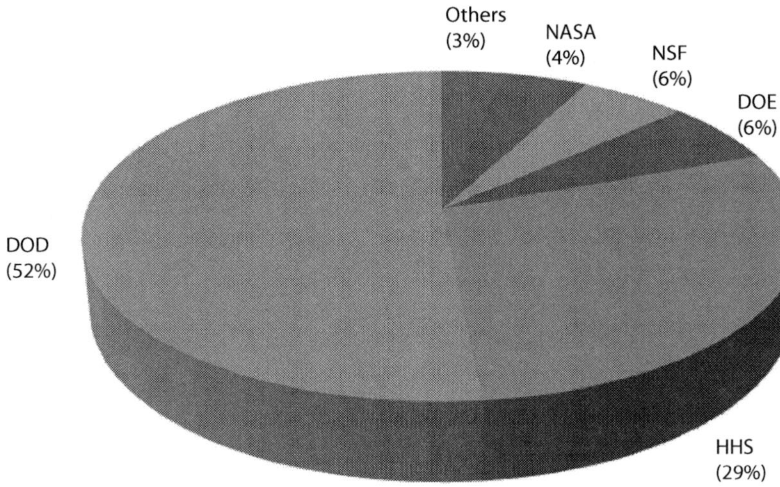

FIGURE 1-1 SBIR/STTR funding by federal agency share, FY2010.
SOURCE: Small Business Administration (SBA) SBIR website. Accessed November 1, 2013.

reauthorization. In fiscal year (FY) 2010, the 11 federal agencies administering the SBIR program disbursed \$2.51 billion, and in FY 2014 they disbursed \$2.08 billion.[6]

Five agencies administer greater than 96 percent of SBIR/STTR funds: the Department of Defense (DoD), the Department of Health and Human Services (HHS) (particularly the National Institutes of Health [NIH]), the Department of Energy (DoE), the National Aeronautics and Space Administration (NASA), and NSF. Figure 1-1 shows the percentages of SBIR and related Small Business Technology Transfer (STTR) funding in FY2010 provided by each of these five federal agencies, as well as the remaining combined percentage provided by all other agencies that administer SBIR/STTR programs. In FY 2010, for example, NSF made 549 SBIR/STTR awards amounting to \$118.7 million. In FY 2014, NSF made 411 SBIR/STTR awards amounting to \$130.2 million.

In December 2011, Congress reauthorized the program for an additional 6 years,[7] with a number of important modifications. Many of these modifications—for example, changes in standard award size—were based on recommendations

[6]Small Business Association SBIR/STTR annual report, accessed April 2015, <*http://www.sbir.gov*>.

[7]Section 5137 of Public Law 112-81.

made in a 2008 National Research Council (NRC)[8] report on the SBIR program.[9] The reauthorization also called for further studies by the Academies.

In a follow-up to the first-round assessment, NSF requested that the Academies provide a subsequent round of assessment, focused on operational questions with a view to identifying further improvements to the program.

This introduction provides a context for the analysis of program developments and transitions described in the remainder of the report. The first section provides an overview of the SBIR program's history across the federal government. This is followed by a summary of the major changes mandated by the 2011 reauthorization and the subsequent Small Business Administration (SBA) Policy Directive; a review of the program's advantages and limitations, in particular the challenges faced by entrepreneurs using (and seeking to use) the program and by agency officials running the program; and a summary of the technical challenges facing this assessment and the committee's solutions to those challenges.

PROGRAM HISTORY AND STRUCTURE[10]

During the 1980s, the perceived challenge of Japanese industrial growth in sectors traditionally dominated by U.S. firms—autos, steel, and semiconductors—led to serious concerns about U.S. competitiveness.[11] A key concern was the perceived failure of American industry "to translate its research prowess into commercial advantage."[12] Although the United States enjoyed dominance in basic research—much of which was federally funded—applying this research to the development of innovative products and technologies remained challenging. As the great corporate laboratories of the post-war period were buffeted by

---

[8]Effective July 1, 2015, the institution is called the National Academies of Sciences, Engineering, and Medicine. References in this report to the National Research Council, or NRC, are used in a historic context identifying programs prior to July 1.

[9]National Research Council, *An Assessment of the SBIR Program*, (Washington, DC: The National Academies Press, 2008). The Academies' first-round assessment of the SBIR program was mandated in the SBIR Reauthorization Act of 2000, Public Law 106-554, Appendix I-H.R. 5667, Section 108.

[10]Parts of this section are based on the Academies' previous report on the NSF SBIR program—National Research Council, *An Assessment of the SBIR Program at the National Science Foundation* (Washington, DC: The National Academies Press, 2008).

[11]See John Alic, "Evaluating Competitiveness at the Office of Technology Assessment," *Technology in Society* 9, no. 1 (1987): 1-17 for a review of how these issues emerged and evolved within the context of a series of analyses at a Congressional agency.

[12]David C. Mowery, "America's Industrial Resurgence (?): An Overview," in *U.S. Industry in 2000: Studies in Competitive Performance*, David C. Mowery, ed., (Washington, DC: National Academy Press, 1999), p. 1. Other studies highlighting poor economic performance in the 1980s include Michael L. Dertouzos et al., *Made in America: The MIT Commission on Industrial Productivity* (Cambridge, MA: The MIT Press, 1989); and Otto Eckstein, *DRI Report on U.S. Manufacturing Industries* (New York: McGraw Hill, 1984).

change, new models such as the cooperative model utilized by some Japanese kieretsu offered new sources of dynamism and more competitive firms.[13]

At the same time, new evidence emerged to indicate that small businesses were an increasingly important source of innovation and job creation. David Birch and others suggested that national policies should promote and build on these developments.[14] This evidence reinforced recommendations from federal commissions dating back to the 1960s, that is, that federal R&D funding should provide more support for innovative small businesses (which was opposed by traditional recipients of federal R&D funding).[15]

Early-stage financial support for high-risk technologies with commercial promise was first advanced by Roland Tibbetts at NSF, who in 1976 advocated for shifting some NSF funding to innovative technology-based small businesses. Following a period of analysis and discussion, and support by the Reagan administration for an expansion of the practice across federal agencies, Congress passed the Small Business Innovation Research Development Act of 1982, establishing the SBIR program.

The scale of the SBIR program was gradually increased. At the start, the SBIR program required agencies with extramural R&D budgets in excess of $100 million[16] to set aside 0.2 percent of their funds for SBIR. Program funding totaled $45 million in the program's first year of operation (1983). Over the next 6 years, the set-aside grew to 1.25 percent.[17]

---

[13]David L. Birch, "Who Creates Jobs?" *National Affairs*, <*http://www.nationalaffairs.com/doclib/20080708_1981651whocreatesjobsdavidlbirch.pdf*>. Accessed August 13, 2014. Birch's work greatly influenced perceptions of the role of small firms. Over the past 20 years, it has been carefully scrutinized, leading to the discovery of some methodological flaws, namely making dynamic inferences from static comparisons, confusing gross and net job creation, and admitting biases from chosen regression techniques. See Steven J. Davis, John Haltiwanger, and Scott Schuh, "Small Business and Job Creation: Dissecting the Myth and Reassessing the Facts, Working Paper No. 4492 (Cambridge, MA: National Bureau of Economic Research, 1993). According to Per Davidsson, these methodological fallacies, however, "ha[ve] not had a major influence on the empirically based conclusion that small firms are over-represented in job creation." See Per Davidsson, "Methodological Concerns in the Estimation of Job Creation in Different Firm Size Classes" (working paper, Jönköping International Business School, 1996).

[14]David Birch, "Who Creates Jobs?"

[15]For an overview of the origins and history of the SBIR program, see George Brown and James Turner, "Reworking the Federal Role in Small Business Research," *Issues in Science and Technology*, Summer 1999, pp. 51-58.

[16]That is, those agencies spending more than $100 million on research conducted outside agency labs.

[17]Additional information regarding the SBIR program's legislative history can be accessed from the Library of Congress. See <*http://thomas.loc.gov/cgi-bin/bdquery/z?d097:SN00881:@@@L*>.

BOX 1-1
Commercialization Language from 1992 SBIR Reauthorization

Phase II "awards shall be made based on the scientific and technical merit and feasibility of the proposals, as evidenced by the first phase, considering, among other things, the proposal's commercial potential, as evidenced by

(i) the small business concern's record of successfully commercializing SBIR or other research;
(ii) the existence of second phase funding commitments from private sector or non-SBIR funding sources;
(iii) the existence of third phase, follow-on commitments for the subject of the research; and
(iv) the presence of other indicators of the commercial potential of the idea."

SOURCE: P.L. 102-564-OCT. 28, 1992.

The SBIR Reauthorizations of 1992 and 2000

The SBIR program approached reauthorization in 1992 amidst continued worries about the U.S. economy's capacity to commercialize inventions. Finding that "U.S. technological performance is challenged less in the creation of new technologies than in their commercialization and adoption," the Academies recommended an increase in SBIR funding as a means to improve the economy's ability to adopt and commercialize new technologies.[18]

The Small Business Research and Development Enhancement Act (P.L. 102-564) reauthorized the SBIR program until September 30, 2000, and doubled the set-aside rate to 2.5 percent. The legislation also more strongly emphasized the need for commercialization of SBIR-funded technologies.[19] The 1992 legislative language explicitly highlighted commercial potential as a criterion for awarding SBIR grants, as indicated in Box 1-1.

At the same time, Congress expanded the SBIR program's purposes to "emphasize the program's goal of increasing private sector commercialization developed through Federal research and development and to improve the federal government's dissemination of information concerning the small business

[18]See National Research Council, *The Government Role in Civilian Technology: Building a New Alliance* (Washington, DC: National Academy Press, 1992), p. 29.

[19]See Robert Archibald and David Finifter, "Evaluation of the Department of Defense Small Business Innovation Research Program and the Fast Track Initiative: A Balanced Approach," in National Research Council, *The Small Business Innovation Research Program: An Assessment of the Department of Defense Fast Track Initiative* (Washington, DC: National Academy Press, 2000), pp. 211-250.

innovation, particularly with regard to woman-owned business concerns and by socially and economically disadvantaged small business concerns."

The Small Business Reauthorization Act of 2000 (P.L. 106-554) extended the SBIR program until September 30, 2008. It also called for an NRC assessment of the program's broader impacts, including those on employment, health, national security, and national competitiveness.[20]

THE 2011 REAUTHORIZATION

The anticipated 2008 reauthorization was delayed in large part by a disagreement between long-time program participants and their advocates in the small business community and proponents of expanded access for venture-backed companies, particularly in biotechnology where proponents argued that the standard path to commercial success involves venture funding at some point.[21] Other issues were also difficult to resolve, but the conflict over participation of venture capital-backed companies dominated the process[22] following an administrative decision to exclude these companies more systematically.[23]

After a much extended discussion, the SBIR and STTR programs were reauthorized through FY2017 through passage of the National Defense Act of December 2011. The new law maintained much of the core structure of both programs but made some important changes, which were to be implemented via the SBA's subsequent Policy Guidance.

The eventual compromise on the venture funding issue allowed (but did not require) agencies to set aside 25 percent of SBIR funding (at NIH, DoE, and NSF) or 15 percent (at the other awarding agencies) for participation by firms benefiting from private, venture capital investment. It is too early in the implementation process to gauge the impact of this change.

Several changes to the program made through reauthorization reflected recommendations by the Academies in prior reports, including the following:

- Increased award size limits
- Expanded program size

---

[20]The current assessment is congruent with the Government Performance and Results Act (GPRA) of 1993: <*http://govinfo.library.unt.edu/npr/library/misc/s20.html*>. As characterized by the Government Accountability Office (GAO), GPRA seeks to shift the focus of government decision making and accountability away from a focus on the activities that are undertaken—such as grants dispensed or inspections made—to a focus on the results of those activities. See <*http://www.gao.gov/new.items/gpra/gpra.htm*>.

[21]Damien C. Specht, "Recent SBIR Extension Debate Reveals Venture Capital Influence," *Procurement Law* 45 (2009): 1.

[22]Wendy H. Schacht, "The Small Business Innovation Research (SBIR) Program: Reauthorization Efforts," Congressional Research Service, Library of Congress, 2008.

[23]Aaron Bouchie, "Increasing Number of Companies Found Ineligible for SBIR Funding," *Nature Biotechnology* 21, no. 10 (2003): 1121-1122.

- Enhanced agency flexibility—for example, to utilize Phase I awards from other agencies or to add a second Phase II
- Improved incentives for the utilization of SBIR technologies in agency acquisition programs
- Explicit requirements to better connect prime contractors with the SBIR program
- Substantial emphasis on developing a more data-driven culture. Addressing Academies recommendations in this area has led to several significant reforms, including the following:
  - adding numerous areas of expanded reporting
  - extending the Academies' evaluation
  - adding further evaluation from other expert bodies, such as the Comptroller General
  - tasking the SBA with creating a unified data platform
- Expanded management resources (through provisions permitting use of up to 3 percent of program funds for defined management purposes)
- Expanded commercialization support (through provisions providing companies with direct access to commercialization support funding and through approval of the approaches piloted in the Commercialization Pilot Program)
- Flexibility for agencies to develop other pilot programs—for example, to skip Phase I, or for NIH to support a new Phase 0 pilot program.

In addition, the reauthorization made changes that were not recommended in previous Academies reports, including the following:

- Expansion of the STTR program
- Limitations on some aspects of agency flexibility, particularly caps on the provision of larger awards
- Introduction of commercialization benchmarks, established by agencies, that companies must meet to remain in the program.

Other clauses of the legislation related to operational issues, such as the definition of specific terms (e.g., "Phase III"), continued and expanded evaluation by the Academies and mandated reports from the Comptroller General on combating waste, fraud, and abuse within the program, and measures of the protection of small companies' intellectual property within the program.

PREVIOUS RESEARCH ON SBIR

Previous studies, most notably by the General Accounting Office and the Small Business Administration, have focused on specific aspects or components of the program.[24] Prior to the first round of the assessment by the Academies, there had been few internal assessments of agency programs. The academic literature on SBIR was also limited.[25] Writing in the 1990s, Joshua Lerner of the Harvard Business School positively assessed the program, finding "that SBIR awardees grew significantly faster than a matched set of firms over a ten-year period."[26] These findings were consistent with the corporate finance literature on capital constraints and the growth literature on the importance of localization economies.[27]

To help fill this assessment gap, and to learn about a large, relatively under-evaluated program, the Academies' Committee for Government-Industry Partnerships for the Development of New Technologies (GIP Committee) prepared the first comprehensive discussion of the SBIR program's history and rationale, reviewed existing research, and identified areas for further research and program improvements.[28] It reported the following:

- The SBIR program enjoyed strong support of parts of the federal government, as well as of the country at large.
- The size and significance of the SBIR program underscored the need for more research on its effectiveness.
- The primary emphasis on commercialization within the SBIR program required further clarification.
- Evaluation methodologies required additional work.[29]

---

[24]An important step in the current evaluation of the SBIR program has been to identify already existing evaluations. These include U.S. General Accounting Office, *Federal Research: Small Business Innovation Research Shows Success But Can Be Strengthened* RCED-92-32, (Washington, DC: U.S. General Accounting Office, 1992); and U.S. General Accounting Office, *Federal Research: Evaluation of Small Business Innovation Can Be Strengthened*, T-RCED-99-198 (Washington, DC: U.S. General Accounting Office, 1999). There is also a 1999 unpublished SBA study of the commercialization of SBIR surveys Phase II awards from 1983 to 1993 among non-DoD agencies. Effective July 7, 2004, the GAO's legal name was changed from the General Accounting Office to the Government Accountability Office.

[25]Early examples of evaluations of the SBIR program include Summer Myers, Robert L. Stern, and Marcia L. Rorke, *A Study of the Small Business Innovation Research Program* (Lake Forest, IL: Mohawk Research Corporation, 1983); and Price Waterhouse, *Survey of Small High-tech Businesses Shows Federal SBIR Awards Spurring Job Growth, Commercial Sales* (Washington, DC: Small Business High Technology Institute, 1985).

[26]Joshua Lerner, "The Government as Venture Capitalist: The Long-Run Effects of the SBIR Program," September 1996, *National Bureau of Economic Research*, Working Paper No. 5753, <*http://www.nber.org/papers/w5753.pdf*>. Accessed August 13, 2014.

[27]See Michael Porter, "Clusters and Competition: New Agendas for Government and Institutions," in *On Competition* (Boston, MA: Harvard Business School Press, 1998).

[28]See National Research Council, *The Small Business Innovation Research Program: Challenges and Opportunities* (Washington, DC: National Academy Press, 1999).

[29]National Research Council, *An Assessment of the DoD SBIR Fast Track Initiative* (Chapter III: Recommendations and Findings), p. 32.

In a later, more comprehensive review of the DoD SBIR program, the GIP Committee found that the program contributed to mission goals by funding valuable innovative projects. It also concluded that a significant number of these projects would not have been undertaken absent SBIR funding and that Fast Track (an initiative to expedite decision-making for SBIR awards to companies that have commitments from outside investors) encouraged the commercialization of new technologies and the entry of new firms into the program.[30]

The GIP Committee also found that the SBIR program affected the development and utilization of human capital and the diffusion of technological knowledge. Case studies showed that the knowledge and human capital generated by the SBIR program have positive economic value, which spills over into other firms through the movement of people and ideas. Furthermore, by acting as a "certifier" of promising new technologies, SBIR awards encourage further private-sector investment in an award-winning firm's technology.[31]

THE ROUND-ONE STUDY OF SBIR

Drawing on these findings and recommendations, the 2000 SBIR reauthorization mandated a comprehensive assessment of the nation's SBIR program, which was conducted in three steps. During the first step, the committee charged with carrying out the study developed a research methodology, which was approved by an independent Academies' panel of experts. This committee gathered information about the program by interviewing officials at the relevant federal agencies and by inviting these officials to describe program operations, challenges, and accomplishments at two major conferences. These conferences highlighted the important differences in agency goals, practices, and evaluations, as well as the evaluation challenges that arise from the diverse program objectives and practices.[32]

The research methodology was implemented during the second step of the study. The committee deployed multiple survey instruments and conducted case studies of a wide variety of SBIR companies. The committee then evaluated the results and developed the findings and recommendations for improving the effectiveness of the SBIR program. It is important to stress that the respondents to the survey represent only a subset of all awardees.[33]

During the third step of the study, the committee reported on the program through a series of publications in 2008-2010: five individual volumes on the five major funding agencies and an additional overview volume. Together, these reports provided the first detailed and comprehensive review of the SBIR pro-

---

[30]Ibid, 33.

[31]Ibid, 33.

[32]Adapted from National Research Council, *Small Business Innovation Research Program: Program Diversity and Assessment Challenges* (Washington, DC: The National Academies Press, 2004).

[33]Box A-1 in Appendix A gives and overview of potential sources of bias in survey results.

gram and, as noted above, became an important input into SBIR reauthorization prior to December 2011. Box 1-2 highlights accomplishments of the round-one, mandated assessment of the SBIR.

THE CURRENT, SECOND-ROUND STUDY: CHALLENGES AND OPPORTUNITIES

The set of reports from the Academies' first-round study of the SBIR program established that, overall, the program is "sound in concept and effective in practice." The second-round study seeks to understand how the program could work better. Box 1-3 highlights the Statement of Task of the second-round study.

The current volume is focused on updating the round-one committee's 2007 assessment of the NSF SBIR program, by updating data, providing new descriptions of recent program and developments, providing fresh company case studies. This volume, in particular, focuses on the efforts made at NSF in recent years to improve the SBIR program. Guided by this Statement of Task, we have sought answers to questions such as the following:

- Are there initiatives and programs within NSF that have made a significant difference to outcomes and in particular to commercialization of SBIR-funded technologies?
- Can they be replicated and expanded?
- What are the main barriers to meeting congressional objectives more fully?
- What program adjustments would better support commercialization?
- What tools would expand utilization by woman- and minority-owned firms and participation by female and minority principal investigators?
- Can links with universities be improved?
- Why do some firms simply drop out of the program?
- Are there aspects of the program that make it less attractive? Could they be addressed?
- What can be done to expand access in underserved states while maintaining the competitive character of the program?
- Can the program generate better data on both process and outcomes and use those data to fine-tune program management?

STUDY METHODOLOGY

It is always useful when assessing government programs to identify comparable programs for appropriate benchmarking. However, in the committee's experience, there are no truly comparable programs in the United States, and those in other countries operate in such different ways that their relevance is limited. The SBIR/STTR programs are relatively unique in terms of scale, integrity, and mission focus.

BOX 1-2
First-Round Assessment of the Small Business Innovation Research (SBIR) Program

Mandated by Congress in the 2000 reauthorization of the SBIR program, the National Research Council's (NRC) first-round SBIR assessment reviewed the SBIR programs at the Department of Defense, National Institutes of Health, NASA, the Department of Energy, and the National Science Foundation. In addition to the release of reports focused on the SBIR program at each of these agencies and a program methodology report that guided the committee's review, the study resulted in a summary of a symposium focused on the diversity of the program and challenge of its assessment, a summary of a symposium focused on the challenges in commercializing SBIR-funded technologies, and two additional reports on special topics in addition to the committee's summary report, *An Assessment of the SBIR Program*. In all, eleven study reports[a] were published:

- *An Assessment of the Small Business Innovation Research Program: Project Methodology* (2004)
- *SBIR—Program Diversity and Assessment Challenges: Report of a Symposium* (2004)
- *SBIR and the Phase III Challenge of Commercialization: Report of a Symposium* (2007)
- *An Assessment of the SBIR Program at the National Science Foundation* (2007)
- *An Assessment of the SBIR Program at the Department of Defense* (2009)
- *An Assessment of the SBIR Program at the Department of Energy* (2008)
- *An Assessment of the SBIR Program* (2008)
- *An Assessment of the SBIR Program at the National Aeronautics and Space Administration* (2009)
- *An Assessment of the SBIR Program at the National Institutes of Health* (2009)
- *Venture Funding and the NIH SBIR Program* (2009)
- *Revisiting the Department of Defense SBIR Fast Track Initiative* (2009)

[a]National Research Council, *An Assessment of the Small Business Innovation Research Program: Project Methodology*, Washington, DC: The National Academies Press, 2004; National Research Council, *SBIR—Program Diversity and Assessment Challenges: Report of a Symposium*, Washington, DC: The National Academies Press, 2004; National Research Council, *SBIR and the Phase III Challenge of Commercialization: Report of a Symposium*, Washington, DC: The National Academies Press, 2007; National Research Council, *An Assessment of the SBIR Program at the National Science Foundation*, Washington, DC: The National Academies Press, 2007; National Research Council, *An Assessment of the SBIR Program at the Department of Defense*, Washington, DC: The National Academies Press, 2009; National Research Council, *An Assessment of the SBIR Program at the Department of Energy*, Washington, DC: The National Academies Press, 2008; National Research Council, *An Assessment of the SBIR Program*, Washington, DC: The National Academies Press, 2008; National Research Council, *An Assessment of the SBIR Program at the National Aeronautics and Space Administration*, Washington, DC: The National Academies Press, 2009; National Research Council, *An Assessment of the SBIR Program at the National Institutes of Health*, Washington, DC: The National Academies Press, 2009; National Research Council, *Venture Funding and the NIH SBIR Program*, Washington, DC: The National Academies Press, 2009; and National Research Council, *Revisiting the Department of Defense SBIR Fast Track Initiative*, Washington, DC: The National Academies Press, 2009.

BOX 1-3
Statement of Task

In accordance with H.R. 5667, Sec. 108, enacted in Public Law 106-554, as amended by H.R. 1540, Sec. 5137, enacted in Public Law 112-81, the National Research Council is to review the Small Business Innovation Research and Small Business Technology Transfer (SBIR/STTR) programs at the Department of Defense, the National Institutes of Health, the National Aeronautics and Space Administration, the Department of Energy, and the National Science Foundation. Building on the outcomes from the Phase I study, this second study is to examine both topics of general policy interest that emerged during the first-phase study and topics of specific interest to individual agencies.

Drawing on the methodology developed in the previous study, an ad hoc committee will issue a revised survey, revisit case studies, and develop additional cases, thereby providing a second snapshot to measure the program's progress against its legislative goals. The committee will prepare one consensus report on the SBIR program at each of the five agencies, providing a second review of the operation of the program, analyzing new topics, and identifying accomplishments, emerging challenges, and possible policy solutions. The committee will prepare an additional consensus report focused on the STTR program at all five agencies. The agency reports will include agency-specific and program-wide findings on the SBIR and STTR programs to submit to the contracting agencies and the Congress.

Although each agency report will be tailored to the needs of that agency, all reports will, where appropriate,

1. review institutional initiatives and structural elements contributing to programmatic success, including gap funding mechanisms such as applying Phase II-plus awards more broadly to address agency needs and operations and streamlining the application process,
2. explore methods to encourage the participation of minorities and women in SBIR and STTR,
3. identify best practice in university-industry partnering and synergies with the two programs,
4. document the role of complementary state and federal programs, and
5. assess the efficacy of post-award commercialization programs.

In partial fulfillment of this Statement of Task, this volume presents the committee's second review of the operation of the SBIR program at the National Science Foundation.

The SBIR program's diversity offers other challenges to evaluation and comparison. "The SBIR program" is in fact a multiplicity of agency-specific programs; it is important to ensure that research focuses on the SBIR at the NSF, not on a generic multiagency conceptualization of the SBIR program.

Focus on Legislative Objectives

It is important to note at the outset that this volume—and this study—does not seek to provide a comprehensive review of the value of the SBIR program, in particular measured against other possible alternative uses of federal funding. This is beyond our scope. Our work is focused on assessing the extent to which the SBIR program at NSF has met the congressional objectives set for the program, to determine in particular whether recent initiatives have improved program outcomes, and to provide recommendations for improving the program further.[34]

Thus, as in the first-round study, the objective of this second round study is "*not* to consider if SBIR should exist or not"—Congress has already decided affirmatively on this question, most recently in the 2011 reauthorization of the program.[35] Rather, we are charged with "providing assessment-based findings of the benefits and costs of SBIR . . . to improve public understanding of the program, as well as recommendations to improve the program's effectiveness." As with the first-round, this study "will *not* seek to compare the value of one area with other areas; this task is the prerogative of the Congress and the Administration acting through the agencies. Instead, the study is concerned with the effective review of each area."[36]

Defining Commercialization

Commercialization offers practical and definitional challenges. As described in Chapter 4, several different definitions of commercialization are used in discussions of the SBIR program. We have concluded that it is important to use more than one simple definition: For example, it is not appropriate to use the number or percentage of projects that reach the marketplace as the sole metric for commercial success. In the private sector, commercial success over the long term requires profitability. But in the short term, commercialization can involve many different aspects of commercial activity, from product rollout to licensing to patenting to acquisition. Even during new product rollout, companies often do

---

[34]These limited objectives are consistent with the methodology developed by the committee. See National Research Council, *An Assessment of the Small Business Innovation Research Program: Project Methodology* (Washington, DC: The National Academies Press, 2004).

[35]National Defense Authorization Act of 2012 (NDAA) HR.1540, Title LI.

[36]National Research Council, *An Assessment of the Small Business Innovation Research Program: Project Methodology*.

not generate immediate profits. In this report we use multiple metrics to address the question of commercialization.

Quantitative Assessment Methods

More practically, several issues relate to the application of quantitative assessment methods, including decisions about which kinds of program participants should be targeted, the number of responses that are appropriate, selection bias, nonresponse bias, the design and implementation of survey questionnaires, and the level of statistical evidence required for drawing conclusions in this case. These and other issues were discussed at a workshop and published in a 2004 report.[37] Also prepared was a peer-reviewed report on study methodology, which provided the baseline for the initial study and for follow-on studies—such as this one.

Survey Development

For the current study, a new survey of SBIR recipients was developed and deployed. This 2011 Survey[38] was based closely on previous surveys, particularly one deployed by the Academies in 2005 as part of its round-one assessment of the SBIR program, but nonetheless included some significant improvements. The description of the survey and improvements are documented in Appendix A of this report. The 2011 survey delves more deeply into the demographics of the program. It addresses in detail the role of agency liaisons. Finally, it provided unique opportunities to collect opinions and recommendations for program improvement from NSF award recipients. The survey generated more than 600 responses from recipients and provided an important pillar to the research conducted for this volume.[39] Appendix A provides a description of the survey methods, including a discussion of the survey outreach and response.

Complementing the 2011 survey was a 2010 Phase IIB Survey of NSF Phase II awards that was carried out on behalf of the committee with the objective of comparing outcomes between awardees receiving standard Phase II awards and those also receiving Phase IIB enhancements.

Issues related to quantitative methodologies are discussed in detail in Chapter 4 and Appendix A. We recognize that the conclusions that can be drawn from this kind of assessment are limited. However, we also conclude that drawing on quantitative analysis is a crucial component of the overall study, given the need

[37] National Research Council, *Program Diversity and Assessment Challenges.*

[38] The survey carried out as part of this study was carried out in 2011, and the survey completed as part of the first-round assessment of the SBIR program was administered in 2005. In this volume all Academies survey references are to the 2011 survey unless noted otherwise.

[39] National Research Council, *An Assessment of the Small Business Innovation Research Program: Project Methodology.*

to identify and assess outcomes that are to be found only at the level of individual projects and participating companies.

A Complement of Approaches

Partly because of these limitations, the 2004 methodology report[40] stressed the importance of utilizing a complement of approaches, which has been adopted here. Although quantitative assessment represents the bedrock of our research and provides insights and evidence that could not be generated through any other modality, it is in and of itself not sufficient to address the multiple questions posed in this analysis. Consequently, we undertook a series of additional activities:

- **Case studies.** We conducted in-depth case studies of 10 NSF SBIR recipients. These companies were geographically and demographically diverse and were at different stages of the company life cycle. Lessons from the case studies are described in Chapter 6, and the cases themselves are included as Appendix E.
- **Workshops.** We conducted workshops to allow stakeholders, agency staff, and academic experts to provide insights into program operations, as well as to identify questions that should be addressed.
- **Analysis of agency data.** As appropriate, we analyzed and included data provided by NSF about the various aspects of its SBIR activities.
- **Open-ended responses from SBIR recipients.** For the first time, we collected textual survey responses. More than 400 recipients provided narrative comments.
- **Agency consultations.** We engaged in discussions with NSF staff about program operations program and the challenges faced.
- **Literature review.** Since the start of our research in this area, a number of papers have been published addressing various aspects of the SBIR program. In addition, other organizations—such as the Government Accountability Office (GAO)—have reviewed specific parts of the SBIR program. We incorporated references to their work, where useful, into its analysis.

In short, within the limitations described, we utilized a complement of tools to ensure that the study reflects a full spectrum of perspectives and expertise. Appendix A provides an overview of the methodological approaches, data sources, and survey tools used in this study.

[40] Ibid.

ORGANIZATION OF THE REPORT

Our analysis and conclusions are organized as follows. Chapter 2 provides background on the mission of the NSF, relates how the SBIR and STTR fit within the NSF's Office of Industrial Innovation and Partnerships, reviews program operations, examining the emergence of what could be called the NSF model of program management, and addresses the congressional mandate to foster the participation of women and minorities and the current status. Chapter 3 reviews NSF data concerning applications and awards to the program, drawing out differences by demographic, geographic, and industrial sector. Chapter 4 analyzes program outcomes related to achieving NSF's SBIR goals. Chapter 5 describes and analyzes in some detail the NSF Phase IIB program, an important initiative that provides matching funds for commercially viable projects at the end of Phase II. Chapter 6 draws on company case studies and on the textual responses from survey respondents to provide a qualitative picture of program operations, issues, and possible solutions. Chapter 7 provides our findings and recommendations.

End-of-chapter Annexes provide additional detail about the chapter topics, and six report appendixes provide additional information. Appendix A sets out an overview of the methodological approaches, data sources, and survey tools used in this assessment. Appendix B describes key changes to the SBIR program from the 2011 Reauthorization. Appendix C reproduces the 2011 Survey instrument. Appendix D shows the 2010 Phase IIB Survey instrument. Appendix E presents the case studies. Appendix F contains a list of references.

2

NSF and Its SBIR Program

NSF was created as an independent federal agency by Congress in 1950, and assigned the mission of supporting research and education for all fields of fundamental science and engineering, except medical sciences, for the purpose of advancing the progress of science while benefitting the nation. Rather than employ its own researchers, NSF supports researchers and educators in their home institutions—colleges, universities, K-12 school systems, businesses, and research and science organizations—throughout the United States through grants and cooperative agreements.

NSF provides funding after receiving proposals requesting specific amounts of support for well-defined, meritorious projects, usually submitted by applicants in response to various NSF funding opportunities announced on the NSF website. A merit review process is used to select proposals for funding from the more than 42,000 proposals currently received yearly by NSF in all its programs. With an annual budget slightly more than $7 billion in FY 2015, NSF funds on the order of 200,000 scientists, engineers, educators and students, and laboratories and field sites yearly.[1]

NSF is organized into seven directorates: Biological Sciences; Computer and Information Science and Engineering; Education and Human Resources; Engineering; Geosciences; Mathematical and Physical Sciences; and Social, Behavioral and Economic Sciences. Within the NSF Directorate for Engineering are five Divisions, one of which is the Division of Industrial Innovation & Partnerships (IIP), which is home to a portfolio of partnership programs—including the SBIR and the STTR

[1]National Science Foundation website, <*http://www.nsf.gov/*>. Website was accessed on April 27, 2015.

programs—which aim at advancing technological innovation across all disciplines for high impact through commercial development, including the SBIR and the STTR programs. The IIP focuses on fostering partnerships across all disciplines to advance technological innovations that have the potential for high impacts through commercial development. In addition to the SBIR and STTR programs, IIP manages several other partnership programs, as well as the Industry/University Cooperative Research Centers (I/UCRC) program, and the Innovation Corps Program (I-Corps). The I-Corps Program helps to equip scientists with the entrepreneurial tools needed to realize commercial potential of innovative technologies, and is considered particularly relevant to the STTR program. (See Box 2-3 and related discussion.) A focus of the set of programs comprising NSF's IIP portfolio is to help startups and other small companies survive what has come to be called "The Valley of Death": the period of difficulty and high failure rates between research on a promising concept and bringing that concept to success as a business.

The Division of Industrial Innovation & Partnerships (IIP) sets the objectives for the NSF SBIR program and works to balance the congressional objectives of the program with mission of the NSF. The NSF SBIR program operates independently from the academic divisions of NSF, but works closely to collaborate with them where possible. NSF program officers that are responsible for managing SBIR are required to remain current on initiatives in their fields across the

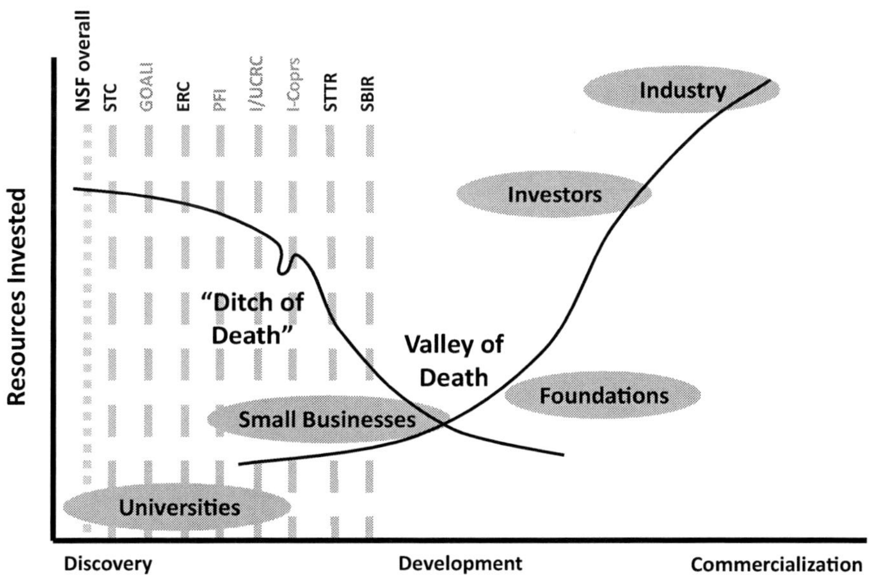

FIGURE 2-1 Portfolio of NSF programs to address the "Valley of Death."
SOURCE: Presentation by Barry Johnson, May 1, 2015 at the Academies workshop on the STTR program, Washington, DC.

foundation, and NSF staff from outside the division can act as reviewers for the SBIR program.

SBIR topics change only incrementally at NSF from solicitation to solicitation and are defined quite broadly. Suggestions for topics from non-SBIR staff at NSF are welcome, but there is no formal approval structure for topics that involves NSF staff from outside the program. Phase I applicants are in addition encouraged to describe how their proposed work links back to academic work funded by NSF.

With this brief overview of the NSF and location of the SBIR program within NSF's Engineering Directorate's IIP Division, this chapter reviews aspects of program management of the SBIR at NSF. A focus is on what can be called the "NSF Model" of SBIR program management. In addition, the chapter discusses important features of the program, feedback about the impact of program directors, and the involvement of women and minorities in the program. Additional support via Phase IIB awards, a distinguishing feature of the program, is discussed separately in Chapter 5.

Sources for this chapter include interviews with staff members, information from the survey and case studies, and documentation from NSF, notably the NSF and SBIR websites and reports of the SBIR Advisory Committee. To our knowledge, there are no formal internal assessments by the National Science Foundation or the National Science Board of the SBIR program at NSF.[2]

FUNDING DECISIONS

The decision to fund some projects and not others is at the heart of competitive awards such as the NSF SBIR program. These decisions are not easy to make—they reflect the interplay among different objectives, the selection process, the backgrounds of key decision makers, and the institutional framework that influences what kinds of projects should be funded.

Topic Selection

The issue of topic selection appears to have evolved at NSF in recent years. One program director (PD) observed that it is "unlikely that a good project would not find a topic." Several case study interviews with companies nonetheless suggested a perception that NSF topics are designed to be quite narrow as a means to pare down the number of applications. However, interviews with NSF staff revealed that topics are primarily used to identify areas of particular interest and are not seen, internally at least, as a means of screening out proposals or of reducing the number of applications to manageable levels.

PDs confirmed in interviews that NSF runs an "open" competition: proposals

[2]Discussion with a senior advisor to the NSF SBIR program on July 7, 2015.

that do not align with a published topic are accepted (unlike at the Department of Defense [DoD] or the Department of Energy [DoE]). As a result, any innovative project has the potential to be funded through the NSF SBIR program, which is especially important given the very wide technical area covered by NSF. Limiting the topics under which companies can apply for SBIR funding would inevitably screen out potentially important innovations and would in effect substitute the market insights of NSF staff for those of company personnel.

Attention goes into the selection of topics Dr. Jesus Soriano (a PD at NSF) observed that in health and medical technologies considerable efforts are made to align SBIR technologies with the academic priorities funded by other NSF programs: "the SBIR program can be seen as one endpoint of the NSF funding pipeline."[3]

Generally, new topics are published every six months. Discussion within the agency begins several months earlier, seeking to identify emerging areas of technology. According to Dr. Prakash Balan, the process provides a catalyst for staff to submit areas on which they will be working. However, he confirmed that topics should be viewed as seeds to spur innovative ideas rather than as a means of restricting applications.[4]

The development process has changed somewhat in recent years. Until 2012, it was organized around topic clusters, under the coordination of cluster leaders. Current practice has abolished the position of cluster leader and has replaced clusters with nine broad topic areas:[5]

- Educational Technologies and Applications (EA)
- Information and Communication Technologies (IC)
- Semiconductors (S) and Photonic (PH) Devices and Materials
- Electronic Hardware, Robotics and Wireless Technologies (EW)
- Advanced Manufacturing and Nanotechnology (MN)
- Advanced Materials and Instrumentation (MI)
- Chemical and Environmental Technologies (CT)
- Biological Technologies (BT)
- Smart Health (SH) and Biomedical (BM) Technologies

Each of these topic areas has further subtopics. The following subtopics for Educational Technologies and Applications were included in the most recent solicitation. The breadth of these topics indicates that NSF is making a concerted effort to provide a very broad platform for the development of new technologies:

- EA1. General Education Applications
- EA2. Global, Distance, and Cyber-learning Education Applications

[3]Dr. Jesus Soriano, program director, interview, May 2, 2014.
[4]Dr. Prakash Balan, program director, interview, May 2, 2014.
[5]See NSF topics webpage, accessed May 19, 2014, <*http://www.nsf.gov/eng/iip/sbir/topicshome.jsp*>.

- EA3. Simulations and Gaming Technology Applications
- EA4. Entrepreneurship Education Applications
- EA5. Tools for Learning and Assessment
- EA6. Information Management and Technology for Education

This broad platform is further expanded by the text of the solicitation itself (see Box 2-1). The wording "The topics *can* include . . ." (emphasis added) indicates that even the wide range of topics published in the solicitation is not meant to be limiting. Though intended to be broad and inclusive, NSF needs to ensure from time to time that this organization and processes are sufficient to keep up with evolving opportunities, and that they are sufficiently timely.

What is not entirely clear is whether the broad pool of applicants and potential applicants understand that the door is open to projects not explicitly covered in the solicitation. Although the language on the topics webpage indicates that NSF is open to such projects, it is probably fair to say that the language is not welcoming: "Certain innovative technologies with high commercial potential may not appear to fit under the nine current solicitation topics or corresponding subtopics. In this case, you may seek advice from the relevant Program Director

BOX 2-1
Details of One NSF Subcomponent

EA4. Entrepreneurship Education Applications
Topics can include

1) entrepreneurship education and training integrating "diverse topics, such as strategic planning, business model development, opportunity recognition, product entry, intellectual property, project management, legal requirements, and business constraints" in innovative ways for success in the contemporary global economy;[a]
2) personal learning environments that allow students to control and experiment with entrepreneurial situations in relation to their personal learning style to acquire knowledge; and
3) innovative tools that enable entrepreneurs and educators to learn or judge the effectiveness and validity of external resources for research, product launch, and effective operations of technological and education related products and services.

[a]National Science Foundation, *<http://www.nsf.gov/eng/iip/sbir/topics/EA.jsp>*. Accessed May 29, 2015.
SOURCE: NSF SBIR/STTR Solicitation, Spring 2014.

(as indicated below) or you may submit the proposal under the topic and subtopic that is the closest match."[6]

Pre-Application Feedback

In recent years, NSF has encouraged potential applicants to submit by email a short executive summary of the project to a relevant program director. The objective is to help the applicant gauge whether a project is likely to meet NSF's selection criteria and hence whether it is a good candidate for funding.

The NSF webpage says that the executive summary should discuss "the following aspects of the project: (1) the company and team, (2) the market opportunity, value proposition, and customers, (3) the technology/innovation, and (4) the competition."

Although not required, potential applicants are encouraged to seek pre-application feedback, because it offers them the opportunity to avoid the cost of application for projects that have little chance of funding. Responses from PDs vary. One PD explained that he has developed an FAQ that covers most topics for most applicants, to which he adds paragraphs specific to the subject solicitation. He stated that applicants appear to like this approach because it provides a good mix of specific and general information.

It was not entirely clear from interviews whether the pre-application process helps companies decide whether or not to apply. Furthermore, NSF does not collect data related to the pre-application process. Still, given the dominant role of PDs in making final selections for funding from recommendations made by a peer review process, companies should seek to take advantage of this feedback mechanism to create links to the PDs.

Selection

NSF takes the peer-review process seriously and includes both technical and commercial reviewers in the peer-review panel: "Groups of technical and commercial experts from around the country participate in confidentially evaluating proposals according to the review criteria below. SBIR/STTR Program Directors use this input to help decide which projects to fund."[7] The merit criteria are provided on the NSF website and in the solicitation instructions (see Box 2-2).

Reviewers are selected for technical and commercial expertise. All are required to sign conflict of interest forms. Commercial reviewers are expected to consider the basic selection criteria as well as additional prompts provided by NSF in the form of a set of questions, which fall into the following four areas

---

[6]NSF SBIR/STTR solicitation topics, accessed May 19, 2014, *<http://www.nsf.gov/eng/iip/sbir/topicshome.jsp>*.

[7]NSF SBIR Information for Applicants, accessed May 1, 2014, *<http://www.nsf.gov/eng/iip/sbir/phase_I.jsp>*.

BOX 2-2
Merit Criteria for NSF SBIR/STTR Awards

"Criterion 1: What is the intellectual merit of the proposed activity?
This criterion addresses the overall quality of the proposed activity to advance science and engineering through research and education.

- Is the proposed plan a sound approach for establishing technical and commercial feasibility?
- To what extent does the proposal suggest and explore unique or ingenious concepts or applications?
- How well qualified is the team (the Principal Investigator, other key staff, consultants, and sub-awardees) to conduct the proposed activity?
- Is there sufficient access to resources (materials and supplies, analytical services, equipment, facilities, etc.)?
- Does the proposal reflect state-of-the-art in the major research activities proposed? (Are advancements in state-of-the-art likely?)
- **For Phase II proposals only**: As a result of Phase I, did the firm succeed in providing a solid foundation for the proposed Phase II activity?

Criterion 2: What are the broader impacts of the proposed activity?
This criterion addresses the overall impact of the proposed activity.

- What may be the commercial and societal benefits of the proposed activity?
- Does the proposal lead to enabling technologies (instrumentation, software, etc.) for further discoveries?
- Does the outcome of the proposed activity lead to a marketable product or process?
- Evaluate the competitive advantage of this technology vs. alternate technologies that can meet the same market needs.
- How well is the proposed activity positioned to attract further funding from non-SBIR sources once the SBIR project ends?
- Can the product or process developed in the project advance NSF´s goals in research and education?
- Does the proposed activity broaden the participation of underrepresented groups (e.g., gender, ethnicity, disability, geography, etc.)?
- Has the proposing firm successfully commercialized SBIR/STTR-supported technology where prior awards have been made? Or has the firm been successful at commercializing technology that has not received SBIR/STTR support?"

SOURCE: National Science Foundation, accessed *<http://www.nsf.gov/eng/iip/sbir/peer_review.jsp>*. Accessed May 19, 2014.

and reflect a considerable understanding of commercialization requirements for innovative products:[8]

- Market opportunity
- Company and team
- Product/technology and competition
- Revenue and finance plan

OTHER ASPECTS OF THE NSF SBIR PROGRAM

Innovative Partnerships

NSF's efforts to support commercialization include funding for a range of supplementary activities beyond the basic Phase II award and related activities. The Phase IIB program is the best known of these and is discussed separately in this report.

NSF also supports other commercial partnerships. It provides up to 20 percent of the Phase II award ($150,000 maximum) for "additional research that goes beyond the Phase II project's objectives to meet the technical specifications or additional proof-of-concept requirements. Additional research is anticipated to enhance the commercial potential and lead to partnerships with industrial partners and venture and angel investors."[9]

NSF also provides funding for direct commercialization assistance, up to $10,000 per Phase II award. In line with the recent reauthorization, this funding is now available for the company to spend on its own selected advisors.

Beyond commercialization, NSF uses supplementary funding to support other program objectives. It funds institutional partnerships between Phase II awardees and the following various other institutions that may offer opportunities to encourage participation by woman and minority students:

- Minority-serving institutions that have a funded Center for Research Excellence in Science and Technology (CREST) or Historically African-American Colleges and University (HBCU) Research Infrastructure for Science and Engineering (RISE) awards ($100,000 per award maximum)
- Community college researchers and students ($40,000 per year maximum)
- Engineering Research Center (ERC) faculty, researchers, and graduate students ($200,000 per year maximum).

---

[8]NSF SBIR/STTR Phase I Commercial Review Form (revised May 2008), n.d.

[9]NSF "Phase II Supplemental Award Opportunities," accessed May 23, 2014, <*http://www.nsf.gov/eng/iip/sbir/Supplement/index.jsp*>.

NSF also funds what it sees as ways of using SBIR to expand educational opportunities. It provides funding for education including the following mechanisms:

- High School Participants (RAHSS), which supports active research participation in SBIR/Small Business Technology Transfer (STTR) projects by high school students interested in science, technology, and mathematics ($6,000 maximum)
- Undergraduate Participants (REU) are funded to participate in SBIR/STTR research ($8,000 per student annually maximum).

NSF has also reached out to the U.S. Patents and Trademark Office (USPTO) to develop a partnership to support companies seeking to protect their intellectual property. This is especially significant because the cost of patenting has traditionally not been covered by SBIR awards. USPTO representatives were available at the recent Phase I awardees boot camp, which is discussed below.[10]

Although Phase I does not provide the opportunities for developing relationships that are apparent in Phase II and beyond, NSF provides some additional channels through which companies can seek support. One PD noted that he makes an effort to contact the Phase I PI by phone 20-30 days before the award starts for a 45- to 60-minute call. He underscores that this is a good opportunity to tell the world that the company received NSF funding both through a press release by the NSF media office listing awards made and by company press releases.

NSF is clearly committed to working during Phase I to ensure that commercial outcomes are maximized. During the most recent award cycle, NSF built on its existing Phase I awardees conference to include a boot camp organized around the principles of the relatively new iCorps program highlighted in Box 2-3. The focus was on ensuring that awardees made the effort necessary to fully connect with potential customers, and to ensure that the eventual product would meet real needs in the marketplace.[11]

The NSF SBIR/STTR Advisory Committee reviewed the I-Corps approach and concluded that "I-Corps has the potential to accelerate the development curve for STTR grantees, because it promises to shorten the time delay (a.k.a. Valley of Death) between research and customer validation/revenue. Another benefit is to capture and capitalize on the value of prior NSF research."[12]

---

[10]Dr. Prakash Balan, interview, May 2, 2014.

[11]Dr. Joseph Hennessy, NSF, interview, March 27, 2014. More information on the NSF Innovation Corps ("iCorps") program is available from NSF at <*http://www.nsf.gov/news/special_reports/i-corps/*>.

[12]NSF SBIR/STTR Advisory Committee, minutes of May 2012 meeting, May 9-12, 2012, p. 8.

BOX 2-3
The NSF I-Corps

Although I-Corps is not designed explicitly for SBIR awardees and is not part of the NSF SBIR program, it is closely aligned and if successful could become an integral part of the program. According to NSF,

"The NSF Innovation Corps (I-Corps) is a set of activities and programs that prepares scientists and engineers to extend their focus beyond the laboratory and broadens the impact of select, NSF-funded, basic-research projects."

"While knowledge gained from NSF-supported basic research frequently advances a particular field of science or engineering, some results also show immediate potential for broader applicability and impact in the commercial world. Such results may be translated through I-Corps into technologies with near-term benefits for the economy and society."

"Combining experience and guidance from established entrepreneurs with a targeted curriculum, I-Corps is a public-private partnership program that teaches grantees to identify valuable product opportunities that can emerge from academic research, and offers entrepreneurship training to student participants."

"I-Corps Teams—composed of academic researchers, student entrepreneurs and business mentors—participate in the I-Corps curriculum administered via online instruction and on-site activities through one of several I-Corps Nodes."

SOURCE: National Science Foundation, *<http://www.nsf.gov/news/special_reports/i-corps/>*.

Funding Gaps

In some cases, the flow of funding from the agency to the awardee can be interrupted between award phases. In some cases, these funding gaps can cause or contribute to added costs, disruption, and even project abandonment. This section addresses the scale and impacts of funding gaps that can develop between Phase I and Phase II of an SBIR award. This problem is especially challenging for small firms, which are less likely to have other funding sources that can be used to keep projects alive until Phase II funding arrives. This section does not address opportunities that were lost to funding gaps, or changes that could be made to reduce them.

NSF has introduced a number of initiatives in recent years to address the problems of gaps in funding between award phases. However, as Table 2-1 shows, about three-quarters of Phase II respondents indicated that their company experienced a gap between the end of Phase I and the start of Phase II for the surveyed award, and the funding gap had a range of consequences for the company. Table 2-2 indicates the types of impact on companies that experienced a funding gap.

TABLE 2-1 Funding Gap Between Phase I and Phase II

Experienced a Gap in Funding Between Phase I and Phase II	Phase II (Percent)
Yes	74
No	26
Total	100
N =	408

SOURCE: 2011 Survey, Question 22.

TABLE 2-2 Effects of Funding Gaps Between Phase I and Phase II

Effects Experienced	Phase II (Percent)
Stopped work on this project during funding gap	29.6
Continued work at reduced pace during funding gap	50.5
Continued work at pace equal to or greater than Phase I pace during funding gap	12.6
Received bridge funding between Phase I and II	5.0
Company ceased all operations during funding gap	0.7
Other	1.7
Total	100
N (Respondents reporting a PI-PII gap) =	301

SOURCE: 2011 Survey, Question 23.

Almost 30 percent of respondents reported that their company stopped work on the project during this period, and more than 50 percent reported a reduced level of effort. However, 12 percent reported that company efforts did not diminish. Only 5 percent reported receipt of bridge funding. Less than 1 percent reported that the company ceased all operations.

Aside from the obvious direct impact of delayed projects, funding gaps can have long-term consequences especially for smaller companies; where in some cases there is insufficient work to retain key project staff during the gap period.

Amount of Funding

Although there are obvious limitations to the utility of asking recipients whether the amount of money provided was sufficient for the project at hand, there is at least some value in determining the extent of affirmative responses.

In this case, about 55 percent of Phase II respondents indicated that the funding was sufficient; about 45 percent indicated that more funding was required. None reported that the funding was more than necessary (see Table 2-3).

Size of Awards

Although awardees often suggest in other contexts (e.g., case study interviews) that the size of awards should be increased (a view especially prevalent before the changes made during the 2011 program reauthorization), the survey asked about the possible trade-off between the size of awards and the number of awards: unless NSF funding for the SBIR program increases, larger awards imply fewer awards. A plurality was in favor of not making that trade-off: 44 percent opposed it while 29 percent were in favor (Table 2-4).

Size of Program

The survey also asked about the possible expansion of the SBIR program itself. About two-thirds of respondents supported an increase in the size of the

TABLE 2-3 Adequacy of Phase II Funding

Amount of Funding Provided for Phase II Award Was	Phase II (Percent)
More than enough	0.0
About the right amount	55.3
Not enough	44.7
Total	100.0
N =	179

SOURCE: 2011 Survey, Question 42.

TABLE 2-4 Tradeoff of Larger Awards for Fewer Awards

Would Prefer More Funding Per Award Even If Fewer Awards	Phase II (Percent)
Yes	28.5
No	43.6
Not sure	27.9
Total	100.0
N =	179

SOURCE: 2011 Survey, Question 43.

TABLE 2-5 Increasing the Size of the SBIR Program

Would recommend that the Program be	Phase I (Percent)	Phase II (Percent)
Expanded (with equivalent funding taken from other federal research programs that you benefit from and value)	66.2	70.4
Kept at about the current level	31.4	29.1
Reduced (with equivalent funding applied to other federal research programs you benefit from and value)	0.5	0.6
Eliminated (with equivalent funding applied to other federal research programs you benefit from and value)	2.0	0.0
Total	100.0	100.0
N =	204	179

SOURCE: 2011 Survey, Question 44.

program even if funding were taken from other federal programs that they value. There were no statistically significant differences between Phase I and Phase II responses (see Table 2-5).

WOMEN AND MINORITIES

One of the four congressional objectives for the SBIR/STTR program is "to foster and encourage participation by minority and disadvantaged persons in technological innovation."[13] Within the SBIR program, the relevant metric is understood to be company ownership,[14] and SBA defines "minority and disadvantaged persons" in the context of the SBIR program as "socially and economically disadvantaged" (SED), including women, designated as "Woman-owned Small Businesses" (WOSB), and people from certain racial/ethnic groups, designated "Minority-owned Small Business (MOSM).[15] MOSBs include the following racial/ethnic groups:[16]

[13]Public Law 97–219, § 2, July 22, 1982, 96 Stat. 217.

[14]Small Business Administration, SBIR/STTR Policy Directive, February 24, 2014, p. 3.

[15]See the SBA description of the SBIR program, <*https://www.sbir.gov/about/about-sbir*> (accessed March 31, 2014) and the SBA definition of Socially and Economically Disadvantaged Individuals (SEDs) <*http://www.sba.com/sba_8%28a%29.htm*>, (accessed on March 31, 2014). Data from NSF rely on self-identification of demographic characteristics by companies during the application process. Self-identification is voluntary, and there is anecdotal evidence that incentives may encourage applicants to over-report or under-report.

[16]See Small Business Administration, "Definition of Socially and Economically Disadvantaged Individuals" at <*http://www.sba.com/sba_8%28a%29.htm*>.

- African-Americans;
- Hispanic Americans (persons with origins from Latin America, South America, Portugal, and Spain);
- Native Americans (American Indians, Eskimos, Aleuts, and Native Hawaiians);
- Asian Pacific Americans (persons with origins from Japan, China, the Philippines, Vietnam, Korea, Samoa, Guam, U.S. Trust Territory of the Pacific Islands [Republic of Palau], Commonwealth of the Northern Mariana Islands, Laos, Cambodia [Kampuchea], Taiwan, Burma, Thailand, Malaysia, Indonesia, Singapore, Brunei, Republic of the Marshall Islands, Federated States of Micronesia, Macao, Hong Kong, Fiji, Tonga, Kiribati, Tuvalu, or Nauru);
- Subcontinent Asian Americans (persons with origins from India, Pakistan, Bangladesh, Sri Lanka, Bhutan, the Maldives Islands, or Nepal).

For ease of reading, the term "minority" or abbreviation "MOSB" will be used to denote membership in at least one of these groups.

Although the WOSB and MOSB formulation has been traditional among SBIR stakeholders, it has several unfortunate consequences:

- It focuses attention entirely on company ownership rather than on "participation" as stated in the statute. There are many different ways to participate in the program, only one of which is ownership.
- It replaces "minority and disadvantaged persons" with "socially and economically disadvantaged small businesses," which aligns the program with SBA definitions of socially and economically disadvantaged rather than with the minority needs at the forefront of the congressional objective.

As a result, agencies disregard all participation other than ownership; for example, no agency maintains data about woman and minority principal investigators. Furthermore, SBA definitions of "socially and economically disadvantaged" have the effect of largely obscuring agency performance in support of the congressional objective.

Analyzing Participation by Minorities and Women

The participation of woman- and minority-owned companies in NSF's SBIR program is informed by data provided by NSF directly, by data obtained by Academies[17] survey, and by an Academies workshop on diversity. In analyzing the data, the following metrics are of particular interest: MOSB and WOSB ap-

---

[17]Effective July 1, 2015, the institution is called the National Academies of Sciences, Engineering, and Medicine. References in this report to the National Research Council, or NRC, are used in a historic context identifying programs prior to July 1.

plications and awards as percentages of overall applications and awards, and the share of MOSBs and WOSBs in the program. Broad conclusions based on the data and workshop evidence are presented here. Supporting tables of the underlying NSF data are provided in detail in Chapter 3. Data tables from the Academies survey are provided in this chapter. Key findings of the Academies workshop on diversity are captured here, but details are in a separate report.[18]

Evidence Provided by NSF Data

NSF data for Phase I provide strong evidence that access to the program by MOSBs is low and not improving; in fact, on almost all relevant metrics, it appears as though access is declining:

- Applications from MOSBs have been flat or have declined slightly since the early 2000s (depending on base year).
- For every year from 2003 to 2012, Phase I success rates (defined as awards as a percentage of applications) were lower for MOSBs than for non-MOSBs—with an average difference of about 6 percentage points.
- The MOSB share of SBIR awards fell steadily from 2003 to 2012, a function of declining numbers of applications and relatively low success rates.

Overall MOSBs have received few awards from NSF.

Although WOSBs have had somewhat more success than MOSBs in Phase I, they have been somewhat less successful than all other applicants:

- Applications from WOSBs in Phase I have broadly mirrored patterns for applications as a whole. However, there was a slight increase in the percentage of applications received from WOSBs during the past 4 years of the study period.
- With one exception, success rates were consistently lower for WOSBs than for non-WOSBs in Phase I. However, the average gap was considerably smaller than that for MOSBs, and in one year (2011) WOSB applications succeeded slightly more often than did non-WOSB applications.
- Yearly Phase I awards to WOSB fluctuated in number, ranging from approximately 19 to 47. Awards to WOSB ranged between approximately 10 and 18 percent, but mostly stayed below 14 percent of total yearly NSF

[18]See National Academies of Sciences, Engineering, and Medicine, *Innovation, Diversity, and the SBIR/STTR Programs* (Washington, DC: The National Academies Press, 2015), the summary report from workshop on "Innovation, Diversity, and Success in the SBIR/STTR Programs," Washington, DC, February 7, 2013. It should also be noted that there are additional sources of information about women and minorities, such as various conferences and books on women, including minority women and men, in STEM.

Phase I awards. NSF data for Phase II show addition evidence of poor performance for MOSBs and WOSBs:

- In 6 of the 10 years of Phase II applications studied, MOSBs had lower success rates than non-MOSBs; and overall had lower average success rates—32 percent for MOSB compared with 44 percent for non-MOSB.
- Over the study period, the MOSB share of Phase II awards declined, and averaged less than 10 percent.
- The numbers of applications from WOSBs increased during the study period, but the reported numbers fluctuated from year-to-year.
- Success rates for WOSBs—though slightly better than those for MOSB—in 6 of 10 years had a lower success rate than did non-WOSBs, and overall had a lower average success rate. Overall, the WOSB share of awards fluctuated from year-to-year, but appears to have trended upward, ranging from near 6 percent at the beginning of the study period, to about 13 percent at the end of the period.

These data would seem to suggest that outreach to woman- and minority-owned companies has not been effective in increasing applications. Furthermore, success rates suggest that neither of these groups has been as successful as other companies in the selection process for awards. At a minimum, NSF should seek an explanation for the observed differences, including gauging the impact of NSF initiatives in this regard.

The data collected and provided by NSF, like those from other agencies, conceal as much as they reveal. The data do not capture other types of participation in the program—notably by PIs. Furthermore, the data are not disaggregated by type of minority, which has the effect of concealing the lack of participation of African-American- and Hispanic-owned companies. These issues are addressed via data drawn from the 2011 Survey, which are the subject of the next section.

Evidence Provided by Academies Survey Data

The 2011 Survey data are especially valuable in understanding the role of women and minorities because they extend beyond company ownership to include an important part of the talent pipeline leading to eventual SBIR awards: the principal investigators (PIs), who in some cases are company founders and in other cases move on to found companies themselves at a later date.

As with the previous Academies survey conducted for the earlier SBIR study, the 2011 survey asked if the PI for the surveyed project belongs to a socially and economically disadvantaged group (SED).[19] About 17 percent of Phase I and 15

---

[19]The Academies survey used SED terminology to focus participation on more than company ownership.

TABLE 2-6 Composition of SED PI Grouping, by Race/Ethnicity, as Percent of SED PIs

Race/Ethnicity of SED PIs	Phase I (Percent)	Phase II (Percent)
Asian-Pacific	45.7	45.2
Asian-Indian	22.9	48.4
Hispanic	22.9	6.5
African-American	8.6	0.0
Native American	0.0	0.0
Other	2.9	3.2
N =	35	31

NOTE: Responses do not add to 100 percent, because respondents were permitted to check more than one category.
SOURCE: 2011 Survey, Question 14C.

percent of Phase II respondents reported that this was the case for their project.[20] The survey then asked respondents to provide details about the PI's racial/ethnic background, using the detailed categories drawn from SBA definitions, with the addition of an "other" category to ensure that all respondents who wished to claim SED status could do so. (Results for women are presented after the results for racial/ethnic groups.)

Responses revealed that more than 90 percent of Phase II projects with an SED PI were Asian-Pacific or Asian-Indian. No PIs were African-American. None were Native American. Moreover, the share of Hispanic PIs declined sharply from 22.9 percent for Phase I to 6.5 percent of Phase II. It must be understood, however, that these percentages are calculated from low absolute numbers—only 35 Phase I and 31 Phase II reported that their PI was SED. Still, the total absence of African-American and Native American PIs in Phase II and the sharp decline in Hispanic PIs across Phases are causes for concern (see Table 2-6).

These data can be placed in the context of the survey population as a whole. Overall, of the 386 respondents to this question, 2.6 percent said that the PI on their project was Hispanic (total: 10) and less than 1 percent said that the PI was African-American (total: 3); none reported that the PI was Native American (see Table 2-7).

Turning from the ethnicity of the PIs to the ethnicity of the owners of surveyed companies, approximately 16 percent of Phase I respondents and 12 percent of Phase II respondents indicated that the company was majority owned by members of SED groups at the time of the award (see Table 2-8).

[20] 2011 Survey, Question 14B.

TABLE 2-7 SED PIs, by Race/Ethnicity, as Percent of All Respondents

Race/Ethnicity of SED PIs	Phase I (Percent)	Phase II (Percent)
Asian-Pacific	7.8	7.7
Asian-Indian	3.9	8.3
Hispanic	3.9	1.1
African-American	1.5	0.0
Native American	0.0	0.0
Other	0.5	0.6
N =	205	181

SOURCE: 2011 Survey, Question 14C.

TABLE 2-8 SED Ownership, Percentage of Respondents

Company Ownership Was Majority SED at the Time of Award	Phase I (Percent)	Phase II (Percent)
Yes	16.1	12.3
No	83.9	87.7
Total	100	100
N =	205	408

SOURCE: 2011 Survey, Question 19B.

However, probing more deeply into the ethnic distribution of SED company ownership allows for further identification of issues. Overall, this distribution is quite similar to that for SED PIs, in that more than 70 percent of both Phase I and Phase II respondents reported majority owners of Asian-Indian and Asian-Pacific ethnicity. Both Hispanic and African American owners garnered a higher share of Phase I than Phase II responses (see Table 2-9). Again, the numbers involved are very small: respondents reported three Phase I and one Phase II African-American-owned companies and seven Phase I and three Phase II awards for Hispanic-owned companies, out of 386 awards surveyed. However, this disaggregation of the SED category reveals very low percentages of African-American-owned companies winning SBIR awards and this requires closer examination.

The discrepancy between Phase I and Phase II responses for both African-American- and Hispanic-owned companies is an additional area for further analysis. We might expect to see approximately the same percentage for both Phases, but differences in percentages are also seen across phases for other groups, such as the Asia-Pacific group.

TABLE 2-9 SED Company Ownership, by Race/Ethnicity and Award Phase

Race/Ethnicity of Company Ownership at Time of Award	Phase I (Percent)	Phase II (Percent)
Asian-Indian	33.3	54.2
Asian-Pacific	39.4	20.8
Hispanic	21.2	12.5
African-American	9.1	4.2
Native American	0.0	0.0
Other	3.0	8.3
N =	33	24

NOTE: Responses do not add to 100 percent because respondents were permitted to check more than one category.
SOURCE: 2011 Survey, Question 19C.

The 2011 Survey also sought information about the gender of PIs. Respondents reported that about 13 percent of PIs were women, for both Phase I and Phase II awards.[21]

The survey also addressed the extent to which SBIR awards are made to woman-owned businesses. These data are provided in Table 2-10.

The difference in percentage woman-owned company between Phase I and Phase II awards may be sufficient to warrant additional research and analysis by the agency.

The Challenge of Improving Diversity[22]

Recognizing that small businesses often introduce the radical ideas that can transform industries and markets, and the need to mobilize all skilled individuals, regardless of race/ethnicity or gender, strengthens the economy and the nation, the committee convened a workshop to draw attention on participation of women, minorities, and both older and younger scientists, engineers, and entrepreneurs in the SBIR program and to identify mechanisms for improving their participation rates.[23] The workshop also drew attention to the fact that improving the participation of women and minorities in the SBIR program is a part of a broader national challenge. See Box 2-4.

---

[21]2011 Survey, Question 14A.

[22]National Research Council, Workshop on "Innovation, Diversity, and Success in the SBIR/STTR Programs," Washington, DC, February 7, 2013.

[23]Ibid, p. 5.

TABLE 2-10 Woman-Owned Businesses

Company Was Majority Woman-Owned at the Time of Award	Phase I (Percent)	Phase II (Percent)
Yes	17.1	11.3
No	82.9	88.7
Total	100.0	100.0
N =	205	408

SOURCE: 2011 Survey, Question 19A.

BOX 2-4
Expanding Participation of Women and Minorities in STEM

The 2011 publication by the National Research Council, *Expanding Underrepresented Minority Participation: America's Science and Technology Talent at a Crossroads*, notes that underrepresented minorities, defined here as Hispanics, African Americans, Native Americans/Alaska Natives, comprise a small percentage at each step of the science, technology, engineering, and mathematics (STEM) education process. The percentage of African Americans and Hispanics interested in STEM undergraduate majors is similar to those of white and Asian Americans, but their completion rates are much lower. At the graduate school level for science and engineering, underrepresented minorities receive only 14.6 master's degrees and 5.4 percent of doctoral degrees. Data from the National Science Board indicates that women earn roughly half of S&E degrees at the bachelor's, master's, and Ph.D. levels, but they earn "fewer than one-third of the doctorates awarded in physical sciences, mathematics and computer sciences, and engineering" and less than a quarter of engineering master's degrees.

Participants at the workshop examined broad demographic trends in the science and engineering workforce and statistical measures from the SBIR program for women and minorities, and searched for pragmatic solutions to boost SBIR awards to women and minorities. The workshop highlighted the fact that women comprise 51 percent of the U.S. population and 27 per cent of STEM graduates, but woman-owned companies have received only about 6 percent of SBIR awards. Hispanics, African Americans, Asian Americans, and Native Americans together comprise 36 percent of the U.S. population and 26 percent of STEM graduates, but less than 10 percent of all SBIR awards. Beyond NSF's SBIR program, the current participation of women and minorities was found to be low and decreasing, and participation of African Americans and Hispanics is particularly low.

Further, participants at this workshop identified steps needed to stimulate participation by underrepresented populations, with a focus on expanding the applicant pool, eliminating barriers in grant applications and selection, and providing greater education and support for entrepreneurship training and commercialization efforts. Examples of specific ways to broaden participation cited included taking advantage of NSF initiatives to encourage SBIR Phase II grantees to hire high school and college students, graduate students, post-docs, and teachers from underrepresented groups; to provide more outreach to graduate students to make them aware of the program; to make it easier for people to apply; and to increase women and minorities on review panels.[24] (See Box 2-5.) Participants also saw the need to align and leverage resources and programs at the state level that aim at providing access and support to woman and minority owned businesses; and to team with other federal and state/local programs which are addressing this issue.

These efforts at NSF notwithstanding, participation rates for women and minorities remain low. In many ways, efforts by NSF staff to address this issue are limited by regulations that are aimed at protecting the identity, race, and sex of grantees and proposers. These rules are also designed to prevent special treatment for those classes of people based on race or sexual orientation.

NSF's SBIR Advisory Committee has recognized this issue some AdCom members have been working to expand the number of women and minorities participating in NSF SBIR program on their own time. Nonetheless, the data revealed by the committee's assessment requires a careful review and a more broad-based response to address this challenge.

Conclusions: Woman and Minority Participation in the NSF SBIR Program

Evidence from the NSF SBIR program and the Academies survey shows minority participation to be at very low levels in NSF SBIR and getting worse. It shows participation rates a little better for women, with some upward trending, but still below the non-woman group. Both are still well below potential. The previous lack of data about African-American, Native American, and Hispanic inclusion in the program concealed what is, by any account, a very disappointing outcome that does not meet congressional objectives of the SBIR program.

NSF acknowledges its limited efforts to address this issue in the past. A lack of sufficient administrative funding has made additional support difficult to fund, and perhaps as a result this objective simply has not been addressed head on. The most current minutes from the NSF SBIR Advisory Board reflect this concern: "We recommend that IIP [the NSF Division for Industrial Innovation and Partnerships] *focus on increasing the number of Phase I proposals submitted from*

[24]Ibid, pp. 53-54.

BOX 2-5
NSF Initiatives to Encourage Diversity in the SBIR Program

In her workshop remarks, Grace Wang, then director of the NSF's Industrial Innovation and Partnerships Division, highlighted her division's mission to drive U.S. innovation by investing in technology and its commercialization, an objective that cannot be accomplished without human talent. "The base of innovation capacity is people--that's the innovators and entrepreneurs," said Dr. Wang. That recognition drives the NSF's interest in broadening participation in science and engineering through the SBIR program. To channel more people into STEM careers, "first, we need to expand the talent pool, and second, retain the talent." said Ms. Wang. "We need to increase the pipeline and stop leakage of the pipeline," she said.

Dr. Wang indicated that according to Census Bureau data, 65 percent of the total U.S. population in 2008 was composed of the underrepresented in science and technology, defined as women, African Americans, Hispanics, Native Americans, and people with disabilities. Yet, these groups constituted only 33 percent of science and engineering occupations in 2006, according to NSF data.

To broaden participation, Dr. Wang noted that the NSF runs several initiatives to encourage SBIR Phase II grantees to hire high school and college students, post-docs, and K-12 and community college teachers:

- Research Assistantships for High School Students, where Phase II grantees hire high school students for up to $6,000 per student.
- Research Experience for Teachers, where Phase II grantees hire K-12 teachers for up to $10,000 per teacher to bring the culture of innovation and entrepreneurship back to the classroom.
- Community College Research Teams, where Phase II grantees receive up to $40,000 and spend at least 75 percent of this award to subcontract to a community college, working with at least one faculty member and one student as a team.
- Research Experience for Undergraduates, the most popular program among Phase II grantees, where companies hire college interns for up to $8,000 per student. About 40 percent of Phase II grantees have hired at least one such student.
- The Phase IIA program gives Phase II grantees $100,000 with the requirement to subcontract 70 percent of the grant to minority-serving institutions to conduct research together.
- The Small Business Post-Doc Research Diversity Fellowship enables post-docs to work for Phase II companies for up to $75,000 each.
- The new Veteran's Research Supplement enables Phase II companies to hire veteran high school and college students, teachers, and community college faculty for up to $10,000 per veteran.
- In addition, the NSF attempts to support existing principal investigators by providing networking and mentoring opportunities specifically for them at the annual grantees' conference.

underrepresented groups (emphasis in the original). We request that IIP collect and present the number of Phase I proposals submitted from underrepresented groups in each AdCom meeting."[25]

THE NSF MODEL: DEDICATED RESOURCES FOCUSED ON COMMERCIALIZATION

Most SBIR programs at other agencies are highly decentralized. At agencies funding a large number of awards, for example the National Institutes of Health and DoD, it is fair to say that individual components run their own SBIR programs with direction but not control from the agency SBIR office. Funding and selection decisions in particular are in the hands of components, not the central office. Even at DoE and NASA, individual program offices have considerable say in selection decisions and in program operations.

The NSF model has a number of innovative features and components. These include the following:

- Investment in a number of high-quality program directors (PDs), so that each PD is responsible for approximately 30 awards at any given point in time
- Close and ongoing connection between PDs and companies, especially for Phase II and beyond
- Selection for innovation and commercialization, with a particular emphasis on the latter
- Additional bridge funding through the Phase IIB program
- Additional commercialization assistance via third-party providers LARTA and Foresight
- Tracking of outcomes at specified points post-award, via the telephone surveys managed by a long-time consultant, George Vermont
- A range of partnership activities to leverage resources and capabilities

Individual PDs are effectively in charge of the entire grant stream from initial draft topic to the conclusion of Phase IIB, which is as far downstream as NSF SBIR funding can reach. The PDs make funding recommendations, based on results from the independent peer review and input from other program staff. Senior program executives and other PDs provide cross-checks and a critical scrutiny of recommendations. The "NSF Model" revolves around the role of the PD.

[25]Report of the Advisory Committee for Small Business Innovation Research (SBIR) and Small Business Technology Transfer (STTR) Programs, May 9-10, 2012, Box 1-1.

Role and Effectiveness of Program Directors

The survey responses provide considerable insight into the effectiveness of PDs. Given the unique role of the PD at NSF, it is understandable that respondents offered a range of views, although overall they were highly positive about the role and impact of PDs.

Respondents reported how often they engaged with their PD. A majority (58 percent) of Phase II respondents reported quarterly contact, while 15 percent reported monthly contact and 23 percent reported annual contact (see Table 2-11).

The survey also sought to determine how easy it was for companies to reach the PD with questions or concerns. Ninety percent of Phase II recipients found it easy or very easy to reach their PD (see Table 2-12).

Interviews and open-ended comments (see below) revealed that some PDs had very positive effects on their awardee companies, while others were of little help.

TABLE 2-11 Frequency of Contact with PDs

PD Contact Cycle	Phase II (Percent)
Weekly	3.9
Monthly	15.0
Quarterly	57.8
Annually	23.3
Total	100.0
N =	206

SOURCE: 2011 Survey, Question 47.

TABLE 2-12 Ease with Which PI Could Contact PD

	Phase II (Percent)
Very easy	28.8
Easy	61.0
Hard	9.6
Very hard	0.6
Total	100.0
N =	177

SOURCE: 2011 Survey, Question 52.

The survey attempted to gauge the distribution of utility by asking respondents how helpful the PD had been to their project (see Table 2-13). Overall, more than 40 percent of Phase II respondents scored PD usefulness at 4 or 5 on a 5-point scale. Conversely, about 30 percent scored usefulness at 1 or 2 on the scale.

One important role of the PD is to provide technical advice to the awardee about the operations of the SBIR program. The program is complex, so a technically knowledgeable PD can be of great use especially to companies that are new to the program.

Given their other priorities, it is possible that PDs may not have enough time to devote to the projects they are supposed to be managing. However, 77 percent of respondents indicated that the PD's available time was sufficient or more than sufficient (see Table 2-14). However, in the open-ended questions, several respondents shared their concern that the PDs were carrying a heavy workload of companies (30 ongoing Phase II awards at any one time, in addition to Phase I, outreach, and selection responsibilities).

TABLE 2-13 Usefulness of the PD

Value of PD	Phase II (Percent)
Invaluable (5)	18.4
4	23.3
3	28.6
2	18.0
No help (1)	11.7
Total	100.0
N =	206

SOURCE: 2011 Survey, Question 48.

TABLE 2-14 PD Time Availability for Surveyed Project

	Phase II (Percent)
More than sufficient	6.4
Sufficient	70.3
Insufficient	23.3
Total	100.0
N =	202

SOURCE: 2011 Survey, Question 54.

Because a core responsibility for PDs is to explain program operations and to support companies through the paperwork, the survey asked about the technical capacity of the PD with regard to the SBIR program. Overall, respondents appeared satisfied; 75 percent indicated that the PD was extremely or quite knowledgeable about the SBIR program, and only 3 percent indicated that the PD was not at all knowledgeable (see Table 2-15). Evidence from open-ended responses suggests that newcomers to the program were more likely to find PDs helpful in dealing with programmatic issues.

PDs can also provide valuable support in a number of areas. They are hired in part because they are technically knowledgeable about specific science and engineering disciplines and therefore can provide valuable technical insights. However, only about 11 percent of respondents indicated that they received a substantial amount of technical help from the PD (scores of 4 or 5 on a 5-point scale); almost 50 percent reported minimal help (Table 2-16).

TABLE 2-15 PD Knowledge About the SBIR Program

	Phase II (Percent)
Extremely knowledgeable	34.8
Quite knowledgeable	39.2
Somewhat knowledgeable	23.0
Not at all knowledgeable	2.9
Total	100.0
N =	204

SOURCE: 2011 Survey, Question 49.

TABLE 2-16 PD Technical Support for Project During Phase II

Technical Help During Phase II	Phase II (Percent)
Most help (5)	2.9
4	8.2
3	19.3
2	22.2
Least help (1)	47.4
Total	100.0
N =	171

SOURCE: 2011 Survey, Question 50.2

PDs can also support awardees by introducing them to technical staff at universities. However, only about 15 percent of respondents reported receipt of this type of support (scores of 4 or 5 on a 5-point scale), while 53 percent reported no help in this area (see Table 2-17).

PDs are sometimes well positioned to provide useful connections to other companies—either other SBIR awardees or other companies with complementary interests or capabilities. Nineteen percent of surveyed companies indicated substantial support (scores of 4 or 5 on a 5-point scale) and 37 percent reported minimal help in this area (see Table 2-18).

TABLE 2-17 PD Support for University Connections

Introduction to University Staff During Phase II	Phase II (Percent)
Most help (5)	4.6
4	10.3
3	13.7
2	18.9
Least help (1)	52.6
Total	100.0
N =	175

SOURCE: 2011 Survey, Question 50.3.

TABLE 2-18 PD Connections to Other Private Firms

Introduction to Staff at Other Companies During Phase II	Phase II (Percent)
Most help (5)	3.4
4	15.5
3	17.8
2	26.4
Least help (1)	36.8
Total	100.0
N =	174

SOURCE: 2011 Survey, Question 50.4.

BOX 2-6
Positive Comments About NSF SBIR/STTR Program Directors

"My initial Program Officer was an invaluable source of information and direction. He guided us through the application process and throughout the funding period. He was very friendly, courteous, and respectful and a pleasure to work with."

"Dr. ___ is an extraordinary asset to NSF. The breadth of his industrial and governmental expertise was very helpful."

"All the program officers at NSF I have worked with have been extremely helpful and knowledgeable. One even introduced us to another company that had potential collaborative value."

"[Our PD] was very effective at encouraging [our] company to make the right company decisions and made himself available to answer questions even prior to submission of application."

"Having worked on SBIR funded projects from four different agencies, our experience with the Project Manager at NSF was the most comprehensive."

"He was just the greatest Program Officer we could have hoped for."

"Through Phase I and Phase II meetings I was also able to get coaching from NSF program managers. The NSF program managers have significant business experience that can be extremely valuable in advising companies."

SOURCE: 2011 Survey.

Open-Ended Responses Related to PD Activities

Survey respondents were offered the opportunity to provide open-ended comments related to their PD.[26] Although each response is different, it is possible to sort them into positive, negative, and neutral comments. About one-third of the comments were negative, one-half were positive, and one-quarter were highly positive.

Positive comments highlighted the connection between specific PDs and their companies, which was evidence for many respondents that the mentoring model (see NSF model section below) has had a positive impact. A sample of positive comments is provided in Box 2-6.

Many of the more negative comments were written by Phase I respondents. As the data indicate, about one-third of Phase II respondents had negative views of their interactions. Key criticisms included the following:

- Insufficient time. Many respondents (even those with positive comments) mentioned the heavy workload. A number indicated that they did not

[26]2011 Survey, Question 56.

have sufficient time with their PD and therefore did not receive the right guidance.

- Flawed Selection Process. A number of Phase I respondents complained that they received an unsatisfactory debrief or that the debrief indicated a flawed selection process. These complaints are likely normal for any grant program, but one respondent specifically noted a weak appeals process.
- Focus on Commercialization. Several interviewees and a number of survey commentators indicated that the current laser focus on commercialization might be counterproductive, excluding technically demanding projects. In interviews, PDs indicated an awareness of this problem and claimed to be addressing it.
- Lack of program guidance. Several respondents (mostly Phase I) complained that they received unsatisfactory and/or insufficient advice about their proposals and technical issues such as budgets. However, NSF provides substantial documentation on its website, and a more proactive approach from at least some of these companies likely would have addressed these issues.

A sample of negative comments is provided in Box 2-7.

Qualifications of PDs

PDs are not like SBIR staff at other agencies. They are specifically hired to manage a portfolio of SBIR projects. So the best comparison is not with staff at other agencies, but with staff at small venture or hedge funds. PDs are carefully selected, have strong backgrounds in business and technology commercialization, often in startups or other highly innovative firms, and are also high-level subject matter experts in specific science and engineering disciplines (see Box 2-8).

Topic Selection

As mentioned earlier in this report, NSF claims to offer a broad set of topics. One PD observed that it was "unlikely that a good project would not find a topic." Some case study interviews suggested that this is not entirely the case and that NSF topics are actually quite narrow, as a means of reducing the number of applications. However, this criticism may apply more to past solicitations than to current ones. This Committee's initial review of the May 2014 solicitation (which addressed areas related to educational applications) concluded that the topics were broad enough to attract a wide range of proposals. Moreover, each subtopic is phrased so that applications "can include" specific areas (see Box 2-9).

The PD is responsible not only for developing topics but also selecting proposals to be recommended for funding. The PD serves as an important mentor during the entire award cycle, providing in-person support that complements

BOX 2-7
Negative Comments about NSF SBIR Program Directors

"Most of the time I feel they are too busy to guide you through the hoops and loops of this whole process."

"Project Managers are always very busy. They seem to not be able to focus any great effort on the details of what we were developing."

"I didn't expect to be engaged too much with the Program Officer in the Phase I. However, I was disappointed by the fact that he seemed to have too heavy of a load and [was] clearly unable to provide adequate attention to any individual program. The initial kick-off meeting at the workshop was much too short and rushed and had little value."

"I would say that my PO's interest in my company and its product pretty much ended when the award was made. I met with him once at the mandatory SBIR PI conference but received no help or guidance from him."

"NSF Program officers focus too much on commercialization, too little on technical guidance. Despite this they have very few useful business contacts."

"Also, the IP-based business model is pushed too hard. In software it is obsolete and should be focused on a services model. The business people at NSF are confused about alternative business models like open source models."

"Our second project manager did not seem to have enough time or understanding of our products."

"Our technology was broad based, but NSF wanted us to find commercialization too fast without improving technology."

"The NSF SBIR Program Manager tried to help us through the administrative process effectively. However, in terms of help in our specific technology/science/commercialization I do not feel the Program Manager was adequately qualified. I do not think there was a clear understanding of entrepreneurship. I do not think there was any entrepreneur on the review panels."

SOURCE: 2011 Survey.

more general commercialization support provided through third parties. Evidence from interviews and the survey indicated that NSF PDs and funded companies develop closer relationship than those found at other agencies. Of course, this close relationship carries certain risks: one PD explained that he makes a point to not provide multiple Phase II awards to a single company in his domain.[27]

The NSF has invested a substantial amount in the administration of the SBIR/STTR program. The cost of the PDs is estimated at $2 million annually, which is provided from NSF's administrative budget, not the SBIR awards budget. This amount does not appear to be excessive for a program of this size.

[27]Steve Konsek, NSF PD, interview, March 14, 2014.

BOX 2-8
Background and Qualifications of NSF SBIR/STTR Program Directors

At the time of this writing, nine SBIR/STTR PDs are employed at NSF. They have deep knowledge of specific fields—for example, biomedical applications or energy. They are very highly qualified individuals. Academically, seven of the nine PDs have PhDs (three in physics, two in chemical engineering, and one each in genetics and cell biology, and medical sciences). They have published numerous scientific peer-reviewed papers, and several hold patents themselves or directed companies that developed patentable technologies. Two hold MBA degrees, one also holds an MD degree, one is a certified Microsoft systems engineer, and one has published well-received books on computer graphics now translated into five languages.

- All have strong backgrounds in the commercialization of technology. Individually, they have managed the development and commercialization of a number of optical fiber and optoelectronic component technologies for British Telecom, and were instrumental in establishing a commercialization center for specialized optical fibers at the University of Sydney;
- co-founded two successful companies, first an environmental engineering technology company and thereafter a biofuels company, which commercialized the PD's patents for industrial multiphase reactors applicable to chemical and biochemical production, as well as energy-efficient patented oxygenation technologies for biological wastewater treatment;
- served as chief technology officer at Illumitex, a venture-backed company developing light emitting diode chips, packages and fixtures for general illumination, after a period as chief of technical staff at Glo, recognized as one of Europe's top LED startups;
- worked in a range of sectors: developed subdivisions, new home construction, manufacturing computers, and custom programming for specific industry applications;
- founded as CEO a Bluetooth wireless product company, "raising equity capital for worldwide operations in the United States, China, and India. He designed, planned, and implemented the product development cycle and managed the marketing strategy, strategic alliances, and business development processes";
- worked as senior research technologist at Kodak for 26 years covering a broad range of materials science technologies where he received several Kodak achievement awards for scientific merit and commercial impact;
- led research and development at Micro Magnetics, focused on a development effort to commercialize a new family of high-performance magnetic microsensor products;
- founded, as "president and CEO, a successful venture-backed life science company (Gentra Systems, Inc.), that developed, manufactured, and sold products for genetic testing and research to clinical and research laboratories worldwide."

SOURCE: Summary and quotations drawn from discussions with six NSF program directors. NSF also provided biographical summaries of all the PDs for review.

BOX 2-9
Example NSF 2014 SBIR Topic Area

EA5. Tools for Learning and Assessment
Topics can include

1) tools and kits that empower students to become scientists, engineers, and educators; tools that allow them to design and build things and increase participation or demonstration in hands-on learning related to science, technology, engineering, math, and entrepreneurship of technical products and services;
2) adaptive learning environments combined with assessments that provide alternative paths of instruction;
3) gesture-based computing applications, semantic analysis, and tools that enable collaborative work with multiple students interacting on content simultaneously;
4) education tools that benefit from objects having their own IP address or location based services for new types of communications, assistive technologies, and new applications of benefit primarily to education;
5) devices that enable expanded dimensional learning such as 3D modeling, computer aided design (CAD), as well as new materials, technologies, and processes for learning and 3D printing suitable for educational settings;
6) Augmented Reality (AR) and tools that layer information over 2D and 3D spaces to provide new environments for learning;
7) wearable information centers, power sources, flexible displays, jewelry, glasses, output devices, and input tools that allow students to interface with computers and other devices in creative new ways that help overcome natural or physical barriers to learning;
8) sensors and systems that detect student engagement, frustration, or boredom while providing real-time feedback to both students and teachers.

SOURCE: National Science Foundation, 2014 SBIR Solicitation.

Overall, the NSF model is intelligently designed and executed by dedicated and highly credentialed staff that appears to be capable of making the judgments demanded by the system.

Also apparent is that the outcomes reflect the focus on commercialization. NSF reports via its own research that about 70 percent of Phase II awardees reach the market; that figure is largely confirmed by Academies survey data. Yet at the same time few projects are generating large returns, while most are not—as is often the case with high-tech funding or projects with higher than average risks. (See Chapter 4). However, the tight focus on commercialization means that other congressional objectives take on a lower priority, which seems to have detracted attention from the objective of encouraging the participation of women and minorities in the program. It is therefore not yet possible to conclude that the current strategy and focus is generating the results that NSF itself is seeking.

3

SBIR Awards at the National Science Foundation

This chapter describes and analyzes Small Business Innovation Research (SBIR) awards made by the National Science Foundation (NSF) during the period 2003 through 2012. This 10-year interval provides sufficient data to analyze trends and the evolution of the program. Selected tables are included in an annex to this chapter. A review of study data sources, methodological approaches, and potential biases can be found in Appendix A of this report.

This chapter undertakes an evaluation of Phase I and Phase II SBIR awards[1] in turn, and considers awards from a range of perspectives, including yearly trends, distribution by state, the impact of multiple awards to individual companies, applications and success rates, and awards to businesses owned by women and members of socially and economically disadvantaged groups, as defined by the Small Business Administration (SBA). The previous chapter (Chapter 2) drew on the latter data, together with other sources, to access the status of women and minorities in NSF's SBIR program.

PHASE I

Phase I SBIR Awards

Following advice from NSF staff, Committee staff[2] used awards data reported via SBA. In most years, NSF awarded between 200 and 300 Phase I

[1]SBA provided a complete dataset of awards 2003-2012 in response to a FOIA request from the Academies. All references to SBA data in this chapter reference that privately provided data set.

[2]Effective July 1, 2015, the institution is called the National Academies of Sciences, Engineering, and Medicine. References in this report to the National Research Council, or NRC, are used in a historic context identifying programs prior to July 1.

awards, averaging 260 a year, with a range of fewer than 200 to slightly more than in 500. Figure 3-1 shows the number of Phase I SBIR awards made by NSF from 2003 through 2012.

The number of Phase I awards largely drives the amount of funding committed to Phase I, because Phase I SBIR awards at NSF are typically made close to a set amount—$100,000 through 2008 and $150,000 from 2009 onward. Figure 3-2 shows total funding for Phase I SBIR awards at NSF over the same period. It illustrates the close linkage between the number of Phase I awards and total Phase I funding.

Regarding overall award distributions, NSF must keep a pipeline to new ideas open in the form of Phase I awards, while reserving enough funds to make Phase II awards and then Phase IIB awards. Figure 3-3 shows the percentage of total SBIR funding expended on Phase I by year. On average, NSF provided about 40 percent of total funding to Phase I awards, except for the unusually high percentages for Phase I in 2003 and 2010, and the low percentages for 2004-2005.

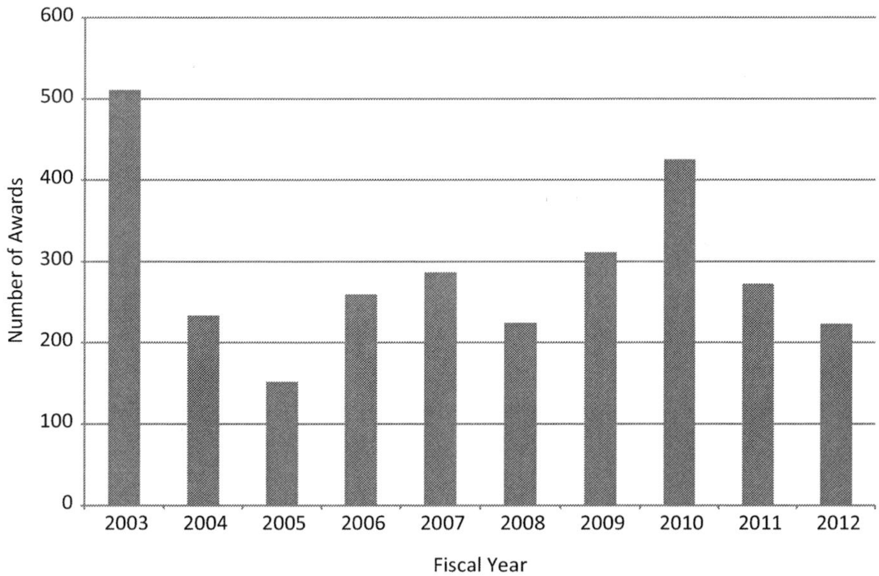

FIGURE 3-1 Number of NSF Phase I SBIR awards, 2003-2012.
SOURCE: Small Business Administration.

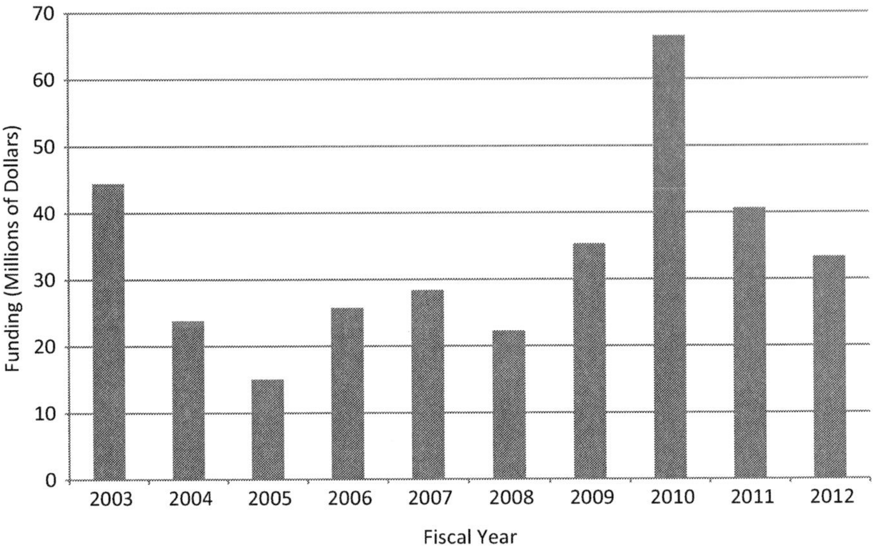

FIGURE 3-2 Funding for NSF Phase I SBIR awards, 2003-2012.
SOURCE: Small Business Administration.

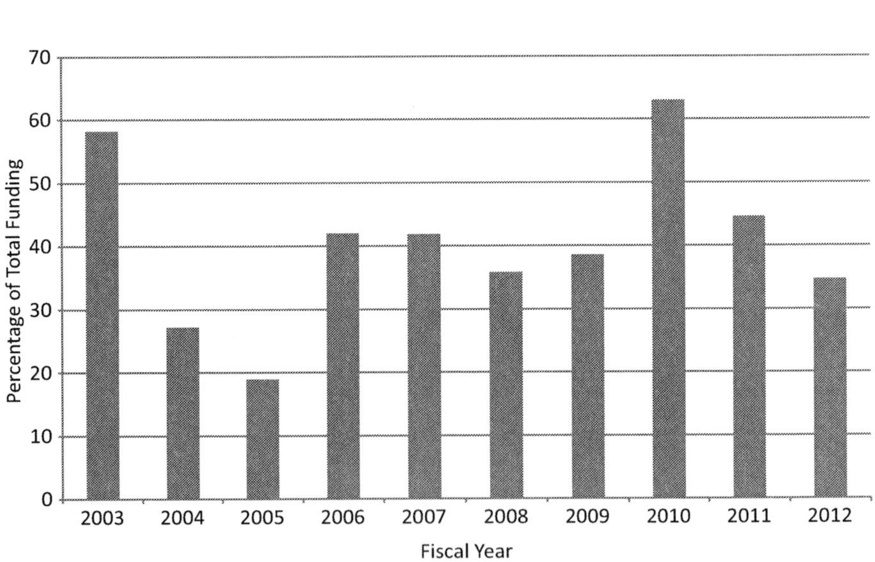

FIGURE 3-3 NSF Phase I SBIR funding as a percentage of total NSF SBIR funding, 2003-2012.
SOURCE: Small Business Administration.

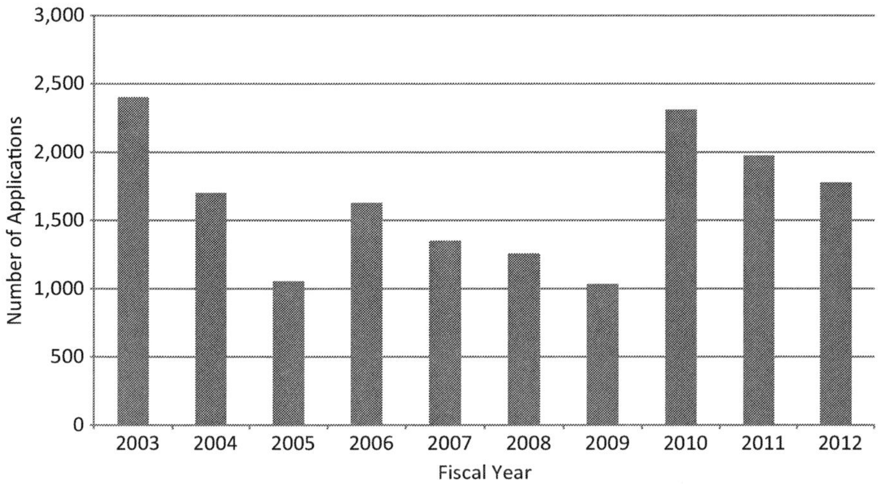

FIGURE 3-4 Number of NSF Phase I SBIR applications, 2003-2012.
SOURCE: National Science Foundation.

Phase I Applications

Figure 3-4 shows the yearly number of applications over this period.[3] Fluctuations in the number of applications do not appear to be correlated with external economic activity: 2003 marks the end of the stock market bottom after the dot.com bust, while 2009 marks the aftermath of the 2008 market crash, which continued well into 2010.

Success Rates of Applicants in Phase I

A fluctuating number of Phase I applications, resulted in a fluctuating success rate, that is, in the percentage of applications that are awarded funding. Figure 3-5 shows the percentage of successful applications during the study period.

In general, success rates ranged from 14 to 21 percent, with the exception of 2009, when a low number of applications was received and the success rate spiked. Overall, 17.3 percent of applications resulted in Phase I SBIR awards at NSF.

[3]All data referenced as NSF data in this section were provided by NSF directly to the study researchers.

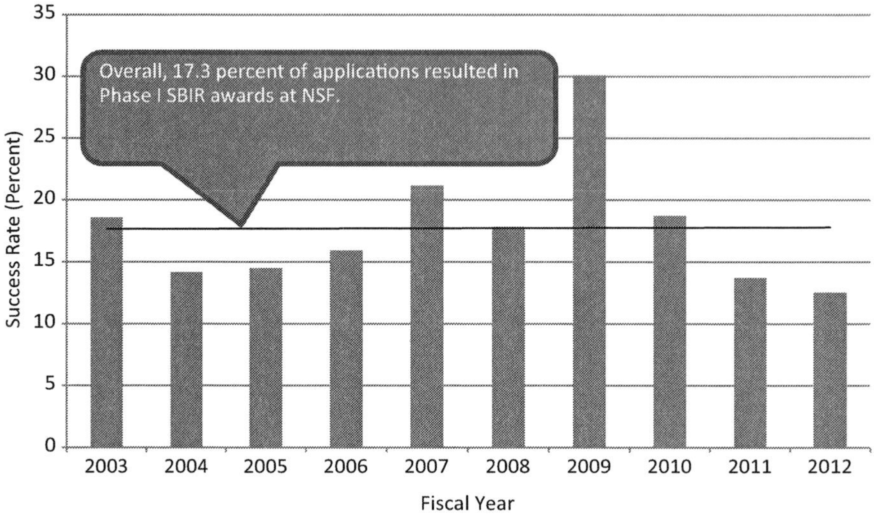

FIGURE 3-5 Success rates for NSF Phase I SBIR applications, 2003-2012.
SOURCE: National Science Foundation.

Phase I Awards and Applications by State

An issue that emerged from the recent reauthorization of the SBIR/STTR programs centers on a concentration of awards in well-known research hubs, and comparatively few awards in less research-intensive regions—a pattern well known in competitively awarded R&D grant programs.

This section reviews awards as well as patterns of applications by state, patterns of ongoing funding via Phase II awards, and contextual information including levels of state gross domestic product (GDP), the amount of research funding deployed in each state, and the distribution of scientists and engineers by state. All of these factors may play a role in the funding distribution. The supporting data tables are presented in the annex of this chapter.

Annex Table 3-1 highlights the wide dispersion in award numbers among the states. States leading in Phase I SBIR awards are California, Massachusetts, New York, Colorado, Pennsylvania, Texas, and Ohio.

The highest average success rate belongs to companies in Wyoming (28.3 percent), which had a low number of applications. Among the states with the highest number of awards, California was near the median with respect to an average success rate of 18.4 percent, as was Massachusetts with an average success rate of 19.2 percent).

The Government Accountability Office (GAO) has noted that the distribution of SBIR awards tends to follow the general distribution of government science and engineering awards.[4] Thus, high-quality projects will not be randomly distributed across the United States.

To examine relationships between some potentially important independent variables and the number of awards, the Committee examined three variables that might explain the differences that remain after accounting for population: research and development (R&D) intensity (percentage of state GDP represented by R&D), the number of scientists and engineers employed per 1000 people, and the population itself. The Committee found modest positive correlations between both R&D variables and the number of awards, as well as the anticipated strong correlation between population size and raw award numbers (see Annex Table 3-5).

Annex Table 3-5 shows that both R&D intensity in the economy and the employment of scientists and engineers are positively correlated with the number of awards—employment of scientists and engineers strongly so. Results were similar when the Pearson r test was applied to rankings of states rather than raw scores for the three variables. Additionally, three high award states with substantially more awards than would be expected given population size (Massachusetts, Maryland, and Colorado) report high levels of scientists and engineering PhDs in their employed workforces.[5]

The Committee therefore concludes that states with higher rates of R&D spending per dollar of GDP and higher rates of employed scientists and engineers per thousand population are more likely to receive higher numbers of NSF Phase I SBIR awards.

From the findings discussed in the next section, the Committee also discounts another possible explanation—the presence of firms with prior awards—as being of limited explanatory power at NSF where a large majority of awards go to first-time applicants.

From a policy perspective, the Committee concludes that the most important variable to explain the distribution of awards by state is simply population, although increasing R&D spending as a share of GDP and increasing the percentage of employed scientists and engineers are both, ceteris paribus, likely to positively affect Phase I award numbers.

---

[4]Government Accountability Office, *Federal Research: Evaluation of Small Business Innovation Research Can Be Strengthened*, GAO/RCED-99-114 (Washington, DC: Government Accountability Office, June 1999), p. 17.

[5]National Science Board, *Science and Engineering Indicators 2014*, (Arlington, VA: National Science Foundation, 2014), Chapter 3.

Phase I Awards by Company[6]

Discussions of the SBIR program often reflect a long-held view that the program is held captive to "SBIR mills" that receive multiple awards and generate little in the way of commercial results. Previous reports reveal this view to be unfounded.[7]

Twenty-two companies received at least six NSF Phase I SBIR awards during the study period (see Annex Table 3-6). The most prolific received fewer than 10 awards. In total, these 22 companies received 157 awards, accounting for 5.4 percent of the 2,896 Phase I awards made by NSF during this period. Overall, 2,009 different companies received at least 1 Phase I award; of these 2,009 companies, 1,504 received only 1 award.

These data indicate that NSF widely distributes Phase I SBIR awards and that, in general, the program does not support companies relying on multiple SBIR awards for survival.

New Companies in the SBIR Program

A positive feature of the SBIR program is that it provides a bridge into the commercial world for new companies as well as early seed funding for companies with few other funding sources. However, these benefits are limited if awards are consistently given to companies that have already benefited from program funding.

At NSF, this is not the case. Figure 3-6 shows the share of Phase I SBIR awards to companies that did not receive a Phase I award during the previous 5 years (the analysis starts at 2008, because the data for the years prior to 2008 are unreliable).[8] About 70 percent of awards went to companies that did not receive an award during the previous 5 years, while 30 percent went to companies that did. This finding underscores the extent to which NSF Phase I funding is widely distributed among different companies.[9]

---

[6]Several difficulties are involved in determining awards by company. SBA data do not include employer identification numbers (EIN) numbers, so companies are identified by name only. Substantial efforts were made to normalize names to account for divergent spelling and punctuation; however, companies also change names, acquire other companies, and spin off subsidiaries among other related activities. These activities cannot be captured using company name alone.

[7]The issue of SBIR mills is discussed in National Research Council, *An Assessment of the SBIR Program* (Washington, DC: The National Academies Press, 2008), p. 88, and the conclusion is that mills do not comprise a serious problem.

[8]"New" companies were identified by taking all company names with awards in a specified area and searching for matches among awards during the previous 5 years. This approach was adopted because NSF does not maintain data on "new" entrants into the program, and simply searching all prior awards (just the previous 5 years) would have introduced bias, because the data set of previous winners continues to increase, which reduces the likelihood that more recent winners would be "new."

[9]This is not to say that having a company receive more than a single Phase I or Phase II SBIR award is necessarily a negative outcome. Some companies are highly inventive and pursue promising opportunities. The more appropriate question is whether the mix between single and multiple award companies is appropriate, and, even more importantly, whether it results from a fair process or whether the process exhibits favoritism or if some companies are exhibiting "troll" behavior.

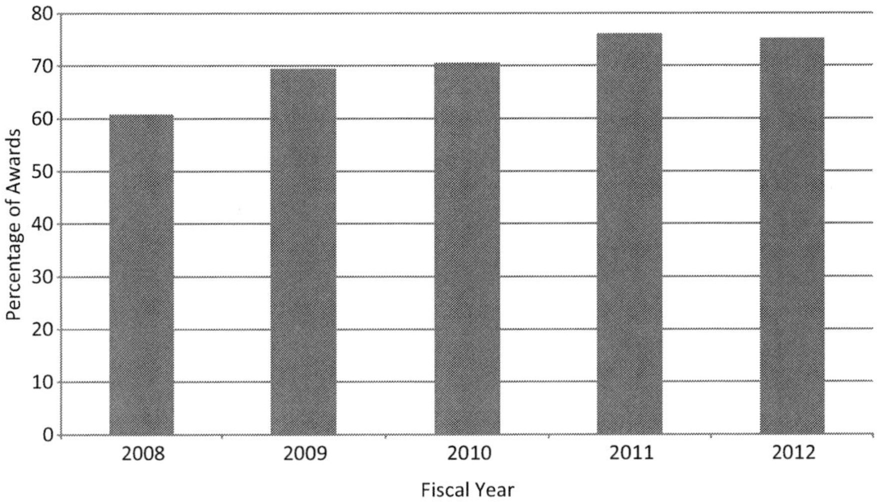

FIGURE 3-6 Percentage of NSF Phase I SBIR awards to non-winners from previous 5 years, 2008-2012.
SOURCE: Small Business Administration.

Interviews with NSF program managers further supported the view that NSF makes a concerted effort to distribute funds widely. For example, one program manager noted that he makes an effort to limit the number of awards to an individual company because success is difficult to predict and therefore developing a portfolio of supported companies is an appropriate approach.

Demographics of Phase I Winning Companies

This section reviews data provided by NSF on awards to and applications from the demographic groups, "Minority-owned Small Business (MOSM) and "Woman-owned Small Businesses" (WOSB).[10] The implications of the data are discussed in Chapter 2.

It should be noted that the NSF/SBA data provides no breakout of the various groups that make up the minority groups. The category contains Asian-Americans, many of whom are prominent—not disadvantaged—in technology

[10]Small Business Administration, Definition of Socially and Economically Disadvantaged Individuals, accessed on March 31, 2014, <*http://www.sdba.com/sba_8%28a%29.htm*>. Data from NSF rely on self-identification of demographic characteristics by companies during the application process. Self-identification is voluntary, and anecdotal evidence indicates incentives both to overreport and underreport.

fields. Their inclusion has tended to obscure the extremely poor performance of the African-American, Native-American, and Hispanic groups. The average performance would likely be lower without the inclusion of the Asian-American groups. This point was revealed by the Academies survey that was able to obtain a breakdown by category.

Minority-owned Small Businesses (MOSBs)

In the context of applications and awards, three data sets are considered: MOSB applications and MOSB awards as a percentage of overall applications and awards, and the share of minority companies in the program.[11]

MOSB Phase I Applications

Figure 3-7 shows the number of applications from self-identified MOSBs during the study period. There is a spike in the number of MOSB in applications most recently in 2010, but this is followed by decreases in the following two years.

MOSB Phase I Success Rates

Figure 3-8 shows the comparative success rates of MOSB and non-MOSB applicants, over the study period. The question is whether MOSBs are less or more successful than non-MOSBs in getting awards.

The success rates of MOSB applicants are strikingly lower than those of non-MOSB applicants in every year. For some agencies, one might consider that winning firms tend to have a stronger track record and that minority firms are less well established. At NSF, however, few companies are "established" in that sense and most awards go to new entrants.

MOSB Phase I Awards

As depicted in Figure 3-9, MOSBs received fewer Phase I SBIR awards from NSF than did non-MOSBs during the entire study period. As shown in Figure 3-10, MOSBs accounted on average for about 10.4 percent of all Phase I awards, and overall this share has declined steadily (see Figure 3-10).

These data show that MOSB participation in Phase I of the NSF SBIR program has declined rather steadily over the study period. The decline has been not only in terms of applications from MOSBs, but in terms of awards to MOSBs.

[11]For a definition of Socially and Economically Disadvantaged Individuals, see Small Business Administration, see <*https://www.sba.gov/content/social-disadvantage-eligibility*>. Accessed on May 29, 2015.

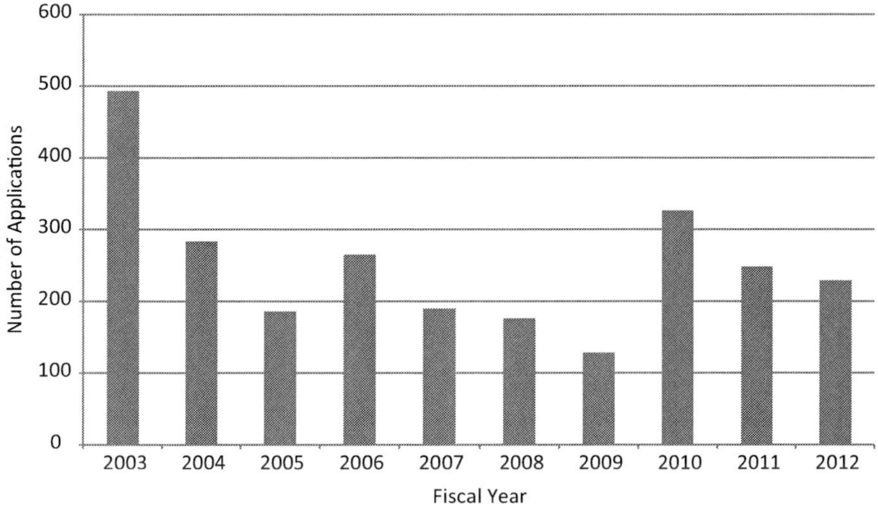

FIGURE 3-7 NSF Phase I SBIR applications from MOSBs, 2003-2012.
SOURCE: National Science Foundation.

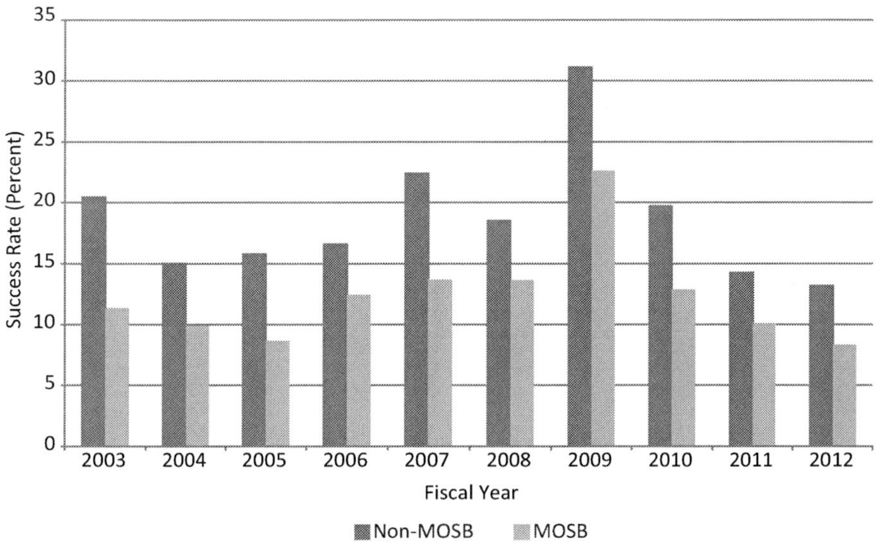

FIGURE 3-8 A Comparison of NSF Phase I SBIR application success rates for MOSB and non-MOSB applicants, 2003-2012.
SOURCE: National Science Foundation.

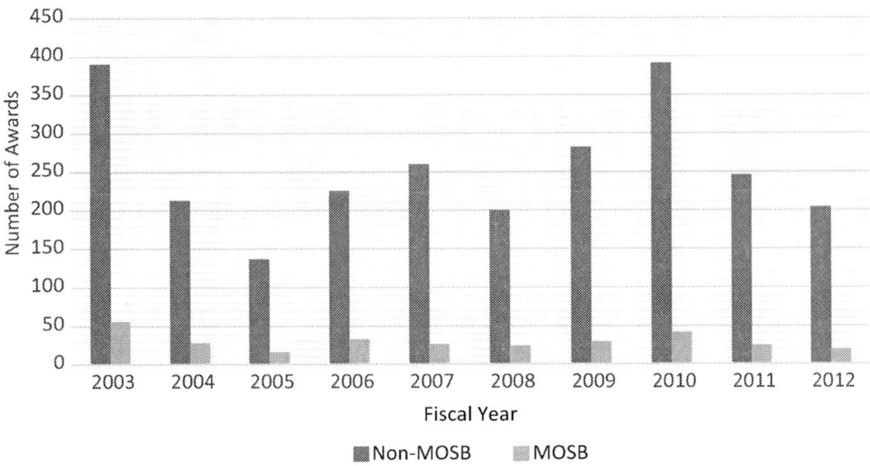

FIGURE 3-9 NSF Phase I SBIR awards for MOSBs and non-MOSBs, 2003-2012.
SOURCE: National Science Foundation.

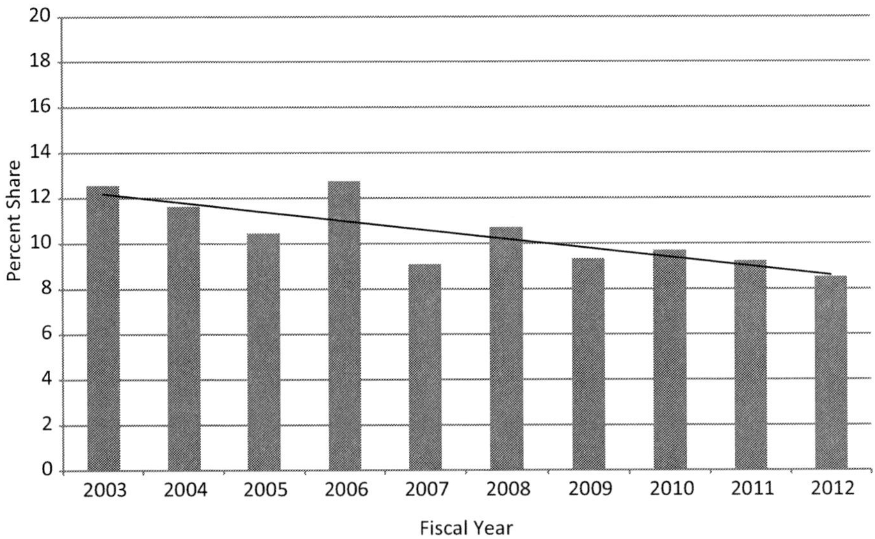

FIGURE 3-10 MOSB share of NSF Phase I SBIR awards, 2003-2012.
SOURCE: National Science Foundation.

Woman-owned Small Businesses (WOSBs)

The congressional mandate to foster participation from socially and economically disadvantaged groups in the SBIR program has historically been interpreted to include WOSBs.

WOSB Phase I Applications

Figure 3-11 appears to show a similar pattern for WOSB Phase 1 applications as was seen for total applications and for MOSB applications (i.e., a repetitive pattern of spiking every three-to-four years followed by declines in the interim). WOSB applications numbered 359 in 2010, compared with 326 MOSB applications in 2010.

Figure 3-12 shows a slight upward trend in the WOSB share of total applications over the study period. During the first 6 years of the study period, the WOSBs averaged 15 percent of total applications, and during the last 4 years of the study period, averaged 17.7 percent.

WOSB Phase I Success Rates

Figure 3-13 shows that WOSBs nearly consistently experienced lower success rates for Phase I compared to other companies. However, the differences were not as dramatic for WOSBs as for MOSBs, and in one year, 2011, the success rates were essentially identical.

WOSB Phase I Awards

The numbers of awards displayed in Figure 3-14 require context. Overall, WOSBs accounted for about 12.9 percent of total Phase I SBIR awards during the study period. The averages for each year are provided in Figure 3-15. The WOSB number of Phase I awards has been increasing slowly, largely reflecting the modest increase in WOSB applications described above.

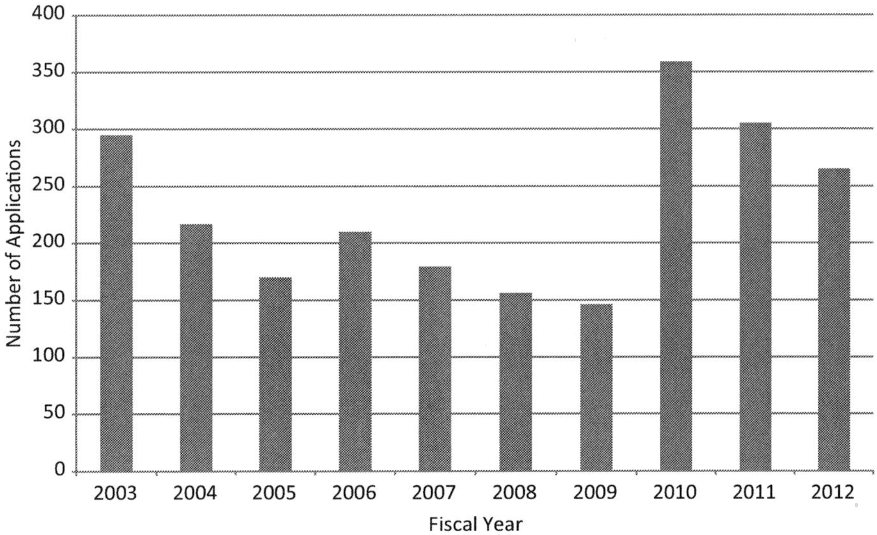

FIGURE 3-11 NSF Phase I SBIR applications from WOSBs, 2003-2012.
SOURCE: National Science Foundation.

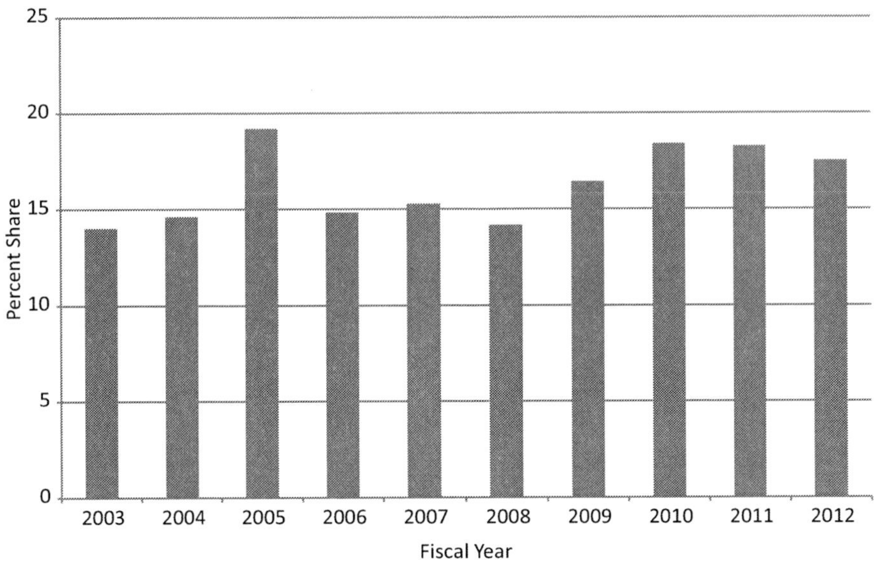

FIGURE 3-12 WOSB share of NSF Phase I SBIR applications, 2003-2012.
SOURCE: National Science Foundation.

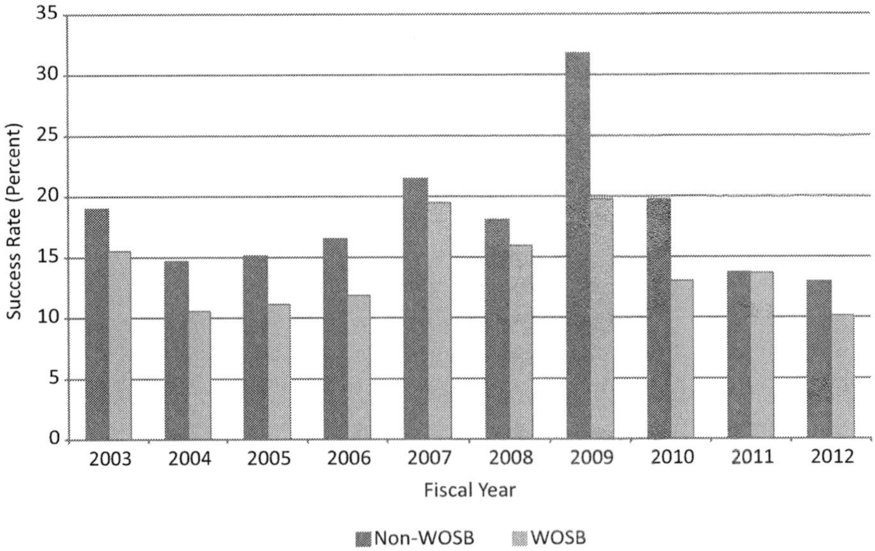

FIGURE 3-13 Comparison of NSF Phase I SBIR success rates for WOSB and non-WOSB applicants, 2003-2012.
SOURCE: National Science Foundation.

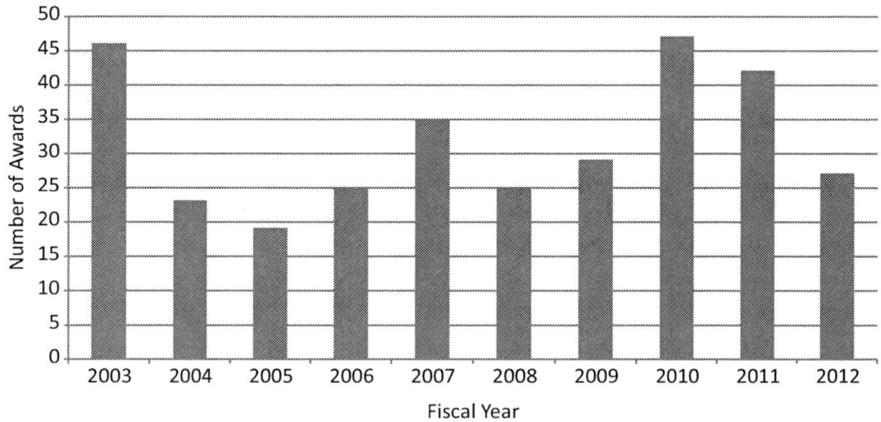

FIGURE 3-14 NSF Phase I SBIR awards to WOSBs, 2003-2012.
SOURCE: Small Business Administration.

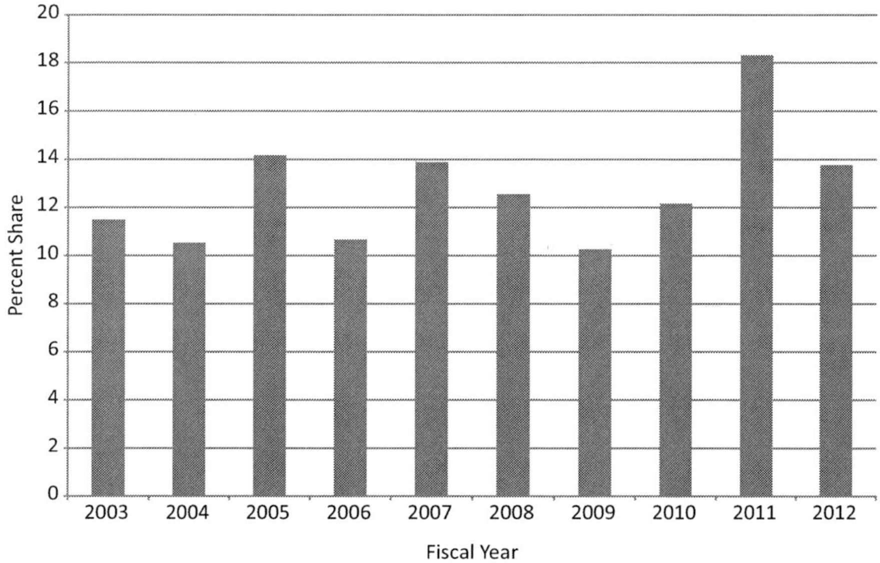

FIGURE 3-15 WOSB share of NSF Phase I SBIR awards, 2003-2012.
SOURCE: National Science Foundation.

PHASE II

Phase II Awards

Figure 3-16 shows the number of Phase II SBIR awards made by NSF during the study period. Following spikes in the number of awards in 2004-2005, the number dropped, but has since increased.

Phase II Applications and Success Rates

Phase II applications, and, in the end awards, have largely been driven by Phase I, because Phase II funding was contingent on completion of a Phase I award until 2011. The number of Phase II applications varied considerably by year, ranging from 133 in 2006 to more than 300 in 2004 and 2011 (see Figure 3-17). This variation is largely explained by the number of Phase I awards made during the previous year (see Figure 3-18).

Analysis using Pearson's r test indicates that the distribution of Phase II awards is strongly correlated with the distribution of Phase I awards, lagged by 1 year (r = 0.73).

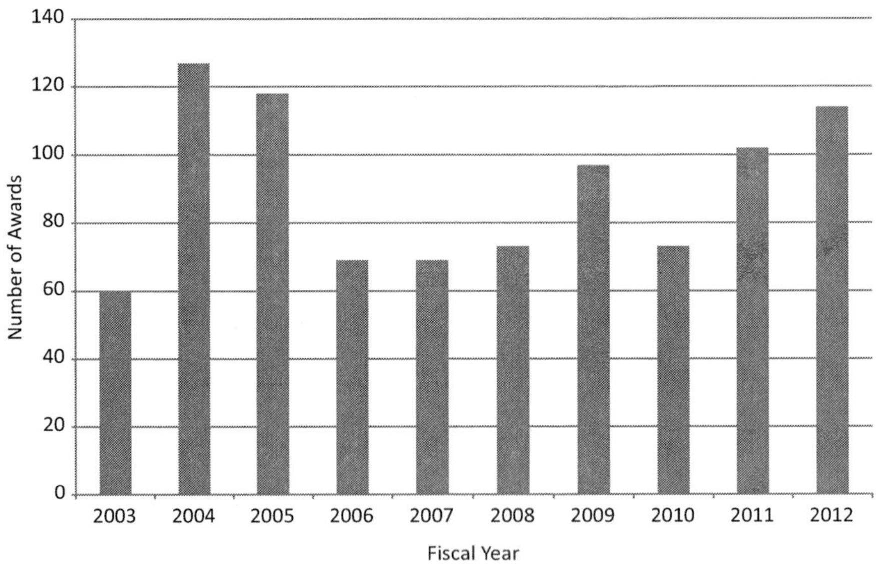

FIGURE 3-16 NSF Phase II SBIR awards, 2003-2012.
SOURCE: National Science Foundation.

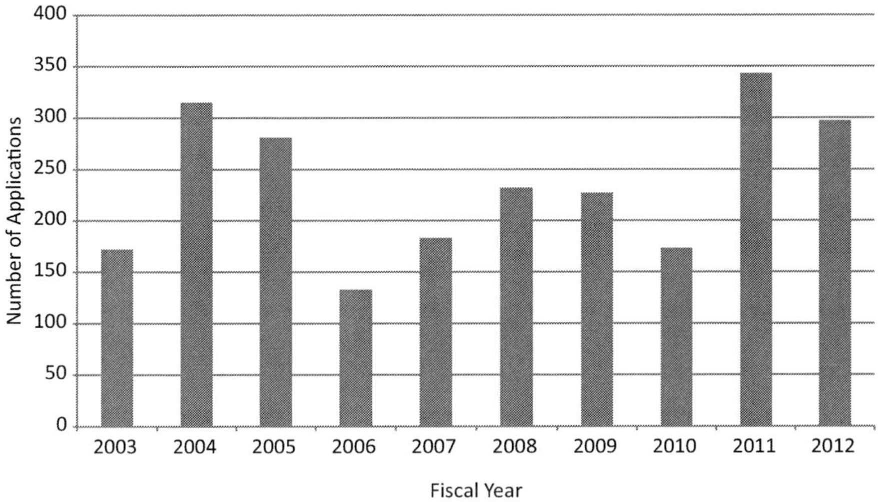

FIGURE 3-17 NSF Phase II SBIR applications, 2003-2012.
SOURCE: National Science Foundation.

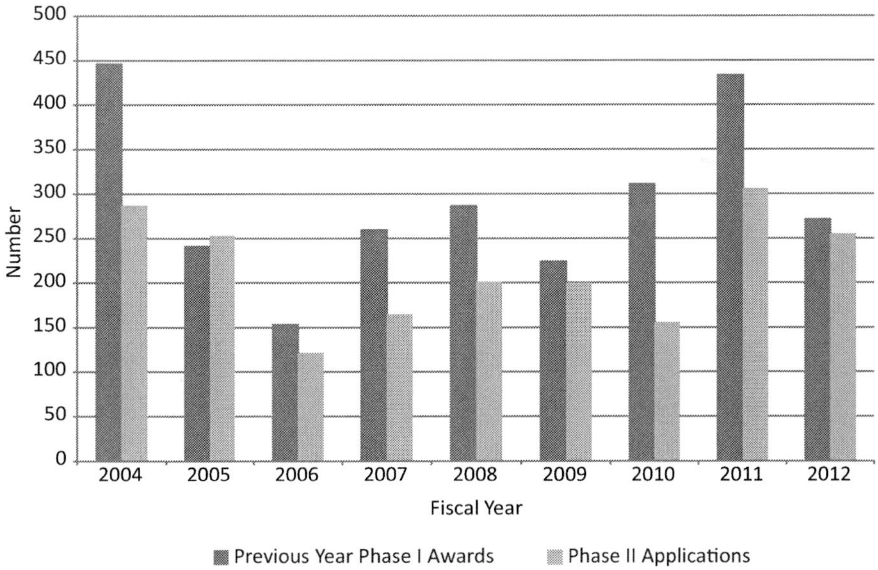

FIGURE 3-18 NSF Phase I SBIR awards made the previous year and Phase II applications, 2003-2012.
SOURCE: National Science Foundation.

As expected, success rates for Phase II were considerably higher than for Phase I (see Figure 3-19). The success rate for Phase II averaged 44 percent. Combining Phase I and Phase II success rates reveals that over the study period, on average, a company applying for a Phase I award had an 8.6 percent chance of receiving both a Phase I and a Phase II award at NSF.

Phase II Awards and Applications by State

The data presented in Annex Table 3-3 show Phase II SBIR applications, awards, and success rates by state, aggregated for the period 2003-2012. As with Phase I (see above), Phase II shows a wide dispersion in award numbers across states.

Not including California where companies experienced an average success rate of 49.8 percent, the average success rate was 41.1 percent. No analysis was conducted to identify specific factors that might account for the higher Phase II success rate in California. As discussed below, however, additional attention by NSF may be warranted to assess why there are differences.

The above section on Phase I awards by state discusses several factors that are related to participation in the SBIR program. This section discusses the extent to which states (such as California) differ in the capacity of their companies to convert Phase I awards into Phase II awards (see Annex Table 3-4).

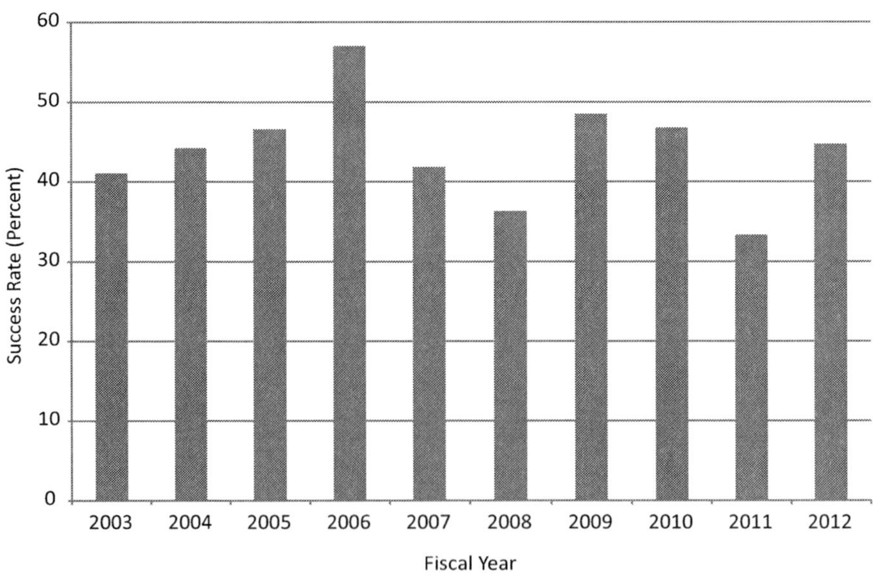

FIGURE 3-19 NSF Phase II SBIR application success rates, 2003-2012.
SOURCE: National Science Foundation.

For states with small numbers of Phase II awards (25 states reported fewer than 10 Phase II awards), the conversion rates are erratic and hence unsuitable for detailed analysis. For the remaining states, 16 had conversion rates greater than 40 percent, two had rates less than 30 percent, and nine had rates between 30 percent and 40 percent. As with Phase I, for all states there is no statistically significant correlation between the number of awards by state and the success rate (Pearson's r = 0.117. However, for the 25 states with at least 10 awards, there is a weak positive relationship (Pearson's r = 0.4997).

At the same time, the correlation between Phase I awards and Phase II awards is very strong (Pearson's r = 0.99), which confirms that the distribution of Phase II awards is strongly dependent on the distribution of Phase I awards.

Phase II Awards by Company

Because Phase I awards were widely distributed among companies it should come as no surprise that Phase II awards were also widely distributed. NSF provided 981 Phase II awards during the study period. Only five companies received five or more awards, claiming 2.8 percent of the total. The remaining awards went to 787 different companies, with 647 receiving only one award. Annex Table 3-7 lists the companies that received five or more Phase II awards.

This finding is not surprising given the preponderance of companies that received only one Phase I award, which automatically made them eligible for only one Phase II award. Moreover, discussions with agency staff revealed an interest within NSF to widely distribute Phase II awards.

Demographics of Phase II Winning Companies

This section relies on data from NSF. Given the limitations of these data, the discussion below does not consider the important question of the composition of MOSB ownership, in particular the share of awards to African-American, Hispanic, and Native-American-owned firms, patterns which are addressed in Chapter 2 of this report.

Minority-owned Small Businesses (MOSB)

This section reviews applications, success rates, and Phase II awards for MOSBs, which are companies that self-certify at least 51 percent ownership by members of a designated minority group as defined by SBA.

MOSB Phase II Applications

The number of Phase II applications from MOSBs during the study period ranged from a high of 41 in 2004, to a low of 12 in 2010 (see Figure 3-20). To a considerable extent, this variation is driven by the number of Phase I awards

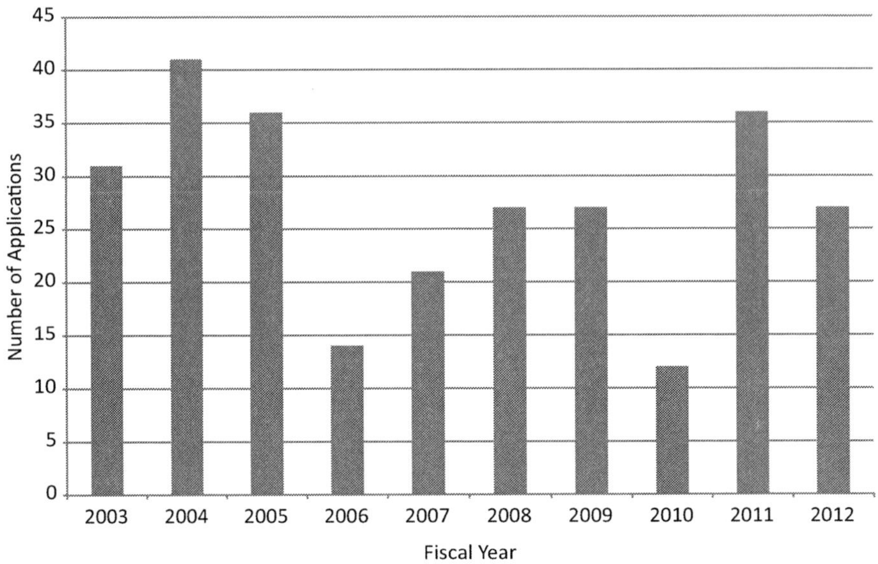

FIGURE 3-20 NSF Phase II SBIR applications from MOSBs, 2003-2012.
SOURCE: National Science Foundation.

BOX 3-1
Analyzing the Participation of Women and Minorities

One of the four primary congressional objectives for the SBIR/STTR program is "to foster and encourage participation by minority and disadvantaged persons in technological innovation."[a] This has in the context of SBIR been taken to mean women and minorities. In practical terms, however, SBA has focused attention in two ways. First, the SBA policy directive lists races/ethnicities which, in its view, qualify as disadvantaged. These include African-American, Hispanic, and Native Americans, as well as Asian Americans. Second, attention is focused solely on firms that are majority owned by women or minorities. As a result, agency data covers only the latter—woman and minority-owned firms—rather than broader measures of "participation."

The implications of these issues are discussed in Chapter 2, where the impact of these narrowed definitions is fully addressed. In the present Chapter 3, we review agency data only, which covers applications and awards to woman-owned firms and firms owned by "socially and economically disadvantaged groups," (as defined in the SBA policy directive).

[a]P. L. 97–219, § 2,July 22, 1982, 96 Stat. 217.

during the preceding year (this is not a 1:1 correspondence, because an NSF Phase I winner may apply for a Phase II award 1 year after the conclusion of its Phase I award, which would be 18 months or even 2 years after the date of the Phase I award). Statistical testing shows a modest positive correlation (Pearson's r = 0.43).

Figure 3-21 compares the number of MOSB Phase I awards lagged one year with MOSB Phase II awards. As might be expected, the two groups follow similar patterns. At the time, the figure shows that many Phase I MOSB awardees do not receive Phase II awards.

MOSB Success Rates

As with Phase I, it is important to compare MOSB Phase II success rates with non-MOSB success rates (see Figure 3-22). For the entire study period, the average success rate for MOSB applicants was 36 percent, 12 percentage points lower than that for non-MOSB applicants (44 percent). The disparity was particularly pronounced in 2008 and 2009 (see Figure 3-22). Unlike Phase I, for which the gaps between MOSB and non-MOSB success rates were almost uniformly negative, in Phase II the gaps were more variable, positive in 4 years and negative in 6 years.

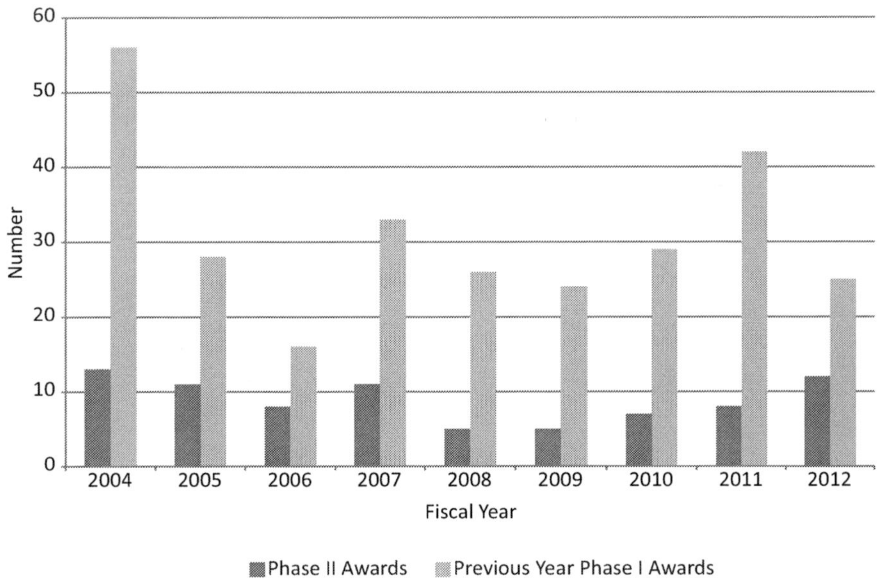

FIGURE 3-21 Comparison of previous year's number of MOSB NSF Phase I awards and MOSB Phase II awards, 2004-2012.
SOURCE: National Science Foundation.

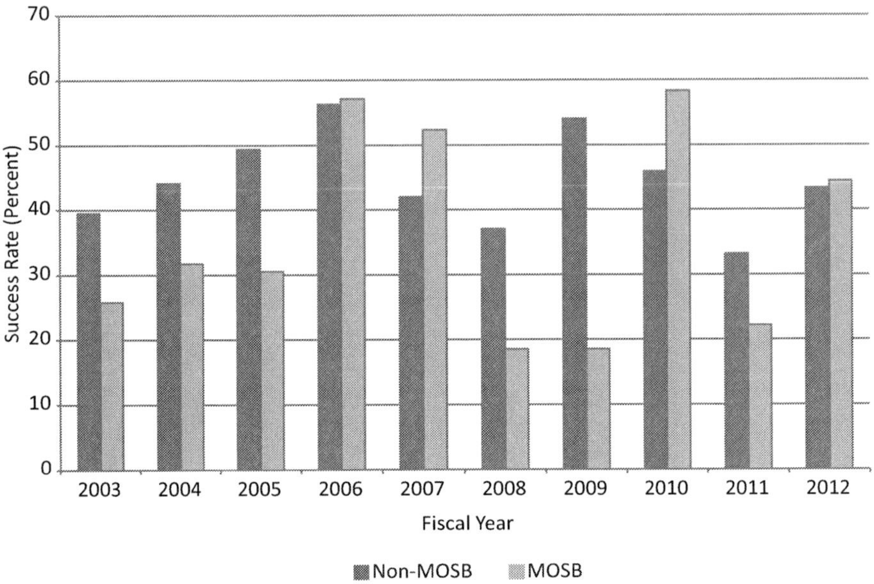

FIGURE 3-22 Comparison of NSF Phase II SBIR success rates for MOSB and non-MOSB applicants, 2003-2012.
SOURCE: National Science Foundation.

It is also interesting to note that when MOSB applications are low, such as in 2006, 2007, and 2010, the MOSB success rate is highest and exceeds the non-MOSB rate. Similarly, the low success years are those with disproportionately more applications.

MOSB Phase II Awards

Figure 3-23 compares the number of Phase II awards to MOSB and non-MOSB applicants. The MOSB share of Phase II awards, and the trend in that share, 2003-2012, can be seen in Figure 3-24.

Woman-owned Small Businesses (WOSB)

This section reviews Phase II applications, success rates, and awards for WOSBs, that is, small business concerns which are at least 51 percent owned by one or more women, or at least 51 percent of the stock is owned by one or more women, and whose management and daily business operations are controlled by one or more women.

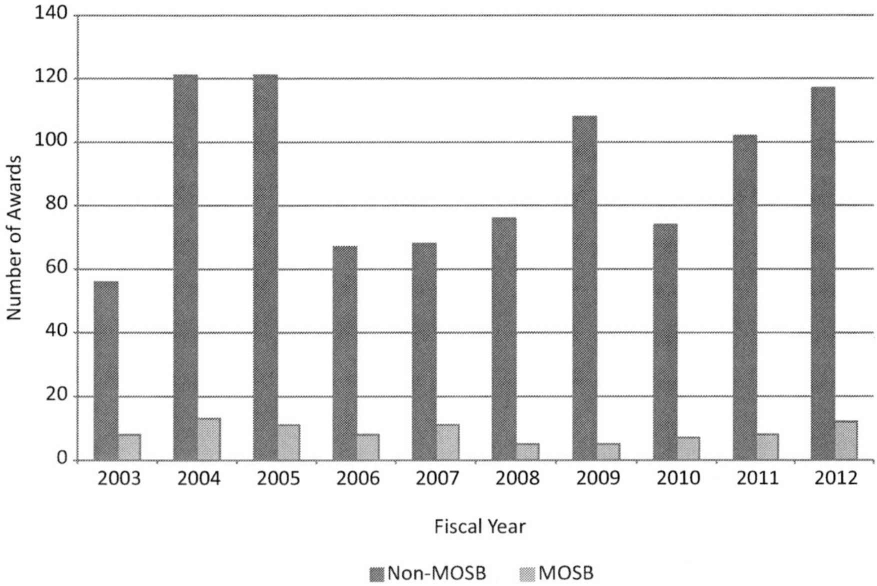

FIGURE 3-23 NSF Phase II SBIR awards to MOSB and non-MOSB firms, 2003-2012.
SOURCE: National Science Foundation.

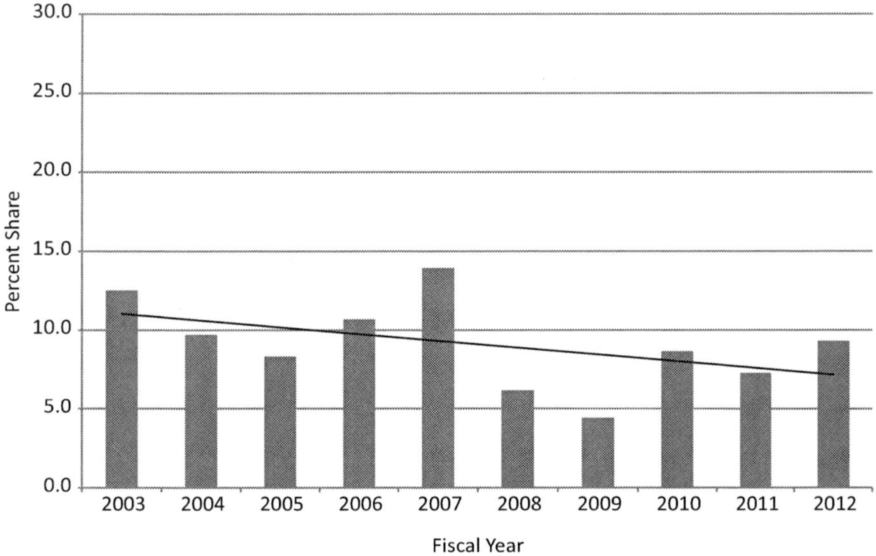

FIGURE 3-24 MOSB share of NSF Phase II SBIR awards and trend line, 2003-2012.
SOURCE: National Science Foundation.

WOSB Phase II Applications

Figure 3-25 shows the number of Phase II applications from WOSBs during the study period. Increases in the number of WOSB applications in 2011 and 2012 helped to drive an apparent overall upward trend across the study period. Note that Phase I applications in Figure 3-17 do not appear to show a similar upward trend.

WOSB Success Rates

Figure 3-26 shows the comparative success rates for WOSB and non-WOSB Phase II SBIR applicants at NSF. Differences between the two populations are captured quite clearly in the "gap" between the rates. For the entire period, WOSBs had an average success rate of 36 percent, lower than non-WOSB applicants (44 percent). For the 5 most recent years, the gap in the average success rates closed to 2 percentage points (38 percent compared to 40 percent).

WOSB Phase II Awards

Figure 3-27 shows the numbers of Phase II awards to WOSB and non-WOSB applicants during the study period. The trend line in Figure 3-28 indicates that the WOSB share of total awards apparently rebounded from relative lows at the beginning of the study period to result in an overall upward trend. From 2005 on, the trend was essentially flat, with award shares ranging from 14.2 percent in 2009 to 7.3 percent in 2011 and averaging 10.5 percent for 2005-2012.

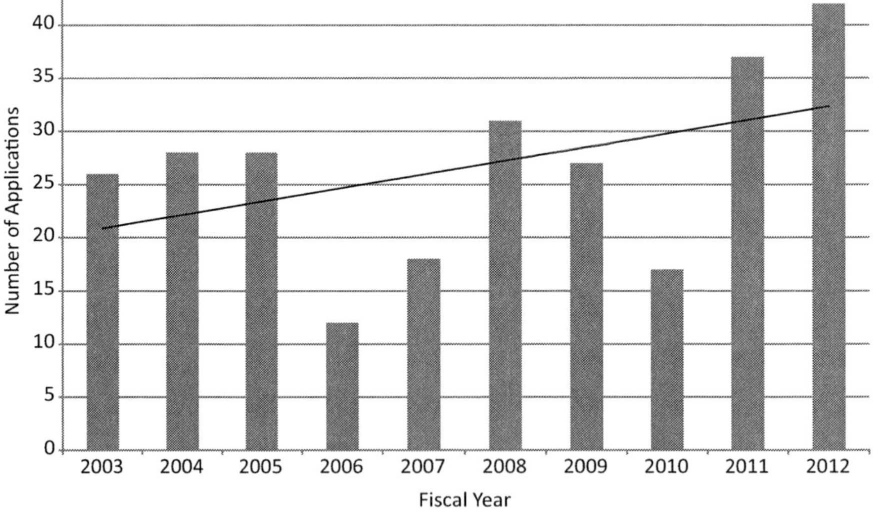

FIGURE 3-25 Number of NSF Phase II SBIR applications from WOSBs, 2003-2012.
SOURCE: National Science Foundation.

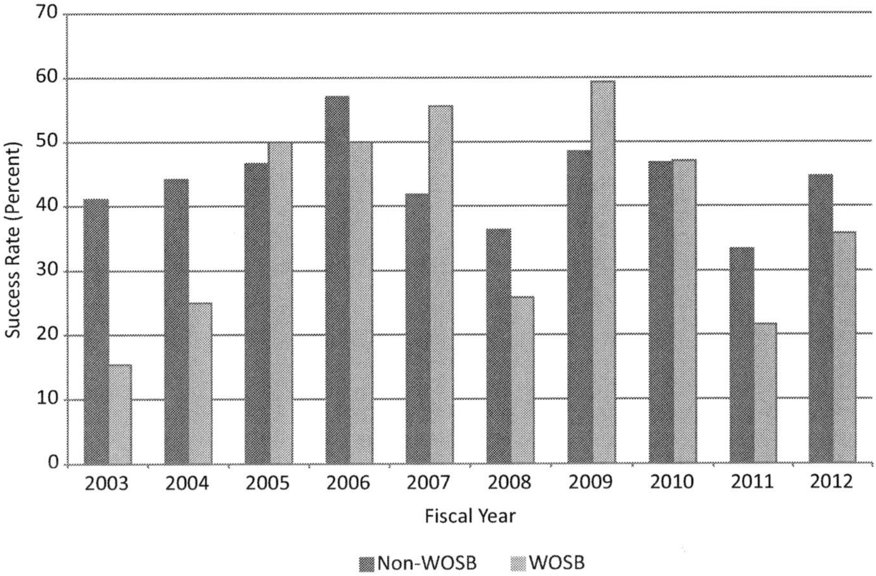

FIGURE 3-26 Comparison of NSF Phase II SBIR success rates for WOSB and non-WOSB applicants, 2003-2012.
SOURCE: National Science Foundation.

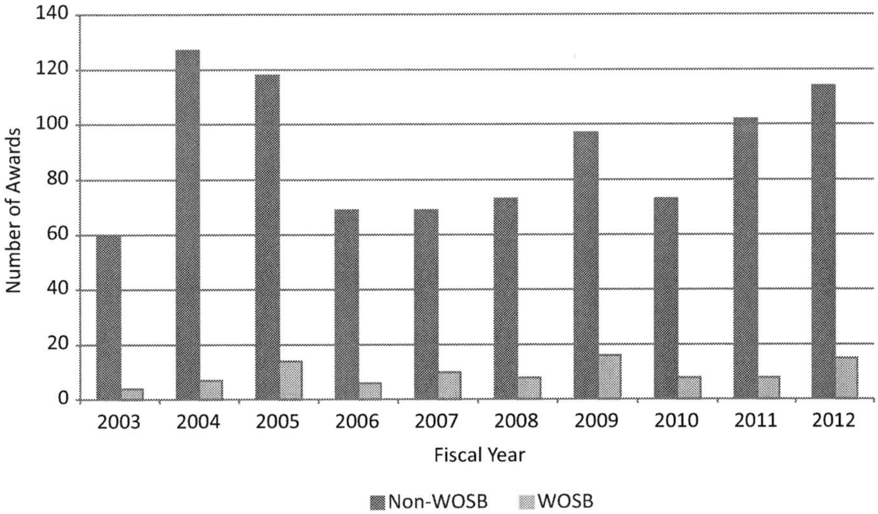

FIGURE 3-27 Comparison of NSF Phase II SBIR awards to WOSB and non-WOSB applicants, 2003-2012.
SOURCE: National Science Foundation.

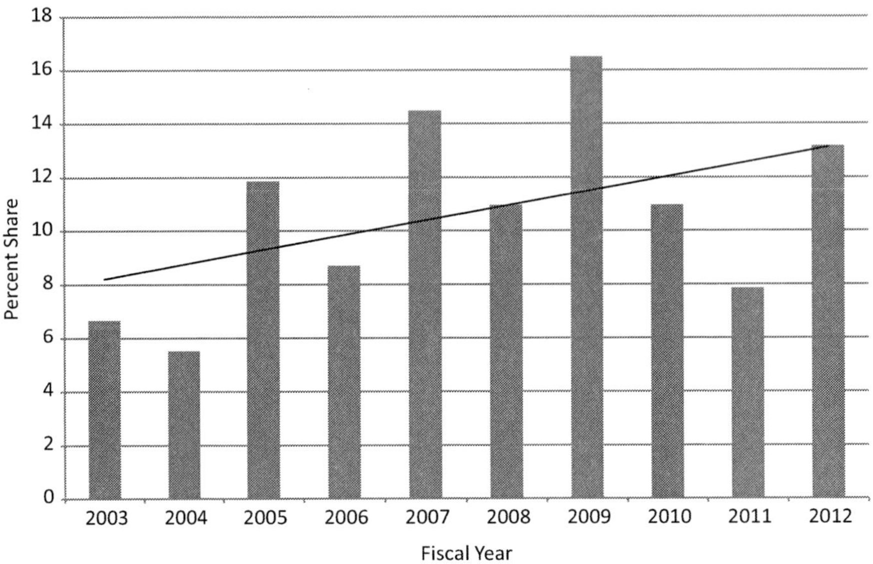

FIGURE 3-28 WOSB share of NSF Phase SBIR II awards, 2003-2012.
SOURCE: National Science Foundation.

ANNEX

DATA TABLES

ANNEX TABLE 3-1 NSF Phase I SBIR Applications, Awards, and Success Rate by State, 2003-2012

State	Number of Awards	Number of Applications	Success Rate (Percent)
CA	545	2,963	18.4
MA	301	1,569	19.2
NY	150	789	19.0
CO	127	731	17.4
PA	122	526	23.2
TX	117	939	12.5
OH	100	646	15.5
MD	91	651	14.0
IL	89	453	19.6
MI	88	523	16.8
NJ	85	632	13.4
NC	69	300	23.0
FL	62	494	12.6
VA	62	727	8.5
GA	61	292	20.9
WI	54	174	31.0
MN	52	341	15.2
AZ	47	434	10.8
CT	41	253	16.2
WA	40	282	14.2
AR	38	168	22.6
NM	37	229	16.2
SC	37	138	26.8
MO	35	188	18.6
UT	35	219	16.0
IN	34	193	17.6
OR	32	196	16.3
MT	31	149	20.8
KY	22	110	20.0
IA	21	73	28.8
ME	20	64	31.3
TN	19	132	14.4
AL	17	152	11.2
KS	14	95	14.7
NE	14	40	35.0
NH	14	84	16.7
OK	13	107	12.1
WY	13	47	27.7
DE	12	125	9.6

continued

ANNEX TABLE 3-1 Continued

State	Number of Awards	Number of Applications	Success Rate (Percent)
ID	12	61	19.7
HI	10	59	16.9
LA	10	48	20.8
MS	9	35	25.7
ND	8	32	25.0
SD	8	55	14.5
VT	8	76	10.5
AK	6	23	26.1
WV	6	50	12.0
PR	4	26	15.4
RI	4	38	10.5
NV	3	74	4.1
DC	2	17	11.8

SOURCE: National Science Foundation.

ANNEX TABLE 3-2 State Rankings: Phase I Awards, Science and Engineering PhDs per 1,000 Populations and R&D as Percentage of State GDP

State	Number of Awards	R&D/ GDP (Percent)	S&E PhDs/ per 1,000 Pop	Population	Rank Awards	Rank R&D/ GDP (Percent)	Rank- S&E PhDs/ per 1,000 Pop	Pop. Rank
CA	545	4.39	0.64	37,253,956	1	5	7	1
MA	301	5.36	1.16	6,547,629	2	3	2	14
NY	150	1.51	0.58	19,378,102	3	33	11	3
CO	127	2.42	0.60	5,029,196	4	19	8	22
PA	122	2.34	0.53	12,702,379	7	21	14	6
TX	117	1.59	0.38	25,145,561	6	30	32	2
OH	100	2.16	0.41	11,536,504	5	25	27	7
MD	91	6.23	1.05	5,773,552	9	2	3	19
IL	89	2.46	0.43	12,830,632	10	18	22	5
MI	88	4.00	0.43	9,883,640	11	6	23	8
NJ	85	3.70	0.56	8,791,894	12	8	13	11
NC	69	2.05	0.50	9,535,483	15	27	20	10
FL	62	1.09	0.25	18,801,310	13	37	49	4
VA	62	2.38	0.57	8,001,024	8	20	12	12
GA	61	1.36	0.36	9,687,653	14	35	35	9
WI	54	2.18	0.38	5,686,986	18	24	33	20
MN	52	2.75	0.50	5,303,925	19	16	19	21
AZ	47	2.22	0.32	6,392,017	17	23	40	16
CT	41	3.35	0.65	3,574,097	16	12	6	29

ANNEX TABLE 3-2 Continued

State	Number of Awards	R&D/ GDP (Percent)	S&E PhDs/ per 1,000 Pop	Population	Rank Awards	Rank R&D/ GDP (Percent)	Rank- S&E PhDs/ per 1,000 Pop	Pop. Rank
WA	40	4.87	0.60	6,724,540	22	4	9	13
AR	38	0.57	0.23	2,915,918	23	49	51	32
NM	37	8.01	0.93	2,059,179	20	1	4	36
SC	37	1.47	0.33	4,625,364	27	34	39	24
MO	35	3.79	0.39	5,988,927	29	7	30	18
UT	35	2.70	0.47	2,763,885	25	17	21	34
IN	34	2.34	0.38	6,483,802	26	22	31	15
OR	32	2.89	0.52	3,831,074	24	15	17	27
MT	31	1.07	0.52	989,415	21	38	16	44
KY	22	0.93	0.28	4,339,367	31	42	45	26
IA	21	2.00	0.36	3,046,355	34	28	36	30
ME	20	0.95	0.37	1,328,361	33	41	34	41
TN	19	1.56	0.41	6,346,105	30	32	28	17
AL	17	2.16	0.34	4,779,736	28	26	37	23
KS	14	1.58	0.29	2,853,118	37	31	44	33
NE	14	1.03	0.33	1,826,341	47	39	38	38
NH	14	3.53	0.43	1,316,470	38	10	24	42
OK	13	0.70	0.30	3,751,351	32	48	42	28
WY	13	0.29	0.28	563,626	35	51	48	51
DE	12	3.70	0.75	897,934	36	9	5	45
ID	12	3.20	0.40	1,567,582	45	13	29	39
HI	10	1.02	0.50	1,360,301	39	40	18	40
LA	10	0.53	0.28	4,533,372	40	50	46	25
MS	9	0.89	0.28	2,967,297	43	44	47	31
ND	8	1.32	0.42	672,591	41	36	26	48
SD	8	0.71	0.31	814,180	44	47	41	46
VT	8	1.75	0.53	625,741	42	29	15	49
AK	6	0.72	0.42	710,231	49	46	25	47
WV	6	0.93	0.30	1,852,994	48	43	43	37
RI	4	2.96	0.59	1,052,567	50	14	10	43
NV	3	0.75	0.25	2,700,551	46	45	50	35
DC	2	3.44	4.79	601,723	51	11	1	50

SOURCES: SBA SBIR awards database (awards data); U.S. Census (population); NSF Science and Engineering Indicators 2014 (R&D%GNP; scientists and engineers per 1,000 populations).

ANNEX TABLE 3-3 NSF Phase II SBIR Awards, Proposals, and Success Rates by State, 2003-2012

State	Number of Awards	Number of Proposals	Success Rate (Percent)
CA	215	432	49.8
MA	119	266	44.7
NY	60	130	46.2
TX	49	102	48.0
CO	47	116	40.5
OH	33	79	41.8
MD	30	77	39.0
PA	30	98	30.6
MI	27	70	38.6
IL	26	62	41.9
FL	25	55	45.5
NC	22	56	39.3
GA	21	51	41.2
NJ	19	69	27.5
WI	19	35	54.3
VA	16	53	30.2
AZ	15	44	34.1
SC	15	25	60.0
CT	14	37	37.8
MN	14	48	29.2
OR	14	26	53.8
WA	14	26	53.8
AR	12	36	33.3
IN	12	30	40.0
MT	11	22	50.0
UT	11	24	45.8
KY	10	19	52.6
NM	9	34	26.5
IA	8	11	72.7
KS	8	15	53.3
MO	8	27	29.6
HI	6	10	60.0
LA	6	7	85.7
ND	6	9	66.7
OK	6	13	46.2
AL	5	15	33.3
VT	5	7	71.4
WY	5	13	38.5
ID	4	11	36.4
NH	4	10	40.0
TN	4	17	23.5
ME	3	16	18.8
DC	2	3	66.7
DE	2	8	25.0
MS	2	9	22.2

ANNEX TABLE 3-3 Continued

State	Number of Awards	Number of Proposals	Success Rate (Percent)
NE	2	10	20.0
NV	2	4	50.0
PR	2	4	50.0
RI	1	4	25.0
SD	1	5	20.0
WV	1	7	14.3
AK	0	3	0

SOURCE: Small Business Administration.

ANNEX TABLE 3-4 NSF Phase II SBIR Awards and Conversion Rates by State, 2003-2012

State	Number of Awards	Phase I-Phase II Conversion Rate (Percent)
CA	215	39.4
MA	119	39.5
NY	60	40.0
TX	49	41.9
CO	47	37.0
OH	33	33.0
MD	30	33.0
PA	30	24.6
MI	27	30.7
IL	26	29.2
FL	25	40.3
NC	22	31.9
GA	21	34.4
NJ	19	22.4
WI	19	35.2
VA	16	25.8
AZ	15	31.9
SC	15	40.5
CT	14	34.1
MN	14	26.9
OR	14	43.8
WA	14	35.0
AR	12	31.6
IN	12	35.3
MT	11	35.5
UT	11	31.4
KY	10	45.5
NM	9	24.3
IA	8	38.1

continued

ANNEX TABLE 3-4 Continued

State	Number of Awards	Phase I-Phase II Conversion Rate (Percent)
KS	8	57.1
MO	8	22.9
HI	6	60.0
LA	6	60.0
ND	6	75.0
OK	6	46.2
AL	5	29.4
VT	5	62.5
WY	5	38.5
ID	4	33.3
NH	4	28.6
TN	4	21.1
ME	3	15.0
DC	2	100.0
DE	2	16.7
MS	2	22.2
NE	2	14.3
NV	2	66.7
PR	2	50.0
RI	1	25.0
SD	1	12.5
WV	1	16.7
AK	0	0.0

SOURCE: National Science Foundation.

ANNEX TABLE 3-5 Explaining Differences in Observed Award Distributions by State, 2003-2012

Variable 1	Variable 2	Pearson r score
Awards	R&D as Percent of GDP	0.373
Awards	S&E PhDs employed per 1,000 pop	0.441
Awards	Population	0.789
Awards-rank	R&D%GDP-rank	0.473
Awards-rank	S&E PhDs-rank	0.449
Awards-rank	Pop-rank	0.818

NOTE: The Pearson r test for awards/employed scientists and engineers excludes the District of Columbia, which is an outlier in opposite directions on both arrays.
SOURCES: Small Business Association SBIR awards database (awards data); U.S. Census (population); NSF Science and Engineering Indicators 2014 (R&D%GDP; scientists and engineers per 1,000 population).

ANNEX TABLE 3-6 Companies Receiving More Than Five NSF Phase I SBIR Awards, 2003-2012

Company Name	Number of Phase I SBIR Awards	Percentage of all Phase I SBIR Awards
Resodyn	9	0.3
Sinmat	9	0.3
TDA Research	9	0.3
Tetramer Technologies	9	0.3
Lynntech	8	0.3
Micro Magnetics	8	0.3
Radiation Monitoring Devices	8	0.3
Sensor Electronic Technology	8	0.3
Advanced Diamond Technologies	7	0.2
BioDetection Instruments	7	0.2
Luna Innovations	7	0.2
Structured Materials Industries	7	0.2
Uncopiers	7	0.2
Advanced Thermal Technologies	6	0.2
Barrett Technology	6	0.2
CCVD dba MicroCoating Technologies (MCT)	6	0.2
Inscent	6	0.2
Materials Modification	6	0.2
NanoSonic	6	0.2
NexTech Materials	6	0.2
Sun Innovations	6	0.2
SVT Associates	6	0.2
Total	157	5.4

SOURCE: Small Business Administration.

ANNEX TABLE 3-7 Companies Receiving Five or More NSF Phase II SBIR Awards, 2003-2012

Company Name	Number of Phase II SBIR Awards	Percent of all Phase II SBIR Awards
Lynntech	7	0.7
Faraday Technology	5	0.5
TDA Research	5	0.5
Tetramer Technologies	5	0.5
WTE	5	0.5
Total	27	2.7

SOURCE: Small Business Administration.

4

Commercial and Knowledge Outcomes

This chapter uses available data from NSF and two award recipient surveys of NSF SBIR winners to analyze outcomes related to achieving three of the four goals of the congressional mandate for the SBIR as was described in Chapter 1. Although the statutory goals of the SBIR program are fourfold, subsequent legislation passed by Congress, as well as administrative policies pursued by NSFs, has focused on the fourth goal, the commercialization of SBIR technologies. Commercialization is among the more measurable outcomes of the SBIR program, and has become the benchmark for program performance. The focus on commercialization, however, should not be allowed to obscure the fact that the program is designed to meet all four congressional mandated objectives. Therefore, this chapter also analyzes knowledge effects, which relate to both the first and second mandated goals, stimulating technological innovation, and helping to meet the NSF mission goals related to expanding scientific and technical knowledge. Outcomes related to the third goal, "to foster and encourage participation by socially and economically disadvantaged small businesses"[1] (defined to include woman-owned small businesses), are treated separately in Chapter 2.

DATA SOURCES AND COMPARISON ANALYSIS

To develop an effective quantitative analysis of the commercial and knowledge outputs of the NSF SBIR program, we have drawn on two main sources: data from the National Science Foundation (NSF), and responses to a large-scale survey of SBIR recipients at NSF. This 2011 survey is based on the 2005 survey

[1]Small Business Administration, SBIR Policy Directive, October 18, 2012, p. 3.

deployed by the Academies,[2] with some additions and modifications.[3] Appendix A provides a detailed description of the survey methodology, including discussion of the response rate and discussions of potential survey bias. Table A-3 shows that the overall number of responses for NSF was 393, reflecting a response rate of 46.2 percent. The 2011 Survey questionnaire is reproduced in Appendix C.

COMMERCIALIZATION

Under previous and current management, the National Science Foundation (NSF) has made a concerted effort to focus on commercialization, the fourth mandated objective. Indicative of this focus, Dr. Kesh Narayanan, a former NSF SBIR program manager, often said that 97.5 percent of NSF funding went to basic science, and that he was in charge of the 2.5 percent of all NSF funding that was aimed at achieving commercial results. Congressional interest in this area is also high.[4] Commercialization is important to the nation's economic prosperity. Commercialization is also among the more quantifiable outcomes, and, therefore, has become a measuring stick for program outcomes.

Some important conceptual challenges must be addressed in tracking commercialization. Like many apparently simple concepts, tracking commercialization becomes progressively more challenging and complex as it is subjected to further scrutiny. A series of questions underscores these challenges:

- Should commercialization include just sales or other kinds of revenue such as licensing fees and funding for further development?
- What is the appropriate benchmark for sales?
  - The fact of any sales whatsoever?
  - Sales equivalent to the cost of awards?
  - Sales that reflect a profit?
  - Sales that drive overall company growth?
  - Very substantial commercial success—which would need to be benchmarked at a specific level of sales (e.g., \$5 million, \$10 million, or some other number)?
- Should commercialization include sales by licensees, which may be many multiples of the revenues provided to the SBIR company through licenses?

---

[2]Effective July 1, 2015, the institution is called the National Academies of Sciences, Engineering, and Medicine. References in this report to the National Research Council, or NRC, are used in a historic context identifying programs prior to July 1.

[3]See National Research Council, *An Assessment of the SBIR Program*, (Washington, DC: The National Academies Press, 2008), Appendix A.

[4]Dr. Narayanan was Division Director of Industrial Innovation and Partnerships from 1994 to 2011 and led the SBIR at NSF in that capacity.

This chapter draws from responses of Phase II SBIR recipients to a large-scale 2011 survey of SBIR recipients at NSF (2011 Survey), as well as previous survey of Phase II and Phase IIB recipients in 2010 (2010 Phase IIB Survey). The 2010 and 2011 surveys are themselves based on a 2005 survey deployed by the Academies, with some additions and modifications.[5] This has permitted the Committee to cast a broad net to capture a wide range of potentially useful data. These data have been analyzed in a variety of ways to provide a range of insights into this complex topic.

Sales and Licensing Revenue

A much used metric for assessing SBIR-type programs is the sum of sales and licensing fees. In previous assessments, the Academies warned extensively against overuse of this metric.[6] The Committee heeded these warnings by adopting a wide range of metrics for use in the current assessment; however, sales and licensing fees are still an important consideration.

The current survey eliminated some questions about licensee activities that have sometimes been collected in the past. Although these activities can be important, the principal investigators leading SBIR-funded projects generally have little insight into the activities of licensees. In addition, descriptions of licensee activities are often subject to nondisclosure provisions. Hence, given the limitations of the data provided, it did not seem appropriate to include questions about licensee activities in the updated survey.

Reaching the Market

The first question addressed by the survey concerns reaching the market: Did the project generate sales, and, if not, are sales expected? The second part of this question is necessary because some projects have long cycle times. Responses are summarized in Table 4-1.

Overall, almost 70 percent of Phase II respondents reported some sales from either the company or a licensee. This percentage is higher than that reported for other agencies (typically 45 percent to 60 percent). A further 19 percent reported that they expected to generate future sales from the surveyed project—a percentage that in part reflects the relatively recent date of some awards in the sample.[7]

[5]The 2011 Survey and 2010 Phase IIB Survey appear in appendixes C and D.

[6]See, for example, National Research Council, *An Assessment of the Small Business Innovation Research Program*, Section 4.2.2.

[7]2011 Survey, Question 35.

TABLE 4-1 Sales Revenue

Outcome	Phase II (Percent)
No sales to date, no sales expected	11.8
No sales to date, sales expected	18.5
Sales to date	69.7
N: TOTAL RESPONDENTS ANSWERING[a]	363

[a] 2010 Phase IIB Survey question was asked only with regard to projects with an active status.
NOTE: 2010 Phase IIB Survey did not break out sales by type.
SOURCE: 2011 Survey, Question 35; and 2010 Phase IIB Survey.

Amount of Sales Revenues

The percentage of projects reaching the market is an important metric, but it is not sufficient. It is also important to understand the dollar distribution of sales revenue. The survey asked respondents who reported some sales from the surveyed project to also report the amount of sales by tier (see Table 4-2).

TABLE 4-2 Distribution of Total Sales Revenue, by Range

Amount	Phase II (Percent)
Under $100,000	21.9
$100,000-$499,999	21.9
$500,000-$999,999	20.2
$1,000,000-$4,999,999	27.2
$5,000,000-$9,999,999	3.5
$10,000,000-$19,999,999	2.6
$20,000,000-$49,999,999	2.6
$50,000,000 or more	—
Total	100.0
N: Projects reporting sales incorporating technology developed by the surveyed project	114
MEAN ($000s)	2,623

SOURCE: 2011 Survey, Question 36, B1.

Slightly more than 90 percent of Phase II respondents reported sales below $5 million. Twenty-two percent reported sales below $100 thousand. Forty-two percent reported sales of more than $100 thousand but less than $1 million. Thirty-six percent reported sales of $1 million or more, and 9 percent were at $5 million or above. None reported sales of $50 million or greater in project related revenues.

Markets by Sector

As shown by Table 4-3, more than one-half of Phase II respondents reported sales to the domestic private sector in the United States. Fifteen percent reported export sales. In addition, overall sales by Phase II respondents to government agencies and prime contractors for DoD or NASA in combination reached 16 percent.

Employment

As with prior surveys, respondents were asked about the size of the company at the time of the award and the size at the time of the survey, in terms of number of employees. As shown by Table 4-4, Phase II funding was more likely to be awarded to firms with 5 or more employees than to those with fewer than 5 employees.

TABLE 4-3 Markets for SBIR Products and Services—Percentage of Total Sales (Mean of All Responses/Category)

Market Sector	Phase II and IIB (Percent)
Domestic private sector	58.5
Export markets	14.9
Department of Defense (DoD)	7.1
State or local governments	2.7
Other federal agencies	2.8
Prime contractors for DoD or NASA	3.2
NASA	0.3
Other (Specify)	10.5
Total	100.0
N =	247

SOURCE: 2011 Survey, Question 37, and 2010 Phase IIB Survey.

TABLE 4-4 Company Size by Number of Employees at Time of Award

Number of Employees	Phase II (Percent)
Under 5	35.7
5 to 9	24.9
10 to 19	18.1
20 to 49	12.3
50 to 99	5.0
100 or more	3.9
Total	100.0
Mean	16
Median	6
N =	371

SOURCE: 2011 Survey, Question 18A.

The mean number of employees was 16 for Phase II respondents, and the median size was 6. The difference between mean and median likely indicates an influence on the mean by outliers, that is, a few companies with high numbers of employees. A striking characteristic of the NSF program is that it makes awards to very small companies—at other agencies the median has been considerably higher.[8]

Respondents were also asked to report the current number of employees. Results are affected by selection bias toward surviving companies. As Table 4-5 shows, the median size grew for Phase II companies from 6 at the time of the award to 10 at the time of the survey.

The percentage of Phase II respondents reporting 50 or more employees about doubled—from 9 percent at the time of award to nearly 17 percent at the time of the survey, which indicates that some NSF SBIR recipients have become relatively large enterprises (within the general universe of small companies).

Further Investment

The ability of SBIR projects and companies to attract further investment has traditionally been an important defining metric for SBIR commercialization outcomes.[9] According to the survey results shown in Table 4-6, 63 percent of

[8]For example, the median number of employees at the Department of Defense is 17. See National Research Council, *SBIR at the Department of Defense* (Washington DC: National Academies Press, 2014), p.62.

[9]See National Research Council, *An Assessment of the SBIR Program.*

TABLE 4-5 Company Size (Employees) at Time of Survey

Number of Employees	Phase II (Percent)
Under 5	23.6
5 to 9	20.6
10 to 19	22.8
20 to 49	16.5
50 to 99	8.1
100 or more	8.4
Total	100.0
Mean	27
Median	10
N=	389

SOURCE: 2011 Survey, Question 18B.

TABLE 4-6 Additional Investment after SBIR Award

Received Additional Investment after SBIR Award?	Phase II (Percent)
Yes	62.8
No	37.2
Total	100.0
N=	409

SOURCE: 2011 Survey, Question 33.

Phase II projects received additional funding (relatively unchanged from the 2009 survey[10]). This indicates that SBIR funded projects are seen to have sufficient value to persuade non-SBIR sources to invest in them.

As with prior surveys, the distribution of additional non-SBIR funding is substantially skewed (see Table 4-7). While 61 percent of Phase II companies reported no funding from non-SBIR sources, 39 percent did. Most reporting funding were in the category of $100 to $500 thousand. Two percent of Phase II respondents reported additional non-SBIR investments of at least $20 million.

[10]See National Research Council, *An Assessment of the SBIR Program at the National Science Foundation* (Washington, DC: The National Academies Press, 2008), Appendix B.

TABLE 4-7 Additional Non-SBIR Federal Funding by Amount

Additional Federal (non-SBIR) Funding	Phase II (Percent)
$0	61.3
Under $100,000	2.4
$100,000-$499,999	17.3
$500,000-$999,999	4.8
$1,000,000-$4,999,999	11.3
$5,000,000-$9,999,999	1.2
$10,000,000-$19,999,999	0.0
$20,000,000-$49,999,999	0.6
$50,000,000 or more	1.2
Total	100
Any funding of this type	38.7
N (those reporting additional funding from some source) =	168

SOURCE: 2011 Survey, Question 34.1.

As shown by Table 4-8, self-funding by one's own company was the prevalent source of additional funding reported by Phase II respondents, while many also reported using personal funds. Direct funding from federal agencies was the second most-cited outside source. Funding by other companies, private equity/angels, and state and local governments were each cited by between 25 and 32 percent of Phase II respondents. Fifteen percent of Phase II companies reported receiving venture funding.

Company-Level Development

Table 4-9 shows company-level changes resulting from the SBIR program. SBIR firms that commercialize their technology often do so through mergers or other company-level activities. Fifteen percent of Phase II respondents indicated that their company had spun off one or more new companies, and 7 percent reported that they had been acquired by or merged with another firm. However, the answers of the majority of respondents indicated that their companies had, as of the time of the survey, not been acquired, had not implemented or planned an initial public offering (IPO), and had not established a spin-off.

Respondents also reported on a range of market-related activities that involved agreements between the surveyed company and other organizations, which can be considered an indication of commercial activity. Table 4-10 shows

TABLE 4-8 Sources of Additional Funding, for Projects Reporting Additional Funding

Type of Funding	Phase II (Percent)
Federal funding	38.7
Venture Capital	14.6
Foreign private	9.2
Private equity/angels	28.1
Other companies	32.2
State/local govt.	25.3
Universities/colleges	8.8
Own company	70.3
Personal funds	31.5
N =	137-185[a]

[a] Actual number depends on type of funding.
NOTE: Additional federal funding was non-SBIR.
SOURCE: 2011 Survey, Question 34.

TABLE 4-9 Company-Level Changes Resulting from SBIR Program

Company-Level Change	Phase II (Percent)
Established one or more spin-off companies	15.0
Been acquired by/merged with another firm	7.0
Made an initial public offering	0.0
Planning to make an initial public offering in 2011-2012	0.0
Any of the above	22.0
N = (company weighted)	83
N = (not company weighted)	149

NOTE: Responses may not sum to 100 percent because respondents could select more than one answer.
SOURCE: 2011 Survey, Question 10.

TABLE 4-10 Market-Oriented Activities—Finalized Agreements with U.S. Companies and Investors

Type of Agreement	Phase II (Percent)
R&D agreement(s)	50.2
Licensing agreement(s)	39.2
Marketing/distribution agreement(s)	28.6
Customer alliance(s)	32.2
Joint venture agreement	9.3
Sale of company	5.7
Sale of technology rights	5.7
Manufacturing agreement(s)	9.7
Partial sale of company	4.4
Company merger	2.2
Other	3.1
N =	227

NOTE: Responses may not sum to 100 percent because respondents could check more than one response.
SOURCE: 2011 Survey, Question 38.1.

finalized agreements between Phase II respondents and U.S. companies and investors by type. Half of Phase II respondents reported completion of at least one R&D agreement with U.S.-based companies or investors. Thirty-nine percent of Phase II respondents reported licensing agreements, and 60 percent reported either marketing/distribution agreements or customer alliances. These four—R&D agreements, licensing agreements, marketing/distribution agreements, and customer alliances—were the most prevalent types of agreements reported by Phase II recipients with U.S. companies and investors.

Table 4-11 shows the percentage of Phase II respondents that entered into various types of agreements with foreign partners. As with domestic partners, the most prevalent types of agreement with foreign partners were R&D agreements reported by 42 percent, licensing agreements reported by 35 percent, marketing/distribution agreements reported by 37.2, and customer alliances reported by 26 percent.

A comparison of forms of agreements of Phase II recipients with U.S. partners and with foreign partners shows the following: R&D agreements, licensing agreements, and customer alliances were reported more often with U.S. partners, while marketing/distribution agreements, joint venture agreements, and sale of company were reported more often with foreign partners.

TABLE 4-11 Finalized Agreements with Foreign Partners

Type of Agreement	Phase II (Percent)
R&D agreement(s)	42.3
Licensing agreement(s)	34.6
Marketing/distribution agreement(s)	37.2
Customer alliance(s)	25.6
Joint venture agreement	11.5
Sale of company	11.5
Sale of technology rights	5.1
Manufacturing agreement(s)	3.8
Partial sale of company	5.1
Company merger	0.0
Other	3.8
N =	227

NOTE: Responses may not sum to 100 percent because respondents could check more than one response.
SOURCE: 2011 Survey, Question 39.1.

Commercialization Training and Marketing

Federal agencies have in recent years increased the amount of commercialization training for SBIR awardees. In some cases, this training has been mandatory. The Committee added a question focused on such training to the 2011 Survey, and the results are reported in Table 4-12.

NSF has focused its commercialization training on Phase II awardees. The responses summarized in Table 4-12 indicate that 61 percent of Phase II respondents had participated in training related to the surveyed award.

The Committee also added a question to the 2011 Survey about the existence of full-time marketing staff at the company, as another metric to gauge the extent to which the company has focused on commercialization. The results are shown in Table 4-13. Forty-seven percent of Phase II respondents reported having one or more full-time marketing staffer. These data suggest that, even with Phase II funding, the very small size of the companies makes it difficult for the majority to support a full-time marketing staffer.

Conclusions: Commercialization Data

Evidence from the 2011 Survey provides useful insight into the commercialization record of SBIR companies at NSF on a number of dimensions. Overall,

TABLE 4-12 Participation in Commercialization Training

Participation in Commercialization Training?	Phase II (Percent)
Yes	60.5
No	39.5
Total	100.0
N =	177

SOURCE: 2011 Survey, Question 17.

TABLE 4-13 Full-Time Marketing Staff

One or More Full-Time Staff for Marketing?	Phase II (Percent)
Yes	46.7
No	53.3
Total	100.0
N = (company weighted)	83
N = (not company weighted)	149

NOTE: Statistical tests were run on responses weighted by company.
SOURCE: 2011 Survey, Q12.

the evidence strongly suggests that NSF is meeting its mandate to encourage the commercialization of federally funded research.

Phase II survey respondents reported that almost 70 percent of Phase II projects had already recorded sales of products or services derived from the surveyed award. A further 18 percent expected future sales. Given the relatively short time between the award date for some of the awards and the survey date, these expectations are not unreasonable.

Sales revenue of Phase II projects which had sales were highly skewed. At the upper end of the distribution, 9 percent reported sales revenue of $5 million or more, and at the lower end of the distribution, 22 percent reported sales revenue under $100 thousand. The remaining nearly 70 percent fell in between, with 42 percent reporting more than $100 thousand, but less than $1 million, and 27 percent reporting $1 million or more but less than $5 million.

Additional non-SBIR investment is another important metric of commercialization. Sixty-two percent of Phase II respondents reported that they had acquired additional funding for the project, which is evidence of the value perceived by investors even before market outcomes are achieved.

The range of market-related activities with other organizations—domestic and foreign—is yet another indication of commercial activity of Phase II SBIR award recipients. The most prevalent types of agreements reported were R&D agreements, licensing agreements, marketing and distribution agreements, and customer alliances. Multiple metrics support the idea that NSF Phase II awardees are making progress toward commercialization.

KNOWLEDGE EFFECTS

Among the four congressionally mandated objectives for the SBIR program is to "stimulate technological innovation," an objective for which knowledge effects serve as metrics. Another mandated objective is to help meet the federal agency mission, which in the case of NSF entails expanding and disseminating scientific and technical knowledge. Again, knowledge effects are pertinent metrics for assessing progress.

To measure progress toward stimulating technological innovation, patenting and patent citation analysis are potent metrics. However, in the context of small business, this standard metric of innovation does not capture the entire story. Many companies interviewed for this report indicated a preference to keep their technology secret or to rely on first-mover advantages and other market-based leverage to defend their technologies. Yet, standard metrics do provide a starting point for quantitative analysis. Consequently, the survey included several metrics related to intellectual property (IP) (i.e., patents, trademarks, copyrights, and peer-reviewed papers).

Patents

Patents are to some degree the lifeblood of high-tech firms. Because patents at small companies often result from multiple contracts in multiple projects, it is important to capture patents related to the surveyed SBIR project as well as patents more generally attributable to SBIR-funded research.

Table 4-14 shows the SBIR-related metrics provided in this study. More than 70 percent of Phase II respondents reported at least 1 patent related to an SBIR-funded technology, and 12 percent of Phase II respondents reported at least 10 such patents. The mean number of patents reported by Phase II respondents is 5.

Regarding patents awarded for the specific SBIR-funded project surveyed, Table 4-15 shows that 70 percent had at least one or more patents, while more than 30 percent of Phase II respondents reported no patents.

Respondents expressed limited interest in either trademarks or copyrights, the two other primary forms of legal protection for intellectual property. Details on these questions are provided in Appendix C.

TABLE 4-14 Number of Patents (at Least in Part) Related to All Company SBIR Awards

Number of Patents	Phase II (Percent)
0	
1 or 2	31.2
3 or 4	17.4
5 to 9	10.9
10 or more	11.9
Total	100.0
Mean	4.8
N = (company weighted)	82
N = (not company weighted)	148

SOURCE: 2011 Survey, Question 11.

TABLE 4-15 Patents Related to Surveyed Project

Number of Patents	Phase II (Percent)
0	30.0
1	32.6
2	22.0
3	4.8
More than 3	10.6
Total	100.0
At least 1	70.0
N=	227

SOURCE: 2011 Survey, Question 39.1.2.

Peer-Reviewed Publications

A broader measure of knowledge effects is provided by taking into account peer-reviewed publications. Other potential measures not included here include tacit knowledge advances such as human capital gains of researchers and others, as well as enhancements of knowledge networks.

TABLE 4-16 Peer-Reviewed Scientific Publications Related to the Surveyed Project

Number of Scientific Publications Submitted	Phase II (Percent)
0	18.4
1	21.9
2	16.7
3	11.4
More than 3	31.6
Total	100.0
At least 1	81.6
N=	114

SOURCE: 2011 Survey, Question 39.4.1.

Data from the survey suggest that companies and their staff publish widely, despite the fact that technical knowledge is often the prime source of competitive advantage of a small business. Survey data on publishing is given in Table 4-16. It shows that more than 80 percent of Phase II respondents indicated that an individual at the surveyed company submitted at least one scientific paper related to the surveyed award. Thirty-two percent reported more than three scientific papers submitted.

Links to Universities

An additional metric for knowledge transfer is the development of linkages with universities, as universities are the acknowledged source of many technological innovations in the US. Previous Academies studies have also utilized this metric for program evaluation. Although the STTR program focuses explicitly on these linkages, data from the survey indicate that SBIR projects also often develop close university ties.

The survey asked several questions about the use of university staff and facilities on the surveyed Phase II projects, and reported the results in Table 4-17. Fifty-eight percent of Phase II respondents reported a university connection of some kind, which represents an increase from the 47 percent of Phase II respondents reported in the 2007 survey.[11] The most reported types of linkage were a

[11]National Research Council, *Assessment of the SBIR Program at the National Science Foundation*, p. 265.

TABLE 4-17 Linkages to Universities

Type of Linkage	Phase II (Percent)
Faculty member(s) or adjunct faculty member(s) worked on this project in a role other than PI	35.5
Graduate students worked on this project	30.5
A university or college was a subcontractor on this project	23.2
The technology for this project was originally developed at a university or college by one of the participants in this project	18.9
The technology for this project was licensed from a university or college	12.6
The PI for this project was at the time of the project an adjunct faculty member	2.5
The PI for this project was at the time of the project a faculty member	4.3
Any of the above	57.7
N =	397

SOURCE: 2011 Survey, Question 59.

faculty member working on the project but not as a principal investigator (PI), graduate students employed on the project, and a university or college as a subcontractor on the project. Also notable among the reported types of linkages was the origination of the project technology from a university or college, either by development by one of the project participants or by having been licensed from a university or college.

Respondents identified 114 different universities and colleges with which they worked in various capacities on the surveyed project. Those universities or colleges mentioned by three or more respondents are listed in Table 4-18.

Many of the institutions are large state universities, some of which have in recent years focused on technology transition as well as basic research. At the top of the tabular listing for having been specifically mentioned as playing a role in at least five surveyed NSF SBIR projects are the Universities of Colorado, Florida, Texas, Georgia, and California-San Diego, plus MIT and Columbia. When all mentions are taken into account (including those universities and branches that received fewer than 3 and therefore are not included in Table 4-18), the University of California system had 8 mentions and the University of Texas had 7. Although far from a perfect metric, the data shown in the table provide a preliminary indication of the connections between specific universities, university systems, and the NSF SBIR program.

Also indicative of a strong linkage between academia and the SBIR was the reporting that about 80 percent of companies had at least one founder with an

TABLE 4-18 University Participants Mentioned by Three or More Respondents

University Name	Number of Mentions
University of Colorado	8
University of Florida	8
University of Texas at Austin	7
Georgia Institute of Technology	6
University of California, San Diego	5
Massachusetts Institute of Technology (MIT)	5
Columbia University	5
University of Illinois	4
University of Delaware	4
University of Nebraska	4
University of Minnesota	4
University of Michigan	4
University of Utah	4
Carnegie Mellon University	3
CalTech	3
University of Arkansas	3
Colorado School of Mines	3
University of Maryland	3
Colorado State University	3
University of North Carolina	3
University of Wisconsin-Madison	3
University of California, Los Angeles (UCLA)	3
Montana State University	3
Stanford University	3

SOURCE: 2011 Survey, Question 60.

academic background, and for slightly less than 30 percent of companies, at least one founder was most recently employed at a college or university.

Conclusions: Knowledge Effects

The survey data reveal that SBIR-funded projects generate a substantial amount of technical knowledge that enters the public domain, thus meeting the congressional objectives for the program to stimulate technological innovation, and to help meet NSF's agency mission to advance the nation's scientific and technical knowledge. What emerges from the survey data is a picture of companies that are embedded in the innovation economy and that use the patent system to protect their intellectual property. More than 70 percent of Phase II companies received at least one patent based on their work under SBIR contracts, while 70 percent received at least one patent related to the surveyed project alone.

That SBIR companies help expand the nation's knowledge base is also indicated by their publishing in peer-reviewed journals. More than four-fifths of the NSF Phase II companies published at least one article based on the SBIR-funded work, and about one-quarter published at least three such papers. These data do not include less formal and perhaps more ubiquitous knowledge transmission, such as conference papers and presentations, and tacit forms of knowledge building, such as embodying human capital in researchers and entrepreneurs.

Finally, many SBIR companies are closely connected to universities. Fifty-eight percent of NSF Phase II respondents reported a university connection on the surveyed project, and seven universities were specifically mentioned as playing a role in at least five surveyed NSF SBIR projects.

ASSESSING THE IMPACT OF NSF SBIR AWARDS USING COUNTERFACTUALS

It is difficult to determine the impact of a given SBIR award or program. Not only do many factors affect the success and failure of companies and projects, but also determining whether a specific factor was a *necessary* condition for success is challenging. Furthermore, the large number of factors and the multiple paths to success and failure lessen the ability to state with confidence that a particular intervention—in this case an SBIR award—constitutes a *sufficient* condition for a project's or company's success.

An alternative, generally practicable approach to assessing impact in applied program evaluation when experimental or quasi-experimental design is infeasible is to use a "counterfactual approach." This entails asking recipients for their views of the program's impact on their project or company by asking, "What would have happened to the project absent the SBIR award? Would it have proceeded anyway and, if so, in what form?" This section addresses responses to the counterfactual questions.

Project Go-Ahead Absent SBIR Funding

The survey asked Phase II recipients whether the surveyed project would have been undertaken absent SBIR funding and, if so, whether the scope and timing would have been affected. Table 4-19 summarizes the responses. Fourteen percent of the respondents believed that the project likely would have proceeded without SBIR funding, but only 2 percent were certain. In contrast, more than

TABLE 4-19 Project Undertaken in the Absence of This SBIR Award

In the Absence of Award, Would Company Have Undertaken Project?	Phase II Respondents (Percent)
Yes	13.7
Definitely yes	2.2
Probably yes	11.5
Uncertain	18.1
No	68.2
Probably not	41.8
Definitely not	26.4
Total	100.0
N =	182

SOURCE: 2011 Survey, Question 24.

two-thirds believed the project likely would not have proceeded absent SBIR funding: 26 percent were definite and a further 42 percent thought it unlikely. Data published in the Academies' 2009 report were very similar.[12] In fact, when project funding cannot be secured in a timely way, the assembled team may dissipate to find other sources of income, leaving the applicant unable to continue.

These data have interesting wider implications for debates about early-stage funding: they suggest a weakness in the "crowding out" hypothesis, because it appears that awardees—presumably those with the closest knowledge of funding prospects for the project—overwhelmingly believed it unlikely that funding alternative to SBIR could be found.[13]

Project Scope Absent SBIR Funding

A second area of review concerns the impact of funding on the scope of the project. It seems likely that SBIR funding would lead to an expansion of the project scope. However, the selection of SBIR awards is based on criteria that

[12]National Research Council, *An Assessment of the SBIR Program of at the National Science Foundation*, Appendix B.

[13]"Crowding out effect" refers to the hypothesis that increasing public sector investment in an area will replace (drive down) private sector spending. See review of crowding out literature in Dirk Czarnitzki and Andreas Fier, "Do Innovation Subsidies Crowd Out Private Investment? Evidence from the German Service Sector," ZEW Discussion Papers, No. 02-04, 2002.

can be drawn quite narrowly. It is therefore possible that tailoring a project to the requirements of a particular solicitation could reduce the scope of the project.

The Committee's analysis focused only on those respondents who were certain that the project would have proceeded without SBIR funding. Table 4-20 shows that, although 75 percent of respondents indicated that the absence of SBIR funding would have limited the project's scope, 8 percent indicated that the project's scope had been limited by participation in the program, most likely to address the specific, narrowly drawn requirements of SBIR topics.

Project Delays Absent SBIR Funding

As with project scope, the immediate supposition is that, absent SBIR funding, projects would have been delayed while other funding was identified and acquired. However, as discussed in Chapter 2, SBIR awards carry delays of their own, which can be substantial in some cases. Thus, this survey question seeks to determine a balance between delays imposed by the need for new funding and delays inherent in the SBIR program.

As shown in Table 4-21, a large majority of respondents who were certain that the project would have proceeded absent SBIR funding agreed that the absence of SBIR funding would have delayed the project. Sixty-five percent of NSF Phase II respondents reported that the project would have been delayed by at least 12 months, 26 percent by at least two years, and 9 percent by at least 3 years. Given that gaps and delays can significantly impact the ability of small companies with limited resources to retain their technical teams, this is a potentially important result. However, the small sample size indicates that these results, although important, should be treated with caution.

TABLE 4-20 SBIR Impact on Scope of Project

Project Scope Absent Award?	Phase II Respondents (Percent)
Broader in scope	8.3
Similar in scope	16.7
Narrower in scope	75.0
Total	100.0
N =	24

NOTE: Only includes respondents who were certain that project would proceed absent SBIR funding.
SOURCE: 2011 Survey, Question 25.

TABLE 4-21 SBIR Impact on Project Delay

Delay Absent Award?	Phase II (Percent)
Under 12 months	34.8
12-23 months	39.1
24 to 35 months	17.4
36 months or longer	8.7
Total	100.0
N =	23

SOURCE: 2011 Survey, Question 26B.

SBIR Funding and Project Duration

The survey also asked respondents to determine how the absence of SBIR funding would have affected the duration of the project. Table 4-22 shows the results. Eighty-three percent reported that the project would have taken longer, while 4 percent thought it would have taken less time without the SBIR award. This too is a potentially important finding in that project delays can delay potential market entry, which can be critical because the window for market entry can be narrow. Again, the small sample size indicates that these results, although important, should be treated with caution.

Long-Term Impacts on the Recipient Company

Although SBIR awards have direct effects on specific projects, they can also have a longer term effect on the trajectory of company development, creating capacity and in some cases providing a needed input that transforms long-term outcomes.

The survey asked respondents about the SBIR program's impact on the company's long-term development. Table 4-23 reveals overwhelmingly positive long-term impact on Phase II companies. Overall, 35 percent of Phase II respondents indicated that the SBIR program had a transformative effect on their company, and an additional 54 percent reported a strongly positive effect. Negative effects were negligible.

Key Aspects of SBIR-Driven Transformation

It is not easy to summarize the numerous ways in which NSF SBIR awards helped to transform recipient companies. Therefore, what follows is a limited list of impacts drawn from verbatim comments from respondents who

TABLE 4-22 SBIR Impact on Project Duration

Duration Absent Award?	Phase II (Percent)
Longer	83.3
The same	12.5
Shorter	4.2
Total	100
N =	24

SOURCE: 2011 Survey, Question 26B.

TABLE 4-23 Long-Term Impacts on Recipient Companies

Long-Term Impact	Phase II (Percent)
Had a transformative effect	34.8
Had a substantial positive long-term effect	53.6
Had a small positive effect	9.3
Had no long-term effect	2.4
Had a negative long-term effect	—
Total	100.0
N = (company weighted)	98
N = (not company weighted)	179

SOURCE: 2011 Survey, Question 45.

reported that SBIR awards had transformed their companies. Respondents indicated that SBIR

- provided first dollars,
- funded areas that did not interest venture capitalists and other funders,
- opened doors to many potential stakeholders in specific technologies, including agencies, prime contractors, investors, suppliers, subcontractors, and universities,
- assisted entry into niche markets too small for major players/funders,
- funded technology development,
- enabled projects with high levels of technical risk,
- supported adaptation of technologies to new uses, markets, and industry sectors,

- diversified expertise and allowed for hiring of specialists,
- provided critical funding during downturns in the business cycle,
- attracted and developed young researchers,
- redirected company activities to new opportunities,
- reduced costs,
- addressed needs that require high tech at low volume and relatively low cost,
- provided companies with credibility because SBIR research is considered to be peer reviewed,
- funded researchers to enter business full time,
- transformed company culture to become more market driven,
- drove researchers to focus on technology transition,
- created new companies and kept companies in business (that would not exist without SBIR funding),
- supported feasibility testing for high-risk/high-payoff projects,
- supported projects with longer time horizons/long sales cycles,
- provided the basis for spin-off companies,
- funded proof of concept,
- encouraged R&D companies to transition into manufacturing,
- funded disruptive technologies,
- provided significant mentoring especially for new businesses,
- stimulated international collaboration, and
- reduced technological risk.

Overall, the strongest conclusion to be drawn from these responses is that small innovative companies are highly sensitive to the impact of exogenous variables. The sudden withdrawal of a sponsor can crush a company; a single contract can provide funding for 2 or 3 years of growth. And above all, these small companies are highly path dependent: what happens to them at a given moment can dramatically affect long-term outcomes. The butterfly effect could have been invented specifically to apply to these kinds of companies.

In the end, the SBIR program has often been for many companies a profoundly positive exogenous variable: one that provides funding, validation, and often market access not otherwise available. Even though it seems tenuous to link one SBIR award to the eventual success of a large company, in fact SBIR awards have been pivotal factors in the growth of what are now large and successful companies. The evidence from survey respondents suggests that this powerful positive jolt is not an uncommon effect of SBIR awards.

Key aspects of SBIR-driven transformation are explored in more detail in Chapter 6, which draws extensively on the numerous open-ended comments received in response to a question about the long-term impact of SBIR awards on recipient companies, as well as on the case studies.

COMPANY AND PROJECT CHARACTERISTICS: INDEPENDENT VARIABLES

The outcomes section above strongly suggests that, overall, Phase II SBIR funding at NSF correlates with commercial success. However, a variety of other factors may play a role, and the survey sought to address these.

Founders and Company Foundation

Venture investors focus heavily on the composition of the company team when deciding whether to make an investment. Accordingly, it seemed appropriate to explore the characteristics of company founders.

Table 4-24 shows most Phase II companies reported that they had multiple founders. Only 30 percent reported a single founder; another 30 percent reported two cofounders; the remaining 41 percent (rounding error) reported three or more cofounders.

An analysis of the number of previous companies (of any kind—not just SBIR recipients) founded by given company founders makes it possible to estimate the use of SBIR awards by serial company founders. As may be determined by Table 4-25, more than half of recipient companies had at least one founder who had previously founded another company, while 46 percent had not. More than a quarter of recipient companies had founders who had previously founded multiple companies.

TABLE 4-24 Number of Founders per Responding Company

Number	Phase II (Percent)
1	29.6
2	29.5
3	22.8
4	10.8
5 or more	7.4
Total	100.0
N = (company weighted)	219
N = (not company weighted)	374

NOTE: The survey received responses from multiple respondents per company. To address company-level questions, these responses were weighted equally so that taken collectively responses from each company had equal impact on statistical analysis. This approach was used for tables based on Questions 3-12 inclusive in the survey questionnaire (see Appendix C).
SOURCE: 2011 Survey, Question 4A.

TABLE 4-25 Number of Previous Companies Started by Founders

Number	Phase II (Percent)
0	46.1
1	28.3
2	14.2
3	7.2
4	1.2
5 or more	3.0
Total	100.0
At least 1	53.9
N= (company weighted)	84
N= (not company weighted)	151

SOURCE: 2011 Survey, Question 4B.

It is worthwhile to learn whether founders had a business background or an academic background before founding their current company. Together the following series of tables provides a good overview of founder background: Table 4-26 shows the share of companies that have founders with a business background; Table 4-27 shows the share with an academic background; and Table 4-28 shows the share with founders most recently employed in business, academia, government, or other areas.

Table 4-26 shows that 64 percent of Phase II companies had at least one founder with a business background, and 23 percent had more than one founder with a business background.

Table 4-27 shows that about 80 percent of Phase II respondents reported at least one founder with an academic background. More than 40 percent of Phase II respondents reported at least two founders with academic backgrounds.

Table 4-28 shows that 73 percent of Phase II respondents reported at least one company founder most recently employed at another private company, while 28 percent of Phase II reported at least one founder previously employed at a college or university. Only 9 percent of Phase II respondents reported most recent prior employment in government and 5 percent in other areas.

Previous Academies studies concluded that, for at least some companies, SBIR funding provided opportunities that led directly to company formation. Table 4-29 shows that 45 percent of Phase II respondents indicated that the surveyed award contributed to some degree to company formation. More than

TABLE 4-26 Number of Founders with Business Backgrounds

Number	Phase II (Percent)
0	36.2
1	40.9
2	16.1
3	4.3
4	1.0
5 or more	1.5
Total	100.0
At least 1	63.8
N = (company weighted)	216
N = (not company weighted)	370

SOURCE: 2011 Survey, Question 4C.

TABLE 4-27 Number of Founders with Academic Backgrounds

Number	Phase II (Percent)
0	19.1
1	38.5
2	21.8
3	11.9
4	5.8
5 or more	2.9
Total	100.0
At least 1	80.9
N = (company weighted)	216
N = (not company weighted)	369

SOURCE: 2011 Survey, Question 4D.

TABLE 4-28 Most Recent Employment of Founders

Most Recent Employment	Phase II (Percent)
Other private company	73.1
College or university	27.6
Government	9.2
Other	4.8
N (unique companies) =	391
N (unique respondents) =	671

NOTE: Totals do not sum to 100 percent because respondents were permitted to check multiple responses.
SOURCE: 2011 Survey, Question 5.

TABLE 4-29 Company Founded Because of SBIR Program

Was Company Founded Because of SBIR Program?	Phase II (Percent)
Yes	24.1
In part	21.3
No	54.6
Total	100.0
N = (company weighted)	85
N = (not company weighted)	152

SOURCE: 2011 Survey, Question 6.

20 percent of Phase II respondents indicated that the company had been founded "because of SBIR."

These data on founders and company foundation indicate that SBIR company founders tended to have strong connections to both business and academia. Companies with founders having a business background were prevalent among recipients of Phase II funding, as were those companies having founders with an academic background, and those companies having more than one founder. It may be that more experience in the private sector and a better understanding of how to address multiple needs within company operations are associated with Phase II success. The observation overall is consistent with the dual emphasis of the NSF SBIR program on both technical strength and commercial potential. Consistent with previous survey findings, SBIR funding had a strong positive effect for many of the respondents on the decision to form a company.

Industry Sector

Previous analysis of the SBIR program did not address a potentially important intervening variable: industry sector. To the extent that commercialization outcomes are affected by the average cycle time of product development in different sectors, the industry sectors of SBIR awards are relevant to impact assessment. For example, product cycle time is much shorter in software than in materials or medical devices.

Table 4-30 shows the distribution of responses of Phase II respondents to a question designed to provide an approximate map of activities by industry sector. There is considerable overlap between some categories, and respondents had substantial leeway to define sectors differently, so these results should be viewed as highly preliminary.

Two key points emerge from the responses. First, engineering and materials were the dominant sectors (at 43 percent and 47 percent of all responses, respectively), followed by information technology (IT) at 21 percent. To a considerable degree, the responses reflect the decisions made at NSF to put forward selected topics in specific areas (as explained in Chapter 2), rather than the endogenous interests of companies themselves.

Project Status and Review of Discontinued Projects

Because the 2011 survey covers 10 years of Phase II awards, projects were at different stages of completion at the time of the survey. Therefore, as noted in previous Academies analyses, project outcomes were in aggregate substantially underreported because of the number of projects whose entire life cycle was not yet complete. Table 4-31 shows the status of the projects at the time of the survey.

Phase II respondents reported products or processes in use at a rate of 47 percent. Fifteen percent of respondents reported that the project is continuing post-award technology development, while 22 percent reported that the project had been discontinued by the company recipient of the award.[14]

As shown in Table 4-32, the survey sought reasons for why projects had been discontinued. Phase II companies were most likely to discontinue their SBIR projects for multiple reasons, among which facing a market that was too small dominated at 35 percent. This problem was followed by not enough funding (23 percent) and technical failure or difficulties (also 23 percent). In fact, the Phase II reasons largely focused on a mismatch between the company's product or service and the marketplace. Thus, it may be that helping companies to address this mismatch or better screening out projects without market potential at the selection stage could be a focus of further initiatives to improve commercialization rates.

---

[14]Sometimes SBIR-funded research is picked up later at another company. This survey focused on the original recipient.

TABLE 4-30 Respondent Identification of Phase II Awards by Industry Sector

Industry Sector	Phase II (Percent)
Aerospace	11.2
Defense-specific products and services	10.2
Energy and the environment	20.9
—Sustainable energy production (solar, wind, geothermal, bioenergy, wave)	7.3
—Energy storage and distribution	4.4
—Energy saving	6.8
—Other energy or environmental products and services	3.4
Engineering	42.7
—Engineering services	4.9
—Scientific instruments and measuring equipment	16.0
—Robotics	1.5
—Sensors	15.0
—Other engineering	9.2
Information technology	20.9
—Computers and peripheral equipment	3.4
—Telecommunications equipment and services	2.4
—Business and productivity software	6.3
—Data processing and database software and services	5.3
—Media products (including web-, print- and wireless-delivered content)	2.9
—Other IT	2.4
Materials (including nanotechnology for materials)	46.6
—Medical technologies	4.9
—Pharmaceuticals	6.3
—Medical devices	9.7
—Other biotechnology products	9.7
—Other medical products and services	1.5
Other	17.5
N =	206

NOTE: Answers do not sum to 100 percent because respondents could select more than one response. Answers within categories do not sum to category totals for the same reason.
SOURCE: 2011 Survey, Question 20.

TABLE 4-31 Current Status of Surveyed Projects

Current Status	Phase II (Percent)
Project has not yet completed (SBIR/STTR) funded research	2.2
Efforts at this company have been discontinued	22.0
Discontinued because no sales or additional funding resulted from this project	12.6
Discontinued but the project did result in sales, licensing of technology, or additional funding	9.3
Project is continuing post-award technology development	14.8
Commercialization is under way	13.7
Products/processes/services are in use	47.3
In use by target customers	36.3
In use by customers not anticipated at the time of the award	11.0
Total	100.0
N =	182

SOURCE: 2011 Survey, Question 30.

TABLE 4-32 Reasons for Project Discontinuation

Reason	Phase II (Percent)
Not enough funding	22.5
Failed to receive Phase II award funding	2.5
Market demand too small	35.0
Company shifted priorities	17.5
Technical failure or difficulties	22.5
Product, process, or service not competitive	15.0
Level of technical risk too high	12.5
Project goal was achieved (e.g., prototype delivered for federal agency use)	7.5
Licensed to another company	7.5
Inadequate sales capability	10.0
Another firm got to the market before us	7.5
Principal investigator left	5.0
Other	17.5
Multiple reasons contributed	52.5
N = (discontinued projects only)	40

SOURCE: 2011 Survey, Question 31.

TABLE 4-33 Primary Reason for Project Discontinuation

Primary Reason	Phase II (Percent)
Failed to receive Phase II award funding	2.5[a]
Market demand too small	25.0
Not enough funding	17.5
Technical failure or difficulties	7.5
Company shifted priorities	5.0
Product, process, or service not competitive	12.5
Licensed to another company	5.0
Level of technical risk too high	5.0
Principal investigator left	2.5
Project goal was achieved (e.g., prototype delivered for federal agency use)	2.5
Inadequate sales capability	0.0
Another firm got to the market before us	0.0
Other reason you mentioned	15.0
N = (discontinued projects only)	40

[a]A Phase II project may fail because it did not receive a second related Phase II at a later date.
SOURCE: 2011 Survey, Question 32.

These findings are further supported by responses to an additional question, which asked respondents to indicate which factor was the primary reason for discontinuation. As shown in Table 4-33, a lack of market demand at 25 percent was the single largest reason given for Phase II project discontinuation, followed by not enough funding at 18 percent.

Based on the data provided in these tables, it seems reasonable to conclude that SBIR funding makes a substantive difference to the plans of companies, and that the absence of alternative funding is often a reason for project discontinuation. This finding in turn suggests that the crowding out hypothesis is at best of limited help in explaining company activity in this area.

Company Size by Revenue

The SBIR program is aimed at supporting small firms. As employment data show, most awardee companies are much smaller than the SBA maximum size for small companies (which is typically 500 employees in most sectors).[15]

[15]See Small Business Administration, Table of Small Business Size Standards Matched to North American Industry Classification System Codes, accessed October 20, 2013, <*http://www.naics.com/naicsfiles/Size_Standards_Table.pdf*>.

Table 4-34 shows company size in terms of company revenue as of the most recent fiscal year prior to the survey. Just slightly under half of Phase II respondents reported company revenue below $1 million for the most recent fiscal year, while 13 percent reported company revenue below $100 thousand. More than a third of Phase II companies reported revenue of at least $1 million but less than $5 million, and 15 percent reported revenue of $5 million or more. The median company size as indicated by amount of revenue was $1.1 million for Phase II respondents.

Company Activities and SBIR

An often raised issue about the SBIR program relates to the existence of SBIR "mills," that is, companies that live off a stream of SBIR contracts with little (if any) commercialization of the product or service. Previous Academies reports addressed this issue; it is addressed in Chapters 2 and 3 of this report; and it is addressed in company interviews (see Chapter 6). Many interviewees explained that the SBIR program's role in a company changes over time and that companies generally become less dependent on SBIR funding as products reach the market.

TABLE 4-34 Total Company Revenues, Most Recent Fiscal Year

Revenues	Phase II (Percent)
Less than $100,000	12.8
$100,000-$499,999	18.3
$500,000-$999,999	18.0
$1,000,000-$4,999,999	36.4
$5,000,000-$19,999,999	10.9
$20,000,000-$99,999,999	3.7
$100,000,000 or more	-
Total	100.0
MEAN (Millions of Dollars)	4.9
Grouped Median (Millions of Dollars)	1.1
N = (company weighted)	115
N = (not company weighted)	203

NOTE: Statistical tests were run on responses weighted by company.
SOURCE: 2011 Survey, Question 8.

SBIR Share of R&D Effort

The survey asked respondents to estimate how much of their company's total R&D effort—defined as man-hours of work for scientists and engineers—was devoted to SBIR-funded projects.

As Table 4-35 shows, 45 percent of Phase II respondents indicated that SBIR currently constitutes 10 percent or less of overall company R&D man-hour effort, while 11 percent indicated that it amounted to more than 75 percent. For 76 percent of Phase II respondents, SBIR-funded projects accounted for no more than half the company's R&D effort in the most recent fiscal year.

These responses correspond fairly closely to responses from another survey question about the percentage of company revenues derived from SBIR awards. As shown in Table 4-36, while 34 percent of Phase II respondents reported that their companies received none of their revenue from SBIR sources during the most recent fiscal year, the majority reported that part of their revenue came from SBIR awards. The part of company revenue composed of SBIR funds varied substantially. Fourteen percent of Phase II companies received up to 10 percent of their company revenue in the most recent fiscal year from SBIR awards; 30 percent received up to 25 percent; and 46 percent received as much as 50 percent. Nineteen percent of companies received more than 50 percent of their revenues from SBIR.

TABLE 4-35 Science and Engineering Man-Hours: Percentage Devoted to SBIR Projects, Most Recent Fiscal Year

Science and Engineering Man-Hours Devoted to SBIR Awards	Phase II (Percent)
0	29.4
1-10	15.8
11-25	13.5
26-50	17.8
51-75	12.5
76-100	11.0
Total	100.0
N = (company weighted)	84
N = (not company weighted)	150

NOTE: Statistical tests were run on responses weighted by company.
SOURCE: 2011 Survey, Question 7.

TABLE 4-36 Percentage of Company Revenue from SBIR, Most Recent Fiscal Year

Revenues	Phase II (Percent)
0	34.4
1-10	13.8
11-25	16.6
26-50	16.0
51-75	10.5
76-99	8.7
100	-
Total	100.0
N = (company weighted)	83
N = (not company weighted)	149

SOURCE: 2011 Survey, Question 9.

Prior Use of the SBIR Program

Although standard models of innovation depict a linear sequence whereby ideas are tested in Phase I, prototyped in Phase II, and commercialized in Phase III,[16] there is considerable evidence to suggest that this approach oversimplifies the process. Often, multiple iterations are required, projects must restart with an earlier phase, or multiple efforts are needed to meet specific problems.

The survey asked respondents to indicate how many prior SBIR or STTR Phase I awards had been received by the company that were related to the surveyed award. As Table 4-37 shows, In general, there was little difference between Phase I and Phase II responses. Overall more than 80 percent of companies received at least one prior Phase I award related to the surveyed award. Phase II respondents were more likely to report at least two or more, as well as five or more prior related Phase I awards.

In addition, Phase II respondents were likely to have received at least one prior related Phase II award (see Table 4-38)—almost 85 percent did so, and a third received two or more. Conversely, Phase I respondents were more likely to have not received any related Phase II awards, which was expected because these respondents were explicitly selected to complete the survey because they had not received a Phase II award from the study agencies during the study period.

[16]For a review of the model and its intellectual history, see Benoît Godin, "The Linear Model of Innovation: The Historical Construction of an Analytical Framework," Project on the History and Sociology of S&T Statistics, Working Paper No. 30, 2005.

TABLE 4-37 Prior SBIR or STTR Phase I Awards Related to the Surveyed Project

Number of Prior Related Phase I Awards	Phase I (Percent)	Phase II (Percent)
0	15.7	13.6
1	45.5	38.5
2	24.6	24.3
3	9.4	10.7
4	2.1	4.7
5 or more	2.6	8.3
Total	100.0	100.0
1 or more	84.3	86.4
N =	191	169

SOURCE: 2011 Survey, Question 40.1.1.

TABLE 4-38 SBIR or STTR Phase II Awards Related to the Surveyed Project Technology

Number	Phase II (Percent)
0	15.5
1	51.2
2	19.0
3	8.3
4	3.0
5 or more	3.0
Total	100.0
1 or more	84.5
N =	168

SOURCE: 2011 Survey, Question 40.1.2.

These data suggest that projects require more than one Phase II award before they commercialize, which in turn supports the view that innovative products emerge from clusters of activity rather than from a simple linear development from Phase I to Phase II to commercialization.

Prior Investment

One question surrounding the SBIR program related to its role in the sequence of funding that leads from an idea to a product. Table 4-39 probes the sources of funding for the given technology area prior to company receipt of the Phase II SBIR award. About two-thirds of all Phase II respondents reported at least one source of additional funding for the technology areas prior to the Phase II SBIR award. The most prevalent source of funding prior to the Phase II SBIR award was internal company investment (including borrowed money), reported by half the respondents. Previous SBIR/STTR funding (excluding preceding Phase I) was another prevalent source, reported by 45 percent of Phase II respondents. Private investors (including angel funding) as a source of funding was reported by 21 percent of Phase II respondents. Other private companies, prior non-SBIR/

TABLE 4-39 Sources of Funding, Prior to SBIR Phase II Award, for Related Technology

Source	Phase II (Percent)
Internal company investment (including borrowed money)	50.4
Private investor (including angel funding)	21.3
Prior [SBIR/STTR] (excluding preceding Phase I)[a]	45.4
Other private company	15.6
Prior non-[SBIR/STTR] federal R&D	12.8
State or local government	12.1
Venture capital	7.1
College or university	5.0
Other	6.4
N =	141

[a] Question only asked for Phase II awards.
NOTE: Totals do not sum to 100 percent because respondents could select more than one funding source.
SOURCE: 2011 Survey, Question 21.

STTR federal R&D, and state or local government followed in order of importance as sources of funding, together reported by 40 percent of respondents as funding sources for their companies. Venture capital was reported as a source of funding by only 7 percent of Phase II respondents.

SUMMARY: QUANTITATIVE IMPACT ANALYSIS

The foregoing analysis can be summarized briefly as follows:

1) SBIR-funded projects at NSF tend to reach the market in large numbers—about 70 percent of surveyed Phase II projects reported sales of related products or services, and almost two-thirds of the remainder expected sales in the future.
2) The scale of commercialization is not large for most projects. Of those Phase II projects experiencing sales from the funded technology, only about 9 percent reported sales of $5 million or more. A few projects produce most of the commercial impact.
3) The SBIR program should not be considered an important program for creating jobs. NSF recipient companies are very small, even within the universe of small companies, and although those that survive do add jobs over time, they do not do so at a rate that would make job creation a primary impact of the program.
4) SBIR funding makes a significant difference in the founding of small innovative companies and the decision to proceed with a specific project: 45 percent of respondents reported that the SBIR program played a role in company foundation, and almost 70 percent of Phase II respondents believed that the surveyed project probably or certainly would not have proceeded without Phase II funding.
5) Knowledge effects in the case of NSF serve two mandated congressional goals for SBIR: (1) to stimulate technological innovation, and (2) to help meet federal agency mission, which in the case of NSF relates to advancing scientific and technical knowledge. Patenting and publications are important metrics for measuring contributions to these goals. About 70 percent of Phase II respondents reported receipt of at least one project-related patent. Similarly, over 80 percent of Phase II respondents reported publishing at least one peer-reviewed technical paper related to the surveyed award; about 40 percent had published at least three such papers.
6) NSF SBIR projects have substantial links with universities: about 57 percent of Phase II projects reported some university involvement, 114 different universities and colleges were identified as being involved in the surveyed project, 24 universities were specifically mentioned by at least three respondents, and 7 universities were specifically mentioned by at least five respondents.

7) Most respondents indicated that the project likely would not have proceeded without SBIR funding. And even among those who believed the project would have proceeded without SBIR funding, most expected that it would have been delayed and/or reduced in scope. Overall, respondents reported that SBIR had profoundly positive long-term impacts on their company. More than one-third of Phase II respondents reported that SBIR had a "transformative effect," and 88 percent reported a substantial positive effect or a transformative effect.
8) Given that the SBIR program supports companies and ideas at earlier stages of development than venture capital investment, the case can be made that it would be appropriate to follow SBIR companies that are showing success over a longer period of time for impact assessment.

5

The Phase IIB Program

The Phase IIB program (formally entitled the "Phase IIB Option" or the "Phase IIB Supplement") is designed to bridge the funding gap between the end of SBIR Phase II and the start of commercial revenues or investment (known as Phase III, but not funded through the SBIR program itself).

FEATURES OF THE PHASE IIB PROGRAM

Introduced by the National Science Foundation (NSF) in 1998,[1] the Phase IIB program was a response to the growing body of evidence that small companies encounter a "Valley of Death" between the end of government funding (in many cases for basic research) and the advent of commercial revenues or commercially based investments.[2] This funding gap appears to be especially relevant to innovative small companies, which may encounter substantial research and development (R&D) expenses well before they generate revenue from their products. As NSF states, "The objective of the Phase IIB Option is to extend the R&D efforts beyond a current grant to meet the product/process/software requirements of a third-party investor to accelerate the Phase II project to the commercialization stage and/or enhance the overall strength of the commercial potential of the Phase II project."[3]

[1]Kesh S. Narayanan, "Testimony Before the Small Business Committee U.S. House of Representatives," April 22, 2009, accessed October 14, 2010, *<http://www.nsf.gov/about/congress/111/ksn_tech_090422.jsp>*.

[2]See National Research Council, *An Assessment of the SBIR Program* (Washington, DC: The National Academies Press, 2008).

[3]National Science Foundation, Phase IIB web page, accessed May 21, 2015, <http://www.nsf.gov/pubs/1999/nsf9957/Chapt-10.htm>. Unless otherwise indicated, information about the NSF Phase IIB program is drawn from the referenced web page, which is the official NSF description of the program.

TABLE 5-1 Leveraged External Funding, 2004-2008

Phase IIB Year	Leveraged Investment (Millions of Dollars)
2008	18.5
2007	36.7
2006	57.8
2005	43.5
2004	10.6

SOURCE: Dr. Kesh Narayanan, NSF, Congressional Testimony, April 22, 2009.

The program was designed to leverage NSF investments to accelerate and enhance commercialization. Funding for standard NSF Phase II SBIR awards was, in general, limited to $500,000, which is $250,000 less than the maximum allowed by Small Business Administration (SBA) guidelines prior to reauthorization.[4] The lower limit provided room for NSF to add Phase IIB funding under the overall Phase II award ceiling, focused on projects that NSF believes to have substantial commercial potential. Critically, third-party validation is provided by the requirement that companies match Phase IIB funds at least 2:1, providing two dollars for every dollar provided by NSF. To ensure that this third-party validation is substantive, only certain kinds of match are permitted (see criteria below under subheading, "Funding Criteria"). The match requirement has the effect of leveraging NSF resources: NSF indicates that substantial additional funding has been leveraged as a result (see Table 5-1). (This question is further addressed in the discussion of survey responses later in this chapter.)

Projects that receive less than $250,000 from NSF are extended for 1 year (making a total of 3 years for Phase II and IIB combined); projects that receive more than $250,000 are extended for an additional 2 years. NSF notes that Phase IIB grants, like other grants, may be completed early. The maximum size of an NSF Phase IIB award is $500,000.

In his congressional testimony, Dr. Narayanan, then Director of the Division of Industrial Innovation & Partnerships, observed, "We have found that awardees that are able to secure the outside funding to qualify for Phase IIB have had better success in commercializing their innovations. After 5 years, about 69 percent of companies that received Phase IIB funding were beginning to see success, whereas only 31 percent of those not having a IIB supplement were successful. Many of the Phase IIB companies have grown in both revenue and employ-

[4]All discussion of the NSF Phase IIB program in this chapter should be viewed in the context of the program prior to reauthorization.

ment and some have been acquired by larger companies."[5] Thus by design, the Phase IIB program has a dual impact: it directly provides promising companies with additional SBIR funding, and at the same time, by requiring that recipients attract at least a 2:1 match from other sources, it leverages the NSF investment.

Phase IIB is a one-time opportunity for NSF Phase II recipients. It applies only to Phase II awards that are currently active, and resubmission at a later date is not permitted. Projects that have received a no-cost extension are not eligible for Phase IIB.

Phase IIB Applications

Phase IIB applications can be submitted at any time during the related Phase II project, up to 30 days before the expiration of the Phase II award. Procedures differ for companies that seek NSF funding of less than or more than $250,000, respectively. All companies must do the following:

- Receive permission to submit an investment package for review
- Submit a completed Phase IIB investment package, which includes documentation that specify the source and conditions governing the Phase IIB matching funds
- Pass NSF Phase IIB Committee review
- Receive application approval from the NSF Phase IIB Committee
- Submit a Phase IIB proposal
- Receive Phase IIB approval from NSF
- Companies seeking $250,000-500,000 must in addition provide an oral presentation to the NSF Phase IIB committee. Presenters must include the Principal Investigator (PI), the CEO, and the third-party investor.

Funding Criteria

Only certain kinds of funding meet NSF criteria to count as matching funds. The NSF Phase IIB web page contains two different descriptions of these criteria:

- Third Party Documentation: "The investment package consists of executed third party documentation from the third party investor(s). The documentation must specify the amount of the investment and the method by which the investor will provide the funding to the company. The third party funding can be cash, liquid assets, tangible financial instruments but not in-kind or other 'intangible assets'. Loans and investments with contingency clauses are not acceptable. Self-funding does not qualify for

[5]Kesh S. Narayanan, "Testimony Before the Small Business Committee U.S. House of Representatives."

the Phase IIB option. The company provides NSF with a letter from the investor(s) stating that the investment was a direct consequence of the NSF Phase II project."

- Link to Phase II Funding: "The third party funding must be a direct consequence of the Phase II award. This funding can be cash, product sales revenues and/or licensing revenues, liquid assets, tangible financial instruments but not in-kind or other 'intangible assets'. Loans and investments with contingency clauses are not acceptable. Self-funding does not qualify for the Phase IIB option" (emphasis in the original).[6]

While there is some overlap between these two descriptions, they are not identical. In particular, the first definition does not include sales or other market revenues, while the second definition does. In fact, the tone of the first definition—which is reiterated separately for applications of less than and more than $250,000—appears to require that matching funds meet a more standard meaning of "investment." The survey evidence discussed below reveals confusion among Phase IIB awardees on this point.

The second description also mandates that the funding be a "direct consequence of the Phase II award." This has the effect of ruling out any funding generated as a consequence of the Phase I award or other company activities that may be coterminous with the Phase II award—a point made in interviews with NSF Phase IIB awardees, some of whom at least believe that this interpretation is implemented at NSF.[7]

NSF has imposed further requirements on larger Phase IIB awards. When Phase NSF IIB funding will be in excess of $250,000, matching funds must come from nongovernmental, private-sector, third-party sources. This may, in some cases, conflict with the inclusion of sales revenues as matching funds, because some NSF projects (about 10 percent of survey responses) indicated that their product sales were to the Department of Defense.

All proposals are submitted electronically via the NSF FastLane program. NSF has set a deadline of 60 days from application for award notification.

NSF Phase IIB "funds can be used only for advancing the research-related elements of the project. The third-party investor funds can be used for research or other business-related efforts to accelerate commercialization. Market research, advertising, patent applications, and business plan improvement are all examples of uses of third-party investor funds."[8]

---

[6]National Science Foundation, Phase IIB web page, accessed October 18, 2010, *<http://www.nsf.gov/eng/iip/sbir/phase_iib.jsp#REVIEW>*.

[7]The company making this point preferred to remain anonymous.

[8]National Science Foundation, *1999 SBIR/STTR Phase I Program Solicitation and Phase II Instruction Guide*, *<http://www.nsf.gov/pubs/1999/nsf9957/Chapt-10.htm#BM10_3>*. Accessed May 29, 2015.

Review of Phase IIB Proposals and Selection Criteria

Phase IIB proposals are reviewed in house by a minimum of two NSF SBIR/STTR Program Officers, based on the following two review criteria drawn verbatim from the NSF Phase IIB web site:

> **"Criterion: What is the intellectual merit of the proposed activity?** Potential considerations: Will the completion of the proposed activity lead to a solid foundation of the scientific and engineering knowledge and understanding base? Has the company progressed satisfactorily in the Phase II activity to justify a Phase IIB activity? Is the proposed plan a sound approach for establishing technical feasibility that could lead to commercialization?"
>
> **"Criterion: What are the broader impacts of the proposed activity?** Potential considerations: Does the commercialization plan summary in the proposed activity show a clear path to commercial and societal benefits? Does the proposed activity reflect changes to the Phase II commercialization plan that further improves the chances of conversion of research in order to provide societal benefits? What are the expectations of the third party and how effective will the third party funded activity lead to commercial and societal benefit? Evaluate the competitive advantages of this technology vs. alternate technologies that can meet similar market needs."[9]

Through survey responses and interviews, several NSF awardee companies indicated that NSF Phase IIB approval has in recent years become strongly focused on opportunities for rapid commercialization. Commercialization is a stated objective of the SBIR program and has been emphasized by NSF SBIR program management in discussions with the Academies,[10] and this perspective is supported by comments from both recipients and management. In addition, it is worth noting that the commercialization imperative appears in both of the above criteria.

Payment and Release of Funding

The Phase IIB match is released on different schedules for projects that have been extended for 1 and 2 years. One-year extensions, which receive NSF funding of up to $250,000, receive 50 percent as an advance payment at the start of the Phase IIB and the remaining 50 percent on approval of the 6-month interim report. Two-year extensions (NSF funding greater than $250,000) receive 25

[9]National Science Foundation, Phase IIB web page, accessed October 18, 2010, <*http://www.nsf.gov/eng/iip/sbir/phase_iib.jsp#REVIEW*>.

[10]Effective July 1, 2015, the institution is called the National Academies of Sciences, Engineering, and Medicine. References in this report to the National Research Council, or NRC, are used in an historic context identifying programs prior to July 1.

percent as an advance payment, 25 percent on approval of the 6-month interim report, 25 percent on approval of the 12-month interim report, and 25 percent on approval of the 18-month interim report. In both cases, the final payment due under the Phase II award is retained until after approval of the final project report. And in both cases, initial payments will be made only after NSF has received a letter stating that the entire amount of any third-party match been transferred to the company.

Reporting Requirements

Phase IIB awardees are required to provide interim reports every 6 months, along with a final project report on conclusion of the award period. These reports are submitted electronically.

PHASE IIB AWARDS

The NSF Phase IIB program has been in operation since 1998. This section summarizes award patterns, based on data provided to the Academies by NSF.

The total number of awards, total NSF funding, and average funding per project are summarized in Table 5-2.

TABLE 5-2 NSF Phase IIB Awards and Funding, by Year 2001-2014

Fiscal Year NSF Supplement Funds Committed	Number of Phase IIB Supplements	Sum of NSF Matching Funds (Dollars)
2001	30	8,552,875
2002	40	9,638,580
2003	25	6,035,296
2004	28	6,074,745
2005	44	14,147,988
2006	52	19,252,650
2007	38	10,849,629
2008	33	9,903,133
2009	31	9,168,638
2010	51	15,977,187
2011	28	9,657,606
2012	56	18,371,621
2013	52	21,457,193
2014	30	10,536,806

SOURCE: National Science Foundation.

Discussions with NSF staff suggest that the annual variation in the number of awards may, in part, reflect the way that records are kept rather than a genuine shift in NSF priorities, as well as variations in the status of potential applicants and other factors. Awards are not a direct reflection of NSF policies.

NSF reports that Phase IIB funding is associated with more than twice the amount of matching funds (see Table 5-3). This issue is addressed further in the detailed discussion of Phase IIB survey results.

An average NSF investment of $281,000 is associated with average matching investments of $830,000—a 3:1 ratio, according to NSF data. The extent to which there is a causal relationship between NSF funding and matching funds is explored through the Phase IIB survey, described below.

Some companies have successfully sought funding for more than one project. Table 5-4 lists the 14 companies that were awarded more than two Phase IIB awards.

2010 PHASE IIB SURVEY RESULTS

This section is based on responses to the 2010 Phase IIB Survey.[11] This is the first survey to target recipients of Phase IIB funding. Its purpose is to compare Phase IIB with Phase II projects (see Box 5-1)

Company Foundation and Status

Almost all responses related to companies that were still in business. Further research (calls and Internet research) indicated that many of the non-responding companies may have gone out of business or had been acquired and were no longer independent entities.

The mean number of founders for these companies was approximately 2.5. About 80 percent of companies reported at least one founder with an academic background, while about three-quarters reported at least one founder with a business background. Serial entrepreneurs are an important element of the ecosystem for high-tech innovative companies, and slightly more than one-half reported that at least one founder had previously founded a different company.

The survey asked whether the awardee company was woman-owned or minority-owned. A higher percentage of Phase IIB respondents reported their companies as woman-owned compared to Phase II respondents. There was a roughly equal response rate between the two groups with regard to minority ownership (see Table 5-5), although this result may be somewhat unreliable because interview evidence suggests that the definition of "minority-owned" is not consistent across companies.

[11]The 2010 Phase IIB Survey was conducted separately from (and prior to) the broader 2011 Survey referenced elsewhere in this volume. All figures in this chapter are based on this Phase IIB survey unless identified otherwise. A discussion of the methodology utilized is provided in Appendix A.

TABLE 5-3 Matching Investment for Phase IIB

Fiscal Year	Number of Awards	Matching Funds (Dollars)	Average Matching Funds (Dollars)
1998	4	799,888	199,972
1999	24	10,069,060	419,544
2000	9	4,570,392	507,821
2001	0	0	0
2002	43	26,866,345	624,799
2003	30	27,259,498	939,983
2004	28	12,049,692	430,346
2005	44	43,514,150	988,958
2006	49	57,812,521	1,179,847
2007	9	7,898,500	1,128,357
2008	21	22,188,527	1,386,783
2009	23	29,150,904	1,325,041

SOURCE: National Science Foundation, private communication.

TABLE 5-4 Companies Receiving More Than Two Phase IIB Awards, 1999-2009

Company	Number of Phase IIB Awards
Intelligent Fiber Optic Systems	6
CFD Research	4
One Cell Systems	3
T/J Technologies	3
Uncopiers	3
Mendel Biotechnology	3
VCOM3D	3
MICROSTRAIN	3
Physical Optics	3
Immersion Corporation	3
wTe	3
Luna Innovations	3
NGIMAT	3
Workplace Technologies Research Institute	3

SOURCE: National Science Foundation.

BOX 5-1
Comparing Phase IIB- and Phase II-only Projects

In order to understand the Phase IIB program more clearly, the Academies undertook a survey of Phase IIB recipients in 2010. The survey was closely based on the Academies' previous Phase II recipient survey of 2005, and had substantial overlap with the subsequent Phase II recipient survey of 2011. An additional series of questions was asked of Phase IIB recipients, focused on the Phase IIB process itself.

The survey provided an opportunity to benchmark Phase IIB outcomes against those of projects that received only Phase II funding. Given that the objective of the Phase IIB program is to identify and then provide additional funding for more commercially promising projects, the null hypothesis would be that Phase IIB projects should be more successful. It is useful to compare the two populations to determine the extent of differential success, and to address other program outcomes beyond commercialization, as well as other process metrics.

The comparisons are not suited to the application of statistical analysis. They are for descriptive purposes only.

TABLE 5-5 Woman- and Minority-Owned Companies

	Phase II (Percent)	Phase IIB (Percent)
Woman-owned	6	15
Minority-owned	12	11
N =	110	117

SOURCE: 2010 Phase IIB Survey, Question 6.

As expected for a sample of small companies, for which the principal investigator (PI) is often the owner, data for woman- and minority-owned companies are reflected in the data for PI demographics (see Table 5-6). Phase IIB respondents reported a higher percentage of both woman and minority PIs than Phase II respondents.

Project Status

Table 5-7 summarizes the current status of the surveyed project. Almost twice the share of Phase IIB projects had generated products, processes, or services that were in use at the time of the survey.

TABLE 5-6 Female and Minority Principal Investigators (PIs)

Demographics	Phase II (Percent)	Phase IIB (Percent)
Woman	7	18
Minority	11	15
N =	110	117

SOURCE: 2010 Phase IIB Survey, Question 17.

TABLE 5-7 Current Project Status

Status	Phase II (Percent)	Phase IIB (Percent)
Products, processes, or services are in use	25	49
Commercialization is under way	23	22
Project is continuing post Phase II development	23	17
Project has not yet completed Phase II	19	3
Efforts at this company have been discontinued	10	9
Total	100	100
N =	110	117

SOURCE: 2010 Phase IIB Survey, Question 7.

Commercial Outcomes

The survey asked a number of questions focused on commercial outcomes from the surveyed projects, with the initial aim of determining whether the projects had in fact reached the market (see Table 5-8).

Although a majority of responses from both groups indicated that sales had occurred, more Phase IIB projects reported sales. Of the Phase II only projects that had not yet reported sales, 31 percent still expected sales in the future.

However, simply reaching the market is not a sufficient metric for commercialization. It is also important to know the size of the commercialization results. Table 5-9 shows the distribution of responses by sales groupings. Table 5-10 shows the average allocation of sales by customer sector and by group. Sales were heavily concentrated in the U.S. private sector.

TABLE 5-8 Sales Status

Sales Status	Phase II (Percent)	Phase IIB (Percent)
Yes, have already made sales	63.0	80.6
No sales to date, but sales expected	31.0	9.7
No sales to date, no sales expected	6.0	9.7
Total	100.0	100.0
N =	78	103

SOURCE: 2010 Phase IIB Survey, Question 10.

TABLE 5-9 Distribution of Responses by Total Amount of Sales per Project

Reported Sales	Phase II (Percent)	Phase IIB (Percent)
Under $100,000	16.7	20.0
$100,000-$499,999	35.7	22.7
$500,000-$999,999	26.2	10.7
$1 million-$1.9 million	7.1	14.7
$2 million-$2.9 million	4.8	9.3
$3 million or more	9.5	22.7
Total	100.0	100.0
Mean (Thousands of Dollars)	1,262	3,030
Median (Thousands of Dollars)	450	650
N (Projects reporting sales) =	42	75

SOURCE: 2010 Phase IIB Survey, Question 13.1.

Employment Impacts

Survey responses suggest that Phase IIB companies tended to be somewhat larger than Phase II only companies at the time of application. More Phase IIB companies than Phase II only companies reported 20-49 employees and 50-99 employees at the time of application (see Table 5-11). Overall, Phase IIB respondents reported double the number of employees at the time of application (22 vs. 11).

The differences in company size are potentially important. Larger companies tend to be better established and have more resources beyond the SBIR award.

TABLE 5-10 Average Distribution of Sales by Customer Sector and Group

Sales Sector	Phase II (Percent)	Phase IIB (Percent)
Domestic private sector	55.7	60.4
Export markets	15.8	16.0
Department of Defense (DoD)	11.0	9.1
Other type of customers	7.5	5.8
State or local governments	2.8	3.8
Other federal agencies	2.9	2.9
Prime contractors for DoD or NASA	4.1	1.6
NASA	0.2	0.5
NSF	0.0	0.0
N (Projects reporting sales) =	46	80

SOURCE: 2010 Phase IIB Survey, Question 14.

TABLE 5-11 Company Size (Employees) at Time of Application

Number of Employees	Phase II (Percent)	Phase IIB (Percent)
None	1.0	0.0
Under 5	46.5	27.7
5 to 9	20.8	25.7
10 to 19	20.8	12.9
20 to 49	6.9	18.8
50 to 99	2.0	9.9
100 or more	2.0	5.0
Total	100.0	100.0
Mean	11	22
Median	5	8
N =	101	101

SOURCE: 2010 Phase IIB Survey, Question 16.1.

Therefore, companies that were able to raise the matching funds for Phase IIB were already likely to be more commercially advanced.

Respondents were also asked to report the number of employees at the time of the survey (see Table 5-12). More Phase IIB than Phase II only respondents reported that their companies had grown substantially since the time of application—on average, Phase IIB respondents reported 78 employees at the time of the survey, while Phase II companies reported 15.

Thus the data reveal that Phase IIB companies not only had more employees at the time of application, but also grew much more than did Phase II only respondents (see Table 5-13). Phase IIB companies increased by 56, or 254 percent, while Phase II only companies on average increased by employees, or 36 percent. This difference is heavily influenced by the large growth of a few companies. The median employment for the Phase IIB group increased by 113 percent (from 8 to 17), while that of the Phase II-only group increased by 60 percent (from 5 to 8).

Survey data indicated that the number of staff hired directly as a result of the SBIR award was limited.[12] Similar results were obtained for employees retained (rather than hired) as a result of the award. Phase IIB companies reported a mean retention almost two times that of Phase II-only companies (4 vs. 2). To summarize, employment effects were larger across the board for the Phase IIB group (see Table 5-14).

Intellectual Property and Knowledge Effects

Commercial returns are not the only significant outcomes of SBIR-funded projects. One of the mandated congressional goals is to support the extension of scientific and technical knowledge. In past Academies reports, work toward this goal has been measured using patent, trademark, and copyright data as proxies for the development of commercially valuable intellectual property.[13]

Table 5-15 shows that a higher percentage of Phase IIB companies reported receiving one or more related patents than Phase II-only companies (46 percent vs. 37 percent), and more Phase IIB companies reported applying for one or more patents than did Phase II-only companies (76 percent vs. 57 percent). Moreover, a higher percentage of Phase IIB companies applied for and received multiple patents.

Another standard metric for knowledge dissemination is authorship of articles published in peer-reviewed journals. One might hypothesize that companies focused on commercialization would be unlikely to generate peer-reviewed publications, and hence that Phase IIB companies would be less focused than Phase II only companies on publishing, but this turns out not to be the case (see Table 5-16).

---

[12]2010 Phase IIB Survey, Question 16.3.

[13]See National Research Council, *An Assessment of the SBIR Program at the National Science Foundation* (Washington, DC: The National Academies Press, 2008), p. 126.

TABLE 5-12 Company Size (Employees) at Time of Survey

Number of Employees	Phase II (Percent)	Phase IIB (Percent)
None	2.8	2.7
Under 5	25.5	13.5
5 to 9	26.4	16.2
10 to 19	28.3	18.9
20 to 49	12.3	19.8
50 to 99	2.8	15.3
100 or more	1.9	13.5
Total	100.0	100.0
Mean	15	78
Median	8	17
N =	106	111

SOURCE: 2010 Phase IIB Survey, Question 18.

TABLE 5-13 Change in Employment

Number of Employees	Phase II	Phase IIB
Time of application		
Mean	11	22
Median	5	8
Time of Survey		
Mean	15	78
Median	8	17

SOURCE: 2010 Phase IIB Survey, Question 16.

TABLE 5-14 Summary of Employment Effects

	Phase II	Phase IIB
Number of employees at time of application (mean)	11	22
Number of employees at the time of the survey (mean)	15	78
Number hired as a result of the project (mean)	2	4
Number retained as a result of the project (mean)	2	4
N =	110	117

SOURCE: 2010 Phase IIB Survey, Question 16.

TABLE 5-15 Patenting Activity

	Patents Applied for:		Patents Received:	
	Phase II (Percent)	Phase IIB (Percent)	Phase II (Percent)	Phase IIB (Percent)
None	33.3	23.6	62.9	53.6
1 or 2	53.3	45.5	32.4	35.5
3 or 4	7.6	20.0	2.9	7.3
5 or more	5.7	10.9	1.9	3.6
Total	100.0	100.0	100.0	100.0
1 or more	66.7	76.4	37.1	46.4
N =	105	110	105	110

SOURCE: 2010 Phase IIB Survey, Question 18.

TABLE 5-16 Publications in Peer-Reviewed Journals

	Phase II (Percent)	Phase IIB (Percent)
None	50.0	34.3
1 or 2	31.4	31.4
3 or 4	9.8	19.6
5 or more	8.8	14.7
Total	100.0	100.0
1 or more	50.0	65.7
BASE: ALL RESPONDENTS	102	102

SOURCE: 2010 Phase IIB Survey, Question 21.

NSF Phase IIB Application and Process

In its analysis of the operations of the Phase IIB program, the Committee surveyed only projects that received Phase IIB funding (as confirmed by the NSF awards database and the company itself). This was done because non-recipient projects would not have inside knowledge of Phase IIB impacts and procedures. At the same time, for the 74 Phase II projects that did not apply for Phase IIB funding were asked why. Just under 60 percent of responses reported that the company did not apply because they could not raise the required matching funds (see Table 5-17). As anticipated by NSF, the requirement for matching funds is a formidable gate for Phase II recipients to pass through on the way to Phase IIB.

Case study interviews with SBIR awardees indicated that NSF enjoys a strong reputation for working closely with awardees and applicants. This positive reputation was confirmed by responses to a survey question about how closely the company worked with the NSF program officer during the Phase IIB application process (see Table 5-18). Almost 80 percent of respondents indicated a considerable degree of contact and guidance from the NSF program officer.

Developing a winning application requires significant efforts. Given the additional complexity of the Phase IIB matching funds requirements, the burden on respondents could be substantial. Table 5-19 indicates the level of effort reported by respondents.

TABLE 5-17 Reasons for Not Applying for Phase IIB Funding

Reason for Not Applying for Phase IIB	Percent (Percent)
Unable to raise matching funds	35
Could not raise matching funds that qualified	32
Not aware of the program	8
Not enough funding to be worthwhile	5
Process too onerous	5
Other	30
N (Phase II projects that did not apply for Phase IIB funding) =	74

NOTE: Responses do not sum to 100 percent because respondents could select more than one answer.
SOURCE: 2010 Phase IIB NSF Survey, Question 29.

TABLE 5-18 Working Relationship with NSF Program Officer

	Phase IIB (Percent)
Not at all	2
Not much	22
We discussed the application in detail	48
The officer provided a lot of guidance	28
Total	100
N (Phase IIB recipient) =	111

SOURCE: 2010 Phase IIB Survey, Question 44.

TABLE 5-19 Amount of Effort Required to Submit Phase IIB Proposal

	Phase IIB (Percent)
No additional effort needed except paperwork	15
Less than 2 weeks FTE for senior company staff	26
2 to 8 weeks FTE for senior company staff	37
2 to 6 months FTE for senior company staff	15
More than 6 months FTE for senior company staff	7
Total	100
N (Phase IIB recipient) =	110

SOURCE: 2010 Phase IIB Survey, Question 44.

The responses confirmed that Phase IIB applications required considerable effort for some companies, with more than 22 percent of companies recording more than 2 months of full-time work from senior staff. For most, however, the burden was much less: 40 percent reported 2 weeks of effort or less.

To provide a comparative benchmark for these data, the survey asked the respondents to compare the effort required to submit a Phase IIB application with that required for other federal R&D projects (see Table 5-20). More than one-third of respondents reported that Phase IIB applications were easier or much easier than those for other federal R&D programs, while 19 percent reported that

TABLE 5-20 Comparison of Phase IIB with Applications to Other Federal R&D Programs

Application Process was:	Phase IIB (Percent)
Much easier than other federal awards	7
Easier	28
About the same	37
More difficult	17
Much more difficult	2
Not sure	9
Total	100
N (Phase IIB recipient) =	110

SOURCE: 2010 Phase IIB Survey, Question 42.

the process was more difficult or much more difficult. These data suggest that the NSF Phase IIB program is, on the whole, not imposing onerous demands on companies in the application process.

Given the different descriptions of NSF matching funds criteria on the NSF Phase IIB web page, the survey sought to determine what successful respondents believed would count as matching funds for NSF Phase IIB purposes. The responses are summarized in Table 5-21

These responses highlight considerable confusion even among companies that were successful in meeting the matching funds criteria. Well over half the responses identified sources as acceptable that are rejected by NSF. Conversely, more than 40 percent of respondents did *not* list equity investment and sales as acceptable matches. Interviews with some successful Phase IIB companies confirmed that sales were not universally understood to be acceptable source of matching funds.

NSF Phase IIB Funding and Matching Funds

One of the primary purposes of the Phase IIB program is to attract additional investment into SBIR awardee companies. Phase IIB requires a minimum of a 2:1 match by the company, and only certain kinds of funding can qualify as matching funds:

"The third party funding must be a direct consequence of the Phase II award. This funding can be cash, product sales revenues and/or licensing revenues, liquid assets, tangible financial instruments but not in-kind or other 'intangible assets'. Loans and investments with contingency clauses are not acceptable. Self-funding

TABLE 5-21 Respondent Beliefs on NSF Phase IIB Matching Funds Criteria

	Phase IIB (Percent)
Equity investments	80
Sales	71
Additional investments from founders	44
In-kind contributions from technical partners	29
In-kind contributions from marketing partners	22
Cash loans	13
Other	15
N (Phase IIB recipient) =	109

NOTE: Responses do not total 100 percent because more than one response was permitted.
SOURCE: 2010 Phase IIB Survey, Question 41.

TABLE 5-22 NSF Phase IIB Funding, by Amount of Funding

	2010 Phase IIB Survey Data (Percent)	NSF Data (Percent)
Under $100,000	10	13
$100,000-$199,999	17	24
$200,000-$299,999	23	24
$300,000-$399,999	10	11
$400,000-$499,999	5	13
$500,000-$999,999	35	15
Total	100	100
Mean (Thousands of Dollars)	315	281
Median (Thousands of Dollars)	281	250
N =	100	285

SOURCE: 2010 Phase IIB Survey, Question 31.1; NSF awards data.

does not qualify for the Phase IIB option." Self-reported data on funding from NSF tracks quite closely with NSF survey data (see Table 5-22).

The self-reported figures in column two track closely with the NSF awards data in column three. The primary discrepancy—in the numbers of awards at $400,000 and above—is explained by the existence of 33 awards that NSF recorded at between $480,000 and $499,999, which awardees likely reported as $500,000.

NSF data and survey data capture the amount of matching investment funds generated in connection with the Phase IIB award. (See Table 5-23). About one-third of projects reported matching funds greater than $1 million, and 25 percent reported funding between $500,000 and $999,999. Both the self-reported and the agency data indicated that NSF achieved its objective of funding projects with at least a 2:1 match from third parties, thus leveraging the Phase IIB NSF investment.

A specific objective of the NSF program is to encourage companies to generate third-party investments. Accordingly it is important to understand the source of the matching funds as well as any prior relationship with the funding source. Tables 5-24 and 5-25 show the distribution of funding by source. Because some respondents answered more than one survey questionnaire, Table 5-25 distinguishes between individual responses (column 3) and individual respondents (column 4). Regardless, more than 20 percent indicated that funding was derived

TABLE 5-23 Matching Funds for Phase IIB Awards, by Amount of Funds

	2010 Phase IIB Survey Data (Percent)	NSF Data (Percent)
Under $100,000	4	0
$100,000-$199,999	8	12
$200,000-$299,999	11	13
$300,000-$399,999	11	7
$400,000-$499,999	4	9
$500,000-$999,999	26	24
$1 million or more	36	35
Total	100	100
Mean (Thousands of Dollars)	651	865
Median (Thousands of Dollars)	500	500
N =	98	275

SOURCE: 2010 Phase IIB Survey, Question 31.2; NSF awards data.

TABLE 5-24 Distribution of Funding by Source of Funding (Number of Responses)

	Responses		
	Count	Percentage of Responses	Percentage of Respondents
Sales	29	22.8	29.9
Another U.S. company	26	20.5	26.8
U.S. angel investment	20	15.7	20.6
State agency funding	15	11.8	15.5
Federal agency funding	13	10.2	13.4
U.S. venture capital	12	9.4	12.4
Foreign funder(s)	9	7.1	9.3
Other internal company resources	3	2.4	3.1
Total responses	127		
N =	97		

NOTES: Includes only respondents who answered questions about both the amount of funding and the source of funding. The percentages in the respondents 'column does not total 100 percent because respondents could select more than one answer.
SOURCE: 2010 Phase IIB Survey, Question 32.

TABLE 5-25 Distribution of Third-Party Funding by Source of Funding (Dollars)

	Percentage of Total Dollars	Total Reported (Dollars)	Average of All Respondents Receiving Funds from This Source (Dollars)
Sales	24.1	15,170,208	523,111
Another U.S. company	21.4	13,369,475	514,211
U.S. angel investment	16.6	10,441,663	522,083
State agency funding	10.7	6,743,034	449,536
Federal agency funding	9.9	6,237,793	479,830
U.S. venture capital	8.3	5,220,832	474,621
Foreign funder(s)	7.5	4,696,157	521,795
Other internal company resources	1.5	952,186	317,395
All sources	100	62,831,349	647,746

N = 97

SOURCE: 2010 Phase IIB Survey, Question 32.

from sales and from other U.S. companies. Nine percent of responses mentioned venture funding.

Conventional wisdom—and previous Academies studies—would lead us to expect that the average amount per project (final column in Table 5-25) from these sources would differ substantially and that venture funding would provide relatively large amounts per project. However, the latter was not the case for these projects. Venture support (for projects with some venture funding) averaged less than $500,000 per funded project; funding from angel investors was more than $500,000, and four sources provided more funding per funded project than did venture sources.[14] The amount of federal agency funding was somewhat surprising given the requirement that Phase IIB awards above $500,000 utilize only private-sector matching funds.

[14]The lower-than-expected venture funding may reflect the fact that VC firms now manage much larger capital bases that they did in the 1980s, and as a result now need to make much larger initial investments in individual projects to make the most of the time and effort of the partners. This trend may mean that the desired size of individual venture investments and the amount of third-part funding needed by most SBIR companies align less well now than in the past.

As noted, the NSF SBIR program encourages companies to find investors. Table 5-26 addresses this question. Only 30 percent of respondents reported matching funds were provided by a new investor, while 45 percent of responses reported funding from a long-term partner of some kind. The substantial number of respondents reporting "other" likely include the 30 percent of respondents who reported sales as a source of matching funds. Overall, these findings suggest that the NSF objective of generating new investment is being partly met.

Phase IIB Impacts: Accelerating Commercialization

The NSF Phase IIB program is aimed at accelerating the commercialization of products and services. The next set of questions focused on these impacts. Table 5-27 summarizes the impact of Phase IIB funding in accelerated commercialization:

TABLE 5-26 Company Relationship to Investor

Source of Funding	Phase IIB (Percent)
A new financial partner or investor	30
A long-time sponsor or vendor relationship	18
A long-time purely financial partner or investor	21
A long-time technical partner	15
Other	36
N (Received Phase IIB funding) =	107

NOTE: Responses do not total 100 percent because respondents could provide more than one answer.
SOURCE: 2010 Phase IIB Survey, Question 33.

TABLE 5-27 Phase IIB Impacts: Accelerating Commercialization

	Phase IIB (Percent)
Yes	80
No	5
Too early to tell	11
Other	4
Total	100
N =	112

SOURCE: 2010 Phase IIB Survey, Question 35.

Eighty percent of respondents indicated that the funding had accelerated commercialization. Five percent thought it had not. Respondents also reported strongly positive views on the long-term impact of Phase IIB on the progress of the company (see Table 5-28).

Nine percent of respondents reported that Phase IIB had a transformative effect on the company, while a further two-thirds reported a "substantial long-term positive effect."

Phase IIB Impacts: Acquiring Additional Investment

As one program objective is to bring in fresh investment, the survey asked respondents whether the Phase IIB match requirements brought in funding that would not otherwise have been acquired by the company. The response was overwhelmingly positive, as described in Table 5-29.

TABLE 5-28 Phase IIB: Long-term Impacts on Company

	Phase IIB (Percent)
Had a transformative effect	9
Had a substantial positive long-term effect	66
Had a small positive effect	21
Had no long-term effect	4
Total	100
N=	112

SOURCE: 2010 Phase IIB Survey, Question 37.

TABLE 5-29 Brought New Funding into the Company that Otherwise Would Not Have Been Acquired

	Phase IIB (Percent)
Yes	61
No	17
Too early to tell	16
Other	6
Total	100
N=	112

SOURCE: 2010 Phase IIB Survey, Question 38.

Sixty-one percent of projects reported that the program had brought additional funding into the company that would not otherwise have been acquired.

Delving a little deeper, respondents who reported receiving Phase IIB funding were also asked whether the availability of Phase IIB funding had added incentives for the company to seek investment funds (see Table 5-30).

Respondents overwhelmingly reported that the Phase IIB funding was at least to some degree an incentive to acquire investment funds. Fifty-two percent indicated that Phase IIB made a great deal of difference or drove the process.

Respondents that indicated that Phase IIB did bring new investment into the company were then asked how much additional funding had been brought in as a result of Phase IIB matching requirements. Responses are described in Table 5-31.

Overall, 20 percent of respondents indicated amounts of less than $200,000, while 59 percent reported amounts of $500,000 or more. The overall average amount was $956,000, and the median was $500,000.

Beyond the additional investment that Phase IIB recipients attribute directly to the Phase IIB program, Phase IIB companies have been more successful in securing additional funding beyond the required match. About three quarters of Phase IIB respondents indicate that their company has received such additional investment funding, compared with about half of Phase II only projects. (See Table 5-32.)

For all projects (Phase II and Phase IIB) reporting additional funding, the survey asked whether they had received any funding (beyond the Phase IIB match) from the following sources (whether as a result of the Phase IIB program or not). These data represent the percentage of respondents indicating that they received additional funding from a given source. Accordingly, the total number of responses adds up to more than 100 percent (see Table 5-33).

A considerably higher percentage of Phase IIB companies reported receiving additional venture funding than did Phase II-only companies (16 percent

TABLE 5-30 Phase IIB Impact: Incentive for Acquiring Investment Fund

	Phase IIB (Percent)
Not at all	8
Somewhat	40
A great deal	42
It drove the process	10
Total	100
N=	108

SOURCE: 2010 Phase IIB Survey, Question 43.

TABLE 5-31 Distribution of Additional Funding Generated Strictly as a Result of Phase IIB Requirements

	Phase IIB (Percent)
Under $100,000	11
$100,000 to $199,999	9
$200,000 to $299,999	7
$300,000 to $399,999	9
$400,000 to $499,999	5
$500,000 to $999,999	20
$1 million or more	39
Total	100
Mean	$956,000
Median	$500,000
N (Phase IIB funding brought additional investment into the company)=	56

SOURCE: 2010 Phase IIB Survey, Question 39.

TABLE 5-32 Additional Investment Reported for the Surveyed Project (Excluding, in the Case of Phase IIB Respondents, Phase IIB Required Matches)

	Phase II (Percent)	Phase IIB (Percent)
Yes	51	74
No	49	26
N=	110	117

SOURCE: 2010 Phase IIB Survey, Question 22.

versus 2 percent). The higher participation of venture capitalists as the SBIR projects move farther out in their development validates that the SBIR program helps companies during a time that is generally too early for extensive attraction of venture capitalists. This finding is evidence that the SBIR is positioned as intended—helping bridge innovative companies across the "Valley of Death."

TABLE 5-33 Sources of Additional Investment

	Phase II (Percent)	Phase IIB (Percent)
Non-SBIR federal funds	27	32
U.S. venture capital	2	16
Foreign sector investment	6	7
Other private equity	18	20
Other domestic private company	18	12
State or local governments	16	16
College or universities	6	3
Your own company (including money borrowed)	65	45
Personal funds	24	16
N=	51	74

NOTE: Responses do not total 100 percent because respondents could provide more than one answer.
SOURCE: 2010 Phase IIB Survey, Question 21.

Phase IIB Impacts: Respondent Perspectives

Table 5-34 summarizes overall respondent views of the Phase IIB program. Overall, 98 percent of respondents have positive or very positive views of the program. None report a negative view. And 99 percent reported that it was worth the effort to apply (even though it costs the equivalent for two months of senior staff time for more than 20 percent of applicants).[15]

More detailed insights into these highly positive responses can be found in additional comments received for survey question 36, which sought to identify ways in which Phase IIB funding made a difference to the company's ability to bring products or services to the market. The comments leave little doubt that for many respondents, Phase IIB funding had a major positive impact on the commercialization of their products and services, and on the long-term success of their company.

Respondent Recommendations for Program Improvement

Respondents were asked two questions about the size of the Phase IIB program. The first focused on overall program size. Responses are summarized in Table 5-35. Three percent of respondents recommended that the program be reduced in size. Conversely, 40 percent recommended that it be expanded.

[15]2010 Phase IIB Survey, Question 46.

TABLE 5-34 Phase IIB Overall Impact: Respondent Views

Views on Phase IIB	Phase IIB (Percent)
Very positive	76
Somewhat positive	22
Neutral	2
Somewhat negative	0
Very negative	0
Total	100
N=	108

SOURCE: 2010 Phase IIB Survey, Question 47.

TABLE 5-35 Should the Size of the Phase IIB Program be Changed?

	Phase IIB (Percent)
Expanded	40
Kept at about the current level	57
Reduced	3
Eliminated	0
Total	100
N=	106

SOURCE: 2010 Phase IIB Survey, Question 49.

This corresponds to respondent views on the adequacy of the individual awards they received: about two thirds of respondents thought they received about the right amount of funding—the remainder almost all thought they had not received enough.[16] However, only about one fifth indicated that the size of awards should be increased if that meant that NSF would award fewer Phase IIB awards (Table 5-36).

One issue that had emerged in the context of case study interviews was whether NSF should relax its prohibition on the use of appropriate in-kind resources as matching funds. No majority view emerged from respondents, although almost half indicated that some relaxation might be appropriate (Table 5-37).

[16]2010 Phase IIB Survey, Question 50.

TABLE 5-36 Should Award Size be Increased Even if that Means Fewer Awards?

	Phase IIB (Percent)
Yes	22
No	46
Not sure	32
Total	100
N=	107

SOURCE: 2010 Phase IIB Survey, Question 51.

TABLE 5-37 Use of Appropriate In-kind Contribution as Matching Funds

	Phase IIB (Percent)
Yes	49
No	30
Not sure	21
Total	100
N=	108

SOURCE: 2010 Phase IIB Survey, Question 52.

CONCLUSIONS

Overall, the NSF Phase IIB program is supporting the accelerated commercialization of SBIR-funded research. It is helping to attract third-party investors, and to accelerate or otherwise enhance the commercialization of SBIR-funded technologies.

The evidence clearly shows that overall there are distinct differences in outcomes for Phase IIB projects compared with Phase II-only projects. In part, this reflects the rigorous selection process—NSF selects Phase IIB projects specifically for their commercial potential, so improved commercial outcomes should be expected. At the same time, evidence from recipients and from case studies strongly suggests that the program also has a catalyzing effect, attracting additional funding, and that it makes a difference to company activities and outcomes.

6

Insights from Case Studies and Extended Survey Responses

Case study and open-ended survey questions offer a way to better understand the numerous ways in which the Small Business Innovation Research (SBIR) program impacts small innovative companies. This chapter addresses a range of impacts and issues that were described by executives interviewed for 12 case studies, which are presented in detail in Appendix E. It supplements the case study perspectives with responses to open-ended questions by all respondents in the 2011 Survey.[1] It is important to note that the viewpoints are those of the respondents, and not necessarily the views of the committee.[2]

Together, the case studies and the response to the open-ended survey questions support a qualitative review that provides considerable context for the data discussed in other chapters and allows company participants to address in their own words those aspects of the SBIR that they believe are working as needed and those that require attention.

ABOUT THE CASE STUDIES

Case studies are an important part of data collected for this report, and need to be read in conjunction with other sources—agency data, the Academies survey, interviews with agency staff and other experts, and workshops on selected topics.

[1]The survey data reported in this chapter are from the 2011 Survey. The 2011 Survey covered NSF Phase I and Phase II SBIR recipients with awards FY1998 to FY2007. The survey included an open comment box where respondents could describe their company's experience with SBIR. Appendix A describes the survey instrument, including the safeguards observed to protect the respondents.

[2]In some cases, these responses may reflect some misunderstandings on the part of executives about the operations of NSF SBIR program.

BOX 6-1
National Science Foundation Company Case Studies

ALD NanoSolutions (ALDN)
Divergence
Intelligent Fiber Optic Systems (IFOS)
Immersion Corporation (Immersion)
Imaging Systems Technology (IST)
Learning in Motion (LIM)
Membrane Technology and Research (MTR)
Mendel (Mendel)
Techno-Sciences Incorporated (TSI)
Touch Graphics (TG)
TRX Systems (TRX)
Work Technology Research Inc. (WTRI)

NOTE: Case studies are included in this volume as Appendix E.

Given that the impact of SBIR funding is complex and often multifaceted, and given that these other data sources provide important insights, case studies allow for an understanding of the narrative and history of recipient firms; in essence, case studies provide context for the data collected elsewhere.

The case studies were selected based on multiple criteria, including number of awards overall, geographic location, firm demographics, industry sector, reported commercialization, and age of firm. Given the multiple variables at play, they are not presented as a quantitative record. Rather they provide qualitative evidence about the individual companies selected, which are, within the limited resources available, as representative as possible of the different component of the awardee population. Box 6-1 lists the 12 NSF case studies from which insights about the range of SBIR impacts are formed. These case studies are presented in full in Appendix E.

ORGANIZATION

The chapter is organized around a series of impacts by topic and a series of issues by topic. Material from the case studies and the survey comments are interwoven throughout the chapter to amplify both the impacts and the issues. Table 6-1 summarizes the principal impact and Issue topics covered qualitatively in this chapter. Within these topic areas are many subtopics. Together, these topical sections on impacts and issues provide the first wide-ranging and publicly available feedback of the NSF SBIR program directly from program recipients.

Together, these sections provide the first wide-ranging and publicly available feedback of the NSF SBIR program from program recipients.

TABLE 6-1 Summary of Impact and Issue Topics Addressed

Impact Topics	Issue Topics
Company Formation and Early-stage Funding	Over-focus on Commercialization and the Venture Capital Model
Funding Otherwise Un-fundable Projects	Application and Selection Procedures
Validation Effects	Funding Gaps
Capacity Building—Human Capital	Funding Levels
Critical Funding	Other Issues
Agency Mission	
Program Operations	

COMPANY FORMATION AND VERY EARLY-STAGE FUNDING

It is often easy to forget that the entire SBIR program provides considerably more funding for very early stage projects than does the entire U.S. venture capital (VC) industry: in 2012, VCs provided $820 million for seed and startup projects compared with the program's overall spending of $2.4 billion.[3] This point is not well known; many are unaware that VC funding goes largely to later stage, less risky projects.

Evidence from case-study interviews and survey responses confirmed that, for many companies, the SBIR program provided the all-important seed funding that allowed the company to get started. During his case-study interview, Dr. Moslehi, President of IFOS, observed, "In its early years, IFOS was a classic Silicon Valley startup—working out of a garage. It did not really gain traction until 2000-2001 when initial Phase II SBIR awards with NASA and NSF allowed me to commit myself full time and for the company to lease its own 4,000-square-foot facility in Sunnyvale, California. SBIR funding was key to equipping the facility and hiring a dedicated founding staff."

Similarly, Mr. Landau, founder of TG, noted during his case-study interview that because the market for assistive devices is too small to interest purely commercial funders, the SBIR program was critical for TG's formation: "SBIR provided initial funding for company formation. The niche market for these assistive devices is not suitable for venture financing given the very small initial size of the market."

During her interview, CEO Carol Politi described the role an NSF SBIR award played in the spin out of TRX from TSI. She explained that the spin out was possible in part because TRX was able to acquire an NSF SBIR award.

Box 6-2 provides verbatim survey responses that pertain to company formation and related seed funding.

[3]PriceWaterHouseCoopers MoneyTree Survey, accessed January 28, 2014, <*https://www.pwcmoneytree.com/MTPublic/ns/nav.jsp?page=historical*>.

BOX 6-2
SBIR and Company Formation: Verbatim Survey Responses

"The funding was key to exploring the concept and allowed the company to be started."

"It's very simple. Without SBIR our company would not exist."

"No SBIR Phase II, no company."

"Our company exists because of the SBIR Award. SBIR awards provided early seed capital to position the company for future investment and growth."

"Company was initiated to win this award."

"The SBIR funding allowed us to generate the first proof of principle data that enabled us to attract more investors and get the company going."

FUNDING OTHERWISE UNFUNDABLE PROJECTS

Commercial funding from investors or lenders is often unavailable to small or newer companies with limited track records that are working to develop products that do not yet exist and hence have no demonstrated existing market. These funding difficulties come in several flavors.

Long Timeline Research

Many survey respondents explained that their research required the investment of considerable time and resources before it would become possible to reach the market. The ongoing support for longer term projects was viewed by many as a particularly important characteristic of NSF SBIR awards. An example of a typically longer term project with the need for sustained funding is a "platform technology," that is, a core technology that could potentially be applied to many different applications if successful. An example was provided by the ALDN case study. Dr. Bueschler (ALDN) noted that the battery technology being developed by her company could have many applications far beyond the initial opportunities being exploited in the auto industry, which itself has a relatively long horizon.

Box 6-3 provides verbatim survey responses pertaining to funding needs of research with long timelines.

Limited Market Size

For many investors, unfavorable risk/return calculations can be especially problematic when rewards are relatively small, that is, when the market served is not large. Although this issue is especially apparent for companies funded by SBIR programs at other agencies, such as NASA, which often procures a single

BOX 6-3
Long-Cycle Research: Verbatim Survey Responses

"The first SBIR funding came at a critical juncture. It was an inflection point and critical to our very existence. We were able to survive and grow a high-risk, long-product cycle Life Science company to develop a platform technology."

"We deliver a new innovative IT solution in the health care space. The sale cycles are very slow, creating a very steep barrier to entry for other small companies. The SBIR enabled us to survive the waiting time and helped us with crucially needed funding. Here typically private investors would not put their money in because of the long/unreasonable ROI time."

"The 10 year gestation period for manufacturing innovations is far too long for venture funding and the investments are often too large for friends and family or other early stage angel sources. The federal government is the only funder of long term, high risk innovation in the U.S. and the SBIR program is the key source of that funding."

item for use in a space flight, it is also true for companies that serve small or specialized market segments.

The case study revealed TG to be such a company. It focuses on developing assistive technology for the blind to support navigation through public places such as museums and schools. Its founder observed, "The company would never have developed any commercial products without the SBIR funding. There is no money in the assistive technologies field for new technologies." In a similar way, case study showed that LIM provides specialized technology with a limited market to the education sector.

Risk

Risk is a key ingredient in private-sector funding algorithms: the greater the risk, the less likely funders are to invest, other factors being the same. And developing new high-tech products is an inherently risky business: the greater the technical challenges, the greater is the technical risk; and the more disruptive the product and the less developed the existing market, the greater the market risk.

Box 6-4 provides verbatim survey responses that help explain how NSF SBIR awards reduce risk.

The SBIR program has the effect of reducing the risk of projects to levels that are acceptable to other investors. Most directly, the SBIR program does move projects further along the technological development curve. In other situations, the SBIR funding allows companies to address projects that would simply be too

BOX 6-4
Risk Reduction: Verbatim Survey Responses

"Reduced technological risk to a level where strategic partners would consider utilizing technology in their products."

"SBIR funding enabled crucial development of the technology to reduce the risk to other investors."

"Much of the proof-of-concept work and high risk work was supported by the SBIR/STTR program, and resulted in new technologies/processes/materials and commercial licenses."

"SBIR funding enabled the company to develop high risk technologies that would never have been funded by traditional VCs."

"SBIR funding provided opportunity to develop IP in very high risk stage when VC funding is very difficult."

"...SBIR funding has proven essential for our company to push forward with game changing ideas that the VC and angel community simply finds too risky to get involved with, despite the fact that there is also large potential reward associated with the risk."

risky to even contemplate. SBIR funding allows companies to move past the go/no go discontinuity by relieving some of the risk involved.

This impact is illustrated by the case study of Divergence, for which the role of SBIR funding has shifted over time. As Mr. Rapp (CEO) observed, "It is now used more for projects that are more speculative and have less data to support them, where VC funding would not be appropriate. This allows a project to mature and prove the design to both company management and reviewers to the point that it could reach product development."

Another example of risk-reducing impact is provided by the case-study illustration of TRX—the recipient of an NSF Phase IIB award. According to Ms. Politi of TRX, NSF support was central in helping the company raise its first angel funding: the ability to point to a federal contribution that leverages the money of investors was "a huge benefit in raising outside money." More generally, as Phase IIB requires matching funds, Ms. Politi observed that "matching programs give you a reason to reach out to people."

TECHNOLOGY AND PRODUCT DEVELOPMENT

Many small companies have limited internal resources for research and development. Often, the SBIR program provides the funding needed to take an idea for an innovative product to the point at which it may enter the market or attract additional funding needed to do so.

Funding Core Technology Development

Box 6-5 gives verbatim survey responses explaining how SBIR awards help fund the development of core technology.

Case studies illustrate how the SBIR program plays a decisive role in the technology development efforts of many small recipient companies, which often lack the cash flow necessary to fund such technology development internally and have very limited access to outside financial resources. A case study of TSI, for example, revealed how a number of SBIR awards from both NSF and other agencies were used to develop the technologies that underpin the company's two core products, SARSAT search and rescue and Trident ship-based monitoring. These technologies have been commercially successful, and this success, according to Dr. Blankenship (CEO), reflects the development of technologies using SBIR funding.

In other case studies, Ms. Wedding of IST notes that "IST could not do the necessary research to develop its innovative products." At Touch Graphics, Mr. Landau observed that "the five or six Phase II awards made it possible to develop most of the TG products now on sale. These came directly from SBIR, although not always by the most direct route." At WTRI, Dr. DiBello said that the SBIR program had been central to the development of WTRI technology and had been used to fund development of each of the WTRI products. SBIR funding helped the company develop an assessment tool that clients could complete on their own—reducing costs and improving quality. Very large scale assessments are now routine at WTRI as a result.

BOX 6-5
Core Technologies: Verbatim Survey Responses

"The SBIR funding supported internal R&D to demonstrate composite processing technology. Our company has taken the technology and applied it to fast adhesive curing and rapid composite assembly processes. These multiple processes are being developed for commercial applications in aerospace, automotive and infrastructure applications."

"This NSF-funded SBIR grant enabled our company to develop an innovative technology that became our first commercial product. This product got us into a lucrative international market in geophysical instrumentation and services."

"Without the SBIR funding from the National Science Foundation, the betavoltaic technology that we are trying to develop would have failed."

"With the SBIR funding our company made significant strides forward in the recurrent selective breeding of corn introgressed with genes from gamagrass. We have discovered many more beneficial traits than just the two that we had funding to work on. We applied for and received 17 US and foreign patents and have products in testing. None of the new discoveries, proof of concept, or research agreements would ever have happened without SBIR funding."

Expansion into New Markets

There is a blurred line between core products and new applications, which often involve core technologies being applied in new ways. Still it seems worth noting that a number of interviewees and many survey respondents indicated that the SBIR program was being used to expand a company's products and offerings beyond its first product and its core product.

In a case study of Mendel Biotechnology, Dr. Gutterson expressed his view that "SBIR is all about leveraging to build off the core Mendel platform into new areas."

At TSI, Dr. Blankenship noted that the early SBIR awards that funded the core technology are now being supplemented by subsequent awards focused on developing new products altogether. For example, the SBIR program is now helping to fund TSI's push into new technologies and new markets such as air-driven technology for aircraft flaps.

At Divergence, Mr. Rapp said that the role of SBIR funding had shifted. Its use has become for more speculative projects with less data to support them, for which VC funding would not be appropriate.

Survey comments also provided examples of using SBIR awards to expand into new markets. One survey respondent emphasized just how important this SBIR role has been to his company: "Over the 30 years the company has been in business the SBIR program has helped on more than one occasion to rejuvenate the company by bringing new technologies that have been the key to our long term sustainability. The sales we have achieved in the past 30 years with the help of SBIR have been in the billions, and this would not have been achieved without SBIR."

Another survey respondent indicated that the SBIR program had funded what had become a major step forward in battery technology: "Company was acquired on basis of SBIR-funded battery technology. . . . Our SBIR-funded cathode material is in high volume production and used in commercial cells."

Spinoffs and Acquisitions

As noted below, SBIR-funded innovations often follow a nonlinear path to the marketplace. In some cases, companies find that their innovations—especially those outside the core technology of the company—are best addressed through a spin-off.

TRX Systems is a spin-off from TSI. Its first CEO—Dr. Carol Teolis—was hired by TSI as a new PhD from the University of Maryland, and then she managed SBIR projects before entering senior TSI management. Her experience at TSI, which included complete management responsibility for a research project for the U.S. Mint, allowed her to develop an understanding of the NSF SBIR program. She successfully applied for SBIR awards for the new company, which currently has 14 employees. Dr. Blankenship believes that three TSI employees

BOX 6-6
Spinoffs: Verbatim Survey Responses

"Led to a spinoff company."

"Two SBIRs allowed us to spin-off a biomedical R&D firm, which then spun off a further firm to commercialize a third SBIR. All told, today those two companies are employing 16 people that would not have been funded without the SBIR program."

"A spin-off company, currently selling products based on SBIR funded technology, was founded by an employee."

"The company would not have survived if not for SBIR funding. Sales related to the work the company did are about $150 million per year. The company was bought by another company, and is now a division with about 75 US employees."

"We were able to explore and then develop software that led to an acquisition of my company into a larger entity, which applied our experience in software construction and marketing to accelerate its own products."

have the potential to follow Dr. Teolis' example and manage spin-offs if the opportunity to do so emerges.

Box 6-6 highlights verbatim survey responses which indicate that spin-offs and other similar ventures have emerged out of SBIR-funded companies.

VALIDATION EFFECTS

A case study of the SBIR-award recipient, Divergence, revealed the importance of validation effects. According to Mr. Rapp and Dr. McCarter of Divergence, the SBIR program had a huge impact on Divergence. It was particularly helpful as the company prepared to offer a B round to venture investors in 2002; SBIR awards were seen by venture investors as important factors in validating the company's research capability. Divergence executives also observed that SBIR awards provided a significant influx of non-dilutive funds (i.e., money an entrepreneur receives that does not affect the ownership of the company), which added to the company's attractiveness to professional investors. Mr. Rapp observed that it made life much easier when talking to investors if he could show that more than 50 percent of income and investment flows came from non-dilutive sources. Investors often find a company more attractive when they find the entrepreneur has tapped into non-dilutive sources, such as SBIR awards.

The previous case study of TRX also indicated important validation effects from receiving SBIR awards from NSF and the Department of Defense (DoD). The validation effects were said to make it easier for the company to raise additional outside capital.

BOX 6-7
Validation Effects: Verbatim Survey Responses

"Having an SBIR funded award provided critical resources and a "stamp of approval" for our research efforts."

"Although the company changed technology and name after this project, an SBIR award helped to find more funding from investors due to the verification of the claimed technology by the NSF scientific council."

"SBIR funding from NSF brought not only funding but credibility for our technology for the investors, especially foreign investors."

"The first SBIR funding gave the company credibility with investors and other grant funding agencies. It also created a sense of can-do within the corporation."

"SBIR funding provided the seed money required to prove out the idea to the point where the VC community became interested in it and provide(d) substantial follow on funding."

Box 6-7 provides verbatim examples from survey responses of NSF SBIR awards providing validation of research efforts and technologies.

CAPACITY BUILDING—HUMAN CAPITAL

It is obvious that SBIR funding can be used in part to provide small companies with necessary equipment, but interviews and survey responses revealed that the human capital effects can be more important.

Most directly, the SBIR program allows companies to hire staff—typically approximately 2-4 full-time staff at the PhD level for a Phase II project. But SBIR has other capacity building effects as well. Box 6-8 presents verbatim survey responses to show how NSF SBIR awards contribute to capacity building.

BOX 6-8
Capacity Building: Facilities—Verbatim Survey Response

"The project resulted in a substantial increase in quality for many other products and demonstrated the necessity of performing the manufacturing in a clean room, which now benefits the whole activity of the company. The high quality [technology] on starter wafers is at the base of many other products now commercialized. Characterizing and finding the changes to be made in the fabrication process to comply with low trace element contaminants is essential for all wafer-products applications entering foundries or clean room facilities. This also enabled a series of bio-applications, for which our technology is now an FDA qualified material."

A case study of TSI provided an account of Dr. Blankenship observation that the SBIR program has played a critical role on the human resources side. SBIR awards provided an almost perfect training ground for project managers. TSI typically hires PhD researchers soon after graduation, at which point they are technically trained but have little understanding of how to manage projects, handle clients, or work to fixed schedules. SBIR projects at TSI are treated as standalone projects and are often handed off to staff not yet ready for a major commercial project. Dr. Blankenship strongly believes that, in the course of managing one or two SBIR awards, these staff members acquire critical management skills, which can then be applied to the management of commercial projects and eventually entire product lines.

A case study of Divergence also found SBIR awards to be helpful to building its research team. Mr. Rapp noted that they were powerfully helpful in the recruitment of high-level scientists, because they not only provided funding for projects but also generated excitement within the company. Eight different principal investigators (PIs) have been in charge of projects at Divergence.

CRITICAL FUNDING

Many of the companies interviewed indicated that the SBIR program provided funding at critical junctures—for example, the case study of TSI reported that TSI management observed that "initially, SBIR awards had provided funding for investigator-initiated research and an important funding stream which allowed for the survival of the company during its early years focused on contract research." Similarly, Ms. Cappo of LIM said that the company likely would not be in existence without this funding, even though SBIR awards did not directly support company foundation.

The wide range of ways in which the SBIR program provided critical funding is, however, best captured in the words of survey respondents, because this was the single impact that generated the most textual responses. See Box 6-9 for verbatim survey responses on how NSF SBIR awards constituted critical funding.

Encouraging More Internal Investment

Among the more interesting impacts mentioned by survey respondents was that SBIR awards encouraged more internal investment. One small company established a "skunk works" in a new building purchased in an industrial park in 2010. At the time of the interview, the company estimated that its investment in the skunk works had exceeded the SBIR award tenfold.

BOX 6-9
Critical Funding Through SBIR: Verbatim Survey Responses

"Enabled us to finance the creation of innovations that we could not have afforded to develop otherwise. We have been able to make these available to more than 1000 school districts, serving 20,000 educators and 2 million students."

"The funding was used to complete critical technology development and initial scale-up."

"The principal source of funding for the company. Starting from zero, revenues just passed $5 million this year."

"The 3 SBIR grants (2 phase I, 1 phase 2) kept the company alive through periods of no other funding. Without SBIR, the development of the technology would never have been completed or commercialized."

"Funded our project in an area (AgBiotech) where there is very little private investment."

Nonlinear Development and New Markets

A considerable number of survey responses illustrated the importance of what can be termed a "nonlinear" path for product development. In many cases, companies struggled to find the right fit between their technical ideas and market needs; often the solution required re-engineering their products, adapting existing approaches, or even starting again after discovering a core technical expectation was simply wrong. Box 6-10 provides verbatim survey responses from NSF SBIR award recipients who found that the awards helped them traverse a complex path of innovation.

Case studies showed that the experiences of both Divergence and TSI illustrate this nonlinear path. At Divergence, initial work partly funded by NSF SBIR awards focused on soybeans, in partnership with Monsanto. Knowledge drawn from the project has since been applied more generally to root crop nematodes. Similarly, NSF has funded work that applies Divergence technologies to corn and sugar cane.

At TSI, the SBIR award is viewed in part as a means to acquire technical skills and know-how that are not necessarily directly commercial but can have significant uses downstream on other projects. For example, TSI won an SBIR award to build high-performance gun turrets. As part of the project, TSI built a prototype that required a high-performance gimbal. Commercially available gimbals were not suitable, so TSI learned to build its own. Although the gun turret project was not picked up for acquisition by DoD, the new gimbal know-how was later applied to coastal radar systems. Similarly, TSI learned how to build high-performance cameras, which are sold as part of its integrated systems.

BOX 6-10
Nonlinear Development Paths: Verbatim Survey Responses

"This research allowed the company to expand operations and to diversify into the closely related fields of aquatic disease control and wastewater treatment."

"SBIR project led to future funding, investment and products outside of the original target application."

"We would not have survived the 2008-2009 disaster otherwise. The technology that we developed is being used in the intended industry, but it is also being used in 2 related industries that we did not anticipate until the Fall of 2009, and the Fall of 2010 respectively. We are now growing in sales and adding jobs. We expect to double in size (10 to 20 employees) in 2012. Thanks to the NSF!"

"Albeit in markets unexpected from the start, the SBIR program has given our company a chance to prove the technology and develop several commercially viable prototypes with anticipated revenue in the near future."

AGENCY MISSION

Although it does not use the SBIR program in the same way that DoD and NASA use it to procure technology, NSF also has specific missions, such as supporting science, engineering and mathematics education and adding to the science and engineering knowledge base. The following comments by survey respondents illustrate some of the agency mission-related activities supported by the SBIR program, with an emphasis on social benefits:

- **Education.** "We have been able to develop a next-generation technology that is highly innovative and that offers considerable value to districts, educators, and learners in mathematics and science. We have been able to offer services related to the technology at a reduction by two thirds of the costs that districts pay otherwise."
- **Public health.** "Enabled us to take software built for one client and generalize it. Resulted in important long-term benefits to health care, in cancer and public health/vaccination administration."
- **Public goods.** "There are significant measurement needs to address air pollution and climate change, and the SBIR program is largely the key NSF mechanism to fund such activities. The market isn't large—indeed it is tiny—so venture capital is unrealistic. . . . There is no way this project would have been undertaken without SBIR funding—no other viable mechanisms really exist."
- **Serving disabled populations.** "It appears to be one of the few ways to fund R&D for products for people with disabilities—high societal benefit, potential to provide revenues to keep our company profitable, but not hav-

ing the huge profit margins that attract typical private investors. . . . There would have been no support to develop this technology, which we expect will be of significant use to people with visual disabilities."

- **Linking academia and business.** "The company employed several students who might have gone out of state without the funding. Also, the founders received experience that convinced them of the value of entrepreneurial endeavors, which has informed their teaching since."
- **Open source technologies.** "We are an open source company providing services. The creation of technology via SBIRs builds a foundation of software technology which we service. We eschew product and licensing and focus on the larger part of the software business which is services. We are also able to make huge contributions to the research community and innovation in general because we release our technology with open source licenses. This spawns a multitude of derivative businesses, many of which are other companies."

Universities and the SBIR Program

Although the Small Business Technology Transfer (STTR) program is specifically aimed at increasing connections between universities and small companies, the SBIR program can have similar effects. An example is provided by the case study of IFOS, whose staff worked with practitioners and surgeons at Stanford University medical school, in cardiology, radiology, and oncology. Through its Stanford connections, the company moved on to partner with several local medical devices companies, which will manage U.S. Food and Drug Administration (FDA) market testing and industry-specific marketing.

A connection may also be increased in some cases between NSF SBIR award recipients and non-university types of research institutes. The Divergence case study revealed that Divergence "has had a close relationship with the Donald Danforth Plant Science Center since 2001. Divergence's laboratories are located next to the Danforth Center, and Divergence collaborates with multiple Danforth investigators, utilizes core laboratories in analytical chemistry and microscopy, and has received joint research grants with the Center."

Survey comments, listed in Box 6-11, suggest that such connections can benefit both sides. For universities, the SBIR program offers a pathway into the commercial world that is otherwise difficult to find, given the very early stage of most university-developed technologies. In addition, companies find that access to university facilities and researchers can provide specialized inputs that are otherwise difficult to find, attract, and fund.

BOX 6-11
SBIR and Universities: Verbatim Survey Responses

"SBIR funding is the only current method by which promising 'breakthrough' technologies developed in an academic setting can be transformed into commercial use."

"A direct result of the commercialization of one of these SBIR awards led us into a 10 year collaboration with a large company, which eventually led to their moving the engineering department for their high end calculator software from Japan to Oregon."

"The ability to conduct funded R&D through the SBIR program and to form collaborations with university researchers through the STTR program enables the company to develop unique technologies with minimal risk, which is critical to my ability to do this type of work."

"SBIR funding enabled the company to hire an engineering team to transfer technologies developed by scientists into viable commercial products."

"This project generated NSF SBIR Phase II award and several supplementary awards for two universities and a school located in a rural area. We have identified and are currently working on a Parkinson's disease diagnostics project in collaboration with a university."

PROGRAM OPERATIONS

Program Directors

One the most distinctive features of the NSF SBIR program is the close connection between program directors and awardees. In the Divergence case study, Dr. McCarter said that "some program officers have been very flexible—indeed, NSF has been especially so over the years, and has been very strong on personal contact between funding officers and PIs." Divergence executives stressed how important it was for the funding officer to maintain close relations with the PI and company management so that he or she could fully understand the project and therefore provide active support as needed.

Similarly, ALDN reported working with NSF to be very easy, with close connections to program managers and limited auditing requirements. At LIM, Ms. Cappo viewed NSF grant administration as highly professional. She said that the NSF program manager provided pre-application advice and early feedback on applications and that the NSF team was highly supportive and focused on ensuring the project's success. Dr. DiBello at WTRI said that the initial NSF SBIR award in 2000 was successful in part because the NSF program officer was familiar with the relevant academic work. At TG, Mr. Landau observed that an important partnership with Exceptional Teaching (ET) in California was in fact engineered by the NSF program director, whose hands-on approach generated additional business for the company and led to a successful connection in which neither party was initially interested.

Box 6-12 provides verbatim survey responses on the topic of NSF Program Directors. These comments support the view that the NSF model of professional program management appears to have strong support among recipient companies.

At the same time, not all reported experiences with NSF Program Managers were positive. One interviewee explained that some interactions were positive, but overall the company's experience with NSF program directors varied. Some program directors, such as its first, were very helpful, believing their mission to be aiding the company. Others were significantly less helpful and displayed little understanding of how small companies worked with their customers. The reported negative experience was in some cases a problem, especially in relation to the oral defense part of the application process for Phase IIB.

Business Training and Commercialization Support

Case-study interviewees rarely made comments about training, but a few did. Dr. Bueschler (ALDN) expressed appreciation for the NSF training. Dr. Gutterson (Mendel) said that Mendel found Foresight's support to be useful. Ms. Politi (TRX) observed that through NSF, TRX received commercialization support from LARTA, whose process was especially helpful in relation to a new collaborative mapping initiative.

Mr. Rapp (Divergence) participated in training programs at both Dawnbreaker and LARTA and rated both as useful in helping inexperienced scientists and engineers understand and prepare for the business world. However, in his view, the strong encouragement to participate taxed the company's executive resources. At the time of the interview, Divergence was looking at world markets for corn seed

BOX 6-12
Program Directors: Verbatim Survey Responses

"In particular, our program director and the NSF program were extremely supportive during all stages of our project."

"Advice from Program Officer was very influential and helpful. Funds came at a critical time."

"Commercialization education/coaching was very valuable because we did not have full time business personnel on the team at the time of the phase I."

"In my experience the SBIR program and NSF's program in particular also provide valuable business education and coaching to small companies that help them navigate the difficulties in building successful businesses."

treatment. Generic business plans were of no use—in fact, the company hired an industry insider with more than 30 years of experience as a consultant. Mr. Rapp noted that his management team had more than 120 years of business experience, and hence the pressure to participate in "training" was not helpful.

Two survey respondents indicated that the prohibition on using SBIR funding for marketing remained a significant barrier. One noted that, because marketing expenses are not allowed in government overhead calculations, a company whose main business is reliant on government funding will eventually find that it does not have the financial resources to conduct commercial marketing or advertising.

ISSUES IN PROGRAM MANAGEMENT

Over Focus on Commercialization and the Venture Capital Model

There is an undeniable tension between the program's Congressional dual mandates to support the commercialization of federally funded programs and to stimulate technological innovation. If innovation concerns turning ideas into products for a market, commercialization concerns realizing that market opportunity. All SBIR programs try to balance both these legislative objectives.

At NSF, the SBIR program is heavily focused on the commercialization of funded technologies. As Dr. Kesh Narayanan, a former director of NSF's Industrial Innovation and Partnerships Program has noted, "the "I" in SBIR is all about Innovation: There are several other innovation research models at NSF; however, SBIR is the only one dedicated to 'for profit' small businesses. The majority of NSF investment for advancing fundamental research is via investments in academia."[4]

Yet as the discussion of commercialization and metrics has shown (Chapter 4), commercialization is not a simple concept. It requires an extended view, because many technologies bear fruit only after many years of effort. It also requires a nuanced view, because commercialization for small businesses is often highly nonlinear.

Among the numerous concerns raised by interviewees, those about NSF's approach to commercialization were raised most often and most strenuously. Some case-study interviewees and survey respondents expressed concern that NSF has adopted an overly narrow view of commercialization and one that fails to adapt to the nuances of commercialization in the small business world. As a result, these interviewees and respondents say, NSF is viewing projects through a narrow lens based on commercialization models from the venture capital (VC) world—a world that several interviewees are determined to avoid if at all possible.

Box 6-13 gives verbatim survey responses pertaining to NSF's SBIR commercialization focus.

---

[4]Correspondence, May 26, 2015.

BOX 6-13
Commercialization Focus: Verbatim Survey Responses

"The most disappointing fact was the requirement for obtaining commitments for investment/major-sales at the end of Phase-I in order to be deemed qualified to receive Phase-II funding."

"NSF is also different in the degree of pressure it exerts toward commercialization. This has become even stronger in the past two years, possibly coinciding with the move to LARTA's commercialization support program."

"One of the WORST measures of an SBIR is 'did the project get to market.' The right question is 'is the company still in business.' It is crazy to think that $100k of funding will result in a marketable product. It does, however, open up the world of possibilities. It creates conversations with potential partners and investors that would not otherwise happen. Often, these conversations lead in directions away from the SBIR technology. BUT that should be viewed as a success!"

"NSF business reviewers tend to have unrealistic expectations on the level of maturity for the output of one phase II effort and this hurts our chances of further NSF funding. VC funded companies seem to be the priority."

In the case study of IFOS, Dr. Moslehi of IFOS noted that some NSF program directors do not entirely appreciate the challenges involved in developing highly innovative technologies for emerging markets. Unlike other agencies, which have an end use in mind related to their own needs and are therefore happy with a working system for their use, NSF struggles to demonstrate a wider impact and consequently expects broader commercial success. Dr. Moslehi explained that this is often unrealistic, considering that venture-backed companies are often funded with tens of millions of dollars to succeed commercially. A $1 million program cannot expect the same results.

Similarly, Dr. Wijmans (MTR) said that he thought NSF is too focused on the VC funding model. He recommended the DoE model as being more realistic—anticipating the need for a demonstration phase after Phase II, which is often funded directly by DoE.

Dr. DiBello of WTRI thought that NSF's SBIR program has changed over the years. When WTRI received its initial awards around 2000, the program was focused on very innovative ideas with potential for significant commercial success. Dr. DiBello described a shift in the program focus toward expecting a project to be innovative as well as far along the commercialization path, even before the start of the first Phase I award. It is not much of an exaggeration, according to Dr. DiBello, to suggest that NSF is seeking projects that are more or less ready for Phase IIB at the time of Phase I. NSF now wants a much more completed idea for Phase I than is reasonable.

Dr. DiBello, of also noted that the NSF focus on encouraging firms to acquire VC funding is itself possibly misplaced. WTRI's experience in fixing broken companies indicates that many of them received VC funding too early, at too high a price.

This new focus seems to influence selection in a number of ways. For example, Dr. McCarter (Divergence) claimed that some applications were dismissed simply because they involved the biotech sector, where regulatory timelines impose significant delays, which pushed the projects out of the timeframe for commercialization that was acceptable to NSF. The rejection of these applications led to indefinite delays on projects that the company believes have a potentially powerful range of applications.

While maintaining a balance between commercialization and innovation is challenging, there is also quantitative evidence to suggest that NSF is now focusing on projects that can reach the market quite quickly; internal statistics suggest that more than 70 percent of all projects have reached the market but also that projects with substantial commercial scale remain scarce.

It seems reasonable to suggest that NSF review its objectives and metrics in this area.

Application and Selection Procedures

Many interviewees complimented NSF selection procedures. Most interviewees had no significant complaints, and IST was particularly positive about the feedback received on its applications. Ms. Wedding said that NSF utilized a substantial number of reviewers even for Phase I applications and provided both a summary review and individual reviewer comments.

However, several companies indicated that selection and review continued to generate possibly unnecessary concern. It is always difficult to produce a completely accurate and objective review, and as noted above NSF was commended by some companies for utilizing outside reviewers. However, on the basis of interviewee comments, there appears to be an appetite among the recipient base for mechanisms to address what they consider as inappropriate or inexplicable rejections. Two such suggested mechanisms are resubmission and rebuttal.

Companies such as Divergence and IST see substantial value in allowing applicants to improve their applications in response to review and resubmit them. These companies pointed out that the National Institutes of Health (NIH) allows resubmissions, and noted that such an approach could work at NSF and other agencies. IST also noted that this mechanism could be especially helpful when proposals are rejected for administrative reasons.

Several companies expressed a wish for mechanisms that would allow companies to rebut misinformed comments in reviews. Mr. Rapp (Divergence) said that one recent NSF reviewer completely misunderstood and therefore torpedoed a Divergence proposal. Reviewer comments included five misstatements and one or two complete misunderstandings. In contrast, Mr. Rapp observed that the U.S. Department of Agriculture (USDA) already uses a system whereby the program officer emails Divergence a list of up to 10 questions arising from review. This gives the company an opportunity to make its case in more detail and to eliminate misunderstandings.

Similarly, IST strongly supported the idea of providing companies with the opportunity to respond to reviewer comments within the framework of the selection process. Ms. Wedding observed that the former Advanced Technology Program (ATP) application process provided that opportunity, by providing companies with a preliminary review and follow-up questions.

Interviewees made other comments and suggestions:

- **Splitting commercial and scientific review.** In its case study, Divergence saw the need for this separation on a recent NIH application, for which review comments were ill-informed about commercial opportunities. USDA is implementing a split review structure.
- **Electronic submission through FastLane.** Interviewees were not enthusiastic about FastLane. One survey respondent noted, "The FastLane process is not well designed for SBIR. Formatting can be very time-consuming, and as FastLane is used for all NSF applications, there are additional sections that are not relevant to SBIR applications. At a minimum, FastLane should identify application elements that are mandatory/not permitted for SBIR."
- **Better focus on proposed budgets.** Budgets are often an afterthought, and Mr. Landau (TG) observed that reviewers rarely pay attention to budgets (NSF and other agencies for which he has reviewed proposals).

Funding Gaps

Data regarding funding gaps are considered in more detail in Chapter 4. The survey data and case studies indicate that funding gaps between Phase I and Phase II remain an issue for many companies. Ms. Cappo of LIM explained that the company experienced a significant gap (about 1 year) between Phase I and Phase II, which presented a substantial problem for the company, even though it was able to fund project staff through other work. Ms. Cappo stated that these gaps and lags should be addressed, given the importance of stable funding for small companies.

Dr. Blankenship at TSI confirmed that the gap between Phase I and Phase II is a serious problem for many smaller companies, although TSI had reached a stage of maturity and scale where it could easily manage the gap with other revenue streams.

Funding Levels

Several interviewees commented on the funding levels for awards (interviews were conducted prior to the recent changes through the reauthorization legislation).

Dr. Bueschler (ALDN) said that the one-size-fits-all funding model is not efficient. For Phase I, $100,000-$150,000 is usually more than enough to achieve proof of concept. However, in some cases, this amount could be an order of magnitude too small—for example, at ALDN some projects would have required $80,000-$100,000 in external testing, so they could not be funded through SBIR Phase I and were dropped. However, for other projects the funding provided was twice the amount needed (e.g., for some software projects). More flexibility rather than more money was needed, in her view.

Mr. Landau at TG also advocated for more flexibility in funding and in timelines. Increased flexibility in these areas would permit TG to return to NSF for further SBIR funding.

At IST, Ms. Wedding did not believe it good policy to focus on more funding for a smaller number of awards. She said that $150,000 is reasonable for Phase I, and $500,000 is reasonable for Phase II. She observed that it was probably better to give smaller awards to more projects. Larger awards would lead to additional focus on hot topics.

Other Issues

Interviewees raised several additional issues in relation to the NSF SBIR program. Some of these are noted below:

- **Venture funding participation.** Dr. Moslehi (IFOS) stated that small businesses that have successfully obtained traditional VC funding should not be allowed to participate in the SBIR/STTR programs. He felt that federal funds should not be used to protect the interests of financial players. He also noted that the commercialization prospects of a well-funded company should not be compared to one that is bootstrapped (growing organically) and that it is a need of SBIR funding to reduce technology risks.
- **Partnerships.** Dr. Moslehi (IFOS) also stated that NSF should focus additional attention on industry-university partnerships organized around long-term, high-risk, high-reward projects with the *potential* for substantial commercialization.
- **Multiple annual application deadlines.** Dr. Gutterson (Mendel) strongly endorsed the need for multiple deadlines; he said that a single annual deadline was no longer sufficient given the rapidly accelerating speed of technical change. Although NSF has multiple annual deadlines, each division within NSF has only a single deadline, so for a given technology only a single funding window is likely to be available. This point was also

raised by other interviewees. Ms. Cappo (LIM) said that given the speed with which market conditions change, a single annual deadline seems unnecessarily inflexible; Mr. Rapp (Divergence) recommended that NSF should—like most other agencies—provide more than one annual opportunity or deadline for each topic.

- **Phase I–Phase II transition.** In different ways, two companies noted the need to emphasize the importance of the Phase I–Phase II transition. Mr. Landau (TG) expressed concerned about firms that won numerous Phase I awards but failed to convert them into Phase II awards. Dr. Blankenship (TSI) underscored the importance of focusing closely on converting Phase I awards: TSI typically matches a Phase I award with internal company money equivalent to about 50 percent of the award to ensure a good result and a strong case for a Phase II award.
- **Paperwork.** Mr. Rapp (Divergence) noted that NSF requires more detailed reporting than most agencies, including time reporting by individuals and a financial report that identifies funds spent on each category.
- **Partnership and business development funding.** Dr. Bueschler (ALDN) stated that support for partnering is a somewhat neglected part of the SBIR program. ALDN, for example, partnered with A123 on batteries, but the latter has become more focused on immediate needs rather than on longer term development. The ALDN platform is generic for all battery materials, and Dr. Bueschler is actively seeking partners, for example, at the University of Colorado and from offshoots. Even a small amount of funding to support partnering might make a difference and would have been especially important during the company's years when dollars were scarce.
- **Phase IIB matching funds.** In particular, IST sees the need for more flexibility on when the company must acquire matching funds. Currently, funding has to be obtained during the exact 18-month period from the start of the Phase II award—the last point of reasonable application for Phase IIB. She does not believe that many companies could start on a new project and reach an investable point so quickly. At a minimum, NSF should accept matching funds acquired during the period after the start of Phase I.
- **Reviewers and reviews.** Dr. DiBello (WTRI) expressed concern about the consistency of quality and qualifications of reviewers; some reviewers did not seem to have the background to follow the proposal details but others did an exemplary job and offered important insights, even if critical of the proposal. This was primarily a problem with Phase I review; in her experience, Phase II reviews separated commercial from technical review and, as a result, the reviews were higher quality. In addition, Dr. DiBello expressed concern about possible conflicts of interest because Phase I reviews at NSF do not disclose panel memberships.[5]

[5] It should be noted that all reviewers must sign a non-conflict of interest form.

- **Award timelines.** Mr. Landau (TG) sees an increasingly poor fit between the timeline for a completed SBIR project and the faster moving markets in which TG operates. He pointed out that the time from Phase I application to Phase II completion is on the order of 3-4 years, which is much too long for commercial applications in his market sector.

  Dr. Moslehi (IFOS) noted that NSF sometimes has unrealistic expectations for its awards and tends to treat all technologies as though they can be commercialized on the same timeline and with the same resources. This approach is not viable. As a result, he believes that NSF is no longer a good source of funding for certain businesses working on challenging emerging technologies. In contrast, Dr. Bueschler (ALDN) observed that people take the time allowed, so shorter timelines are preferable because they force companies to think about what could reasonably be accomplished within a more constrained Phase I.

  Divergence offered a slightly different perspective, recommending that program managers have the flexibility to adapt projects to changing technical realities. Mr. Rapp (Divergence) observed that grant applications require the company to look ahead to where the project might be 12 months in the future. Often, several of the specific milestones to be addressed under a proposed award have been completed by the time funding arrives. It is therefore critical that program officers and technical points of contact (TPOCs) have flexibility to work with PIs to adjust objectives, perhaps by adding more advanced milestones.

- **Topics.** Several interviewees expressed concern that the topics at NSF are becoming increasingly narrow and that important and potentially transformative ideas are being excluded by this narrow technical framework. Ms. Cappo (LIM) said that NSF focuses too much on a limited market—NSF SBIR topics in her area, for example, have become increasingly focused on testing and electronic student records (which are not topics of interest to LIM). However, interviews with NSF staff suggested that concern may reflect a misunderstanding of NSF topics. Overall Dr. Moslehi (IFOS) was concerned that SBIR programs do not adequately distinguish between emerging and non-emerging technologies, between areas where incremental improvements could be marketed and others that are potentially transformative, but where there are limited markets in the short term.

Dr. Gutterson (Mendel) said that he would like to see more broad topics so that firms could decide which technologies fit the agency's requirements. Ms. Cappo (LIM) said that broader topics would be very welcome and would support a wider range of innovation. Dr. Wijmans (MTR) said that MTR was adapting to the narrow technical band for NSF awards and was therefore focused on development of novel or improved materials, because that appears to be critical for success in the NSF SBIR competition.

7

Findings and Recommendations

The findings and recommendations in this chapter reflect the congressional objectives for the SBIR program, as reiterated in the recent program reauthorization and in the subsequent SBA policy Directive that guides program implementation at all agencies.[1] Section 1c of the Directive states program objectives as follows:

> *The statutory purpose of the SBIR Program is to strengthen the role of innovative small business concerns (SBCs) in Federally-funded research or research and development (R/R&D). Specific program purposes are to: (1) Stimulate technological innovation; (2) use small business to meet Federal R/R&D needs; (3) foster and encourage participation by socially and economically disadvantaged small businesses (SDBs), and by women-owned small businesses (WOSBs), in technological innovation; and (4) increase private sector commercialization of innovations derived from Federal R/R&D, thereby increasing competition, productivity and economic growth.*[2]

The findings below review the extent to which each of these program objectives is being addressed at NSF, as well as examine some specific aspects of program management.

FINDINGS

The Small Business Innovation Research (SBIR) program at the National Science Foundation is having a very positive overall impact. It is meeting three

[1]See Box 1-2 and the discussion in Chapter 1 of the committee's task.

[2]SBA SBIR Policy Directive, October 18, 2012, p. 3.

of the four legislative objectives of the program; we find that more needs to be done to "foster and encourage participation by socially and economically disadvantaged small businesses (SDBs), and by women-owned small businesses (WOSBs), in technological innovation." Even so, the program is supporting the birth and growth of small innovative companies, which indirectly impacts all of the congressional objectives.

I. Commercialization

Each agency has its own priorities for the program. At NSF, the overwhelming emphasis has been on commercialization, which here means that projects are commercially successful in private sector markets.

A. **SBIR projects at NSF commercialize at a substantial rate.**[3]
   - Substantial Sales Reported: About 70 percent of Phase II respondents reported sales, based on responses to the 2011 Survey.[4] This represents a slight increase from rates reported in the previous Academies report.
   - Sales Anticipated: An additional 19 percent of Phase II respondents reported that they anticipate future sales.[5]
   - Scale of Commercialization: Despite the high rates of commercialization, the scale of commercialization was limited: no projects surveyed reported aggregate sales of $50 million or more, and only 6 projects (out of 114) reported project-related sales of $10-50 million.[6]

B. **Academies survey data show that NSF SBIR projects are primarily commercializing in the domestic private sector.**
   - Just under 60 percent of respondents with sales reported revenues from domestic private sector customers.[7]
   - About 15 percent reported export customers.[8]
   - About 16 percent of responses identified customers in the public sector (primarily defense and defense contractors.)[9]

C. **Further investment in SBIR is additional evidence of commercial activity.** Subsequent investment provides further evidence that SBIR projects are expected to generate significant commercial value even if they have not yet reached the market. Academies survey data show that

[3]NSF declined to share the results of its own internal commercialization tracking system on confidentiality grounds.

[4]See Table 4-1.

[5]See Table 4-1.

[6]See Table 4-2.

[7]See Table 4-3.

[8]See Table 4-3.

[9]See Table 4-3.

- More than 60 percent of Phase II survey respondents reported additional investment funding.[10]
- The most likely source of additional funding for Phase II recipients (other than their own company and personal funds) was non-SBIR federal funding, which is funding provided by from federal agencies from other budgets outside the SBIR programs (39 percent of responses), closely followed by other companies (32 percent), private equity/angels (28 percent). About 15 percent reported venture capital (VC) funding.[11]

D. **SBIR is associated with job growth.** With respect to direct job growth, Academies survey data indicate that the median size of firms with NSF Phase II awards grew from 6 employees at the time of award to 10 employees at the time of survey.[12]

E. **SBIR funding makes an important difference to project outcomes.** SBIR funding makes the difference in determining project initiation, scope, and timing. Academies survey data show that
- 68 percent of Phase II respondents reported that the project probably or definitely would not have proceeded without SBIR funding.[13]
- 75 percent of those who would likely have proceeded anyway reported that the project would have been narrower in scope.[14]
- About two thirds of those who would likely have proceeded anyway reported that the project would have been delayed by at least one year.[15]

F. **Commercialization and venture funding**
- NSF has focused on an expected pattern of product development and commercialization that is modeled on that for projects funded by venture capitalists (VCs).
- This model is not an appropriate model for NSF SBIR project commercialization for several reasons:
    - o VCs seek projects that meet a number of narrow criteria, including timeline to market exit, size of opportunity, amount of funding required, capabilities of the management team, and industry sector. The vast majority of NSF SBIR projects do not and will not meet these criteria as they are typically specified by VCs.
    - o NSF appears to make a point of funding very small companies at or soon after startup. The median size of Phase II company at

---

[10]See Table 4-6.
[11]See Table 4-8.
[12]See Tables 4-4 and 4-5.
[13]See Table 4-19.
[14]See Table 4-20.
[15]See Table 4-21.

time of award is 6 employees.[16] These companies are generally too small to interest VCs.

- o Data from the Academies survey indicates that 63 percent of Phase II respondents received additional investments, but of these only 15 percent received funding from VCs.[17]

II. Fostering the Participation of Women and Other Underserved Groups in the SBIR Program

A. Current outcomes data show that this objective has not been met.

- Levels of participation by underserved groups are low and declining.
  - o Data from NSF indicate that approximately 10 percent of SBIR Phase I awards go to Minority-Owned Small Businesses (MOSBs). These figures have declined over the past decade, as has the MOSB share of Phase II awards.[18] MOSBs also show lower success rates for both Phase I and Phase II applications than non-MOSB applicants, with the success for Phase I MOSB applicants being lower in every year, and success rate of Phase II MOSBs averaging 12 percentage points lower over the period.[19]
  - o NSF data show that about 13 percent of SBIR Phase I awards were to Woman-Owned Small Businesses (WOSBs) and make up just over 10 percent of Phase II awards.[20] The higher share for WOSBs compared to MOSBs is based on the large numbers of awards won by 3 well established WOSBs.
  - o NSF does not maintain data on woman and minority Principal Investigators (PIs). Data from the Academies survey indicates that these numbers are also low and not rising.
  - o The Academies survey indicated that 13 percent of PIs were female.[21]
  - o These discouraging results correspond to similar poor results across many STEM education fields in the United States.
- NSF has no separate data on African-American-, Hispanic-, or Native American-owned small businesses.
  - o The Academies survey, however, indicated that black- and Hispanic-owned small businesses are themselves a very small

---

[16]See Table 4-4.

[17]See Tables 4-6 and 4-8.

[18]See Figure 3-10 for percentages of SBIR Phase I awards going to MOSB, and Figure 3-24 for percentages of SBIR Phase II awards going to MOSB.

[19]See Figure 3-8 for Phase I MOSB comparative success rates for applications receiving awards, and Figure 3-22 for Phase II MOSB comparative success rates.

[20]See Figure 3-15 for percentages of SBIR Phase I awards going to WOSB, and Figure 3-28 for percentage of SBIR Phase II awards going to WOSB.

[21]See Chapter 2, "Evidence provided by Academies Survey Data," and 2011 Survey, Question 14A.

share of MOSBs overall. Black-owned small businesses accounted for approximately 1 percent of all respondents; Hispanic-owned firms, about 3 percent.[22]

- o In the Academies survey, companies reported that 17 percent of Phase I PIs and 15 percent of Phase II PIs were minority. However, further analysis indicates that less than 1 percent overall were African American, with 1.5 percent in Phase I and none in Phase II. Less than 3 percent overall were Hispanic, with 3.9 percent in Phase I and 1.1 percent in Phase II. There were no Native Americans in Phase I or Phase II.[23]

- <u>There are further questions raised by the data regarding the shares of Phase I and Phase II awards accounted for by WOSB</u>.
  - o The Academies survey data show that WOSBs accounted for 17.1 percent of Phase I companies but only 11.3 percent of Phase II companies responding to the Academies survey.[24]

B. NSF's efforts to "foster and encourage" the participation of woman-owned and minority-owned small businesses have not been effective.

1. <u>No substantial program has been implemented for outreach to these communities</u>.
   - o No examples or other evidence was found of persistent efforts by NSF to reach out to WOSB.
   - o The NSF SBIR/STTR Advisory Committee also raised this issue in 2012.
2. <u>NSF does not report on or sufficiently track participation by WOSB and MOSB</u>.
   - o NSF does not provide an annual report that covers either data on participation or efforts to foster and encourage participation.
   - o No evidence was found that NSF tracks MOSBs at a level sufficient to meet congressional intent.
   - o No evidence was found that NSF tracks the participation of woman and minority principal investigators in the program.

C. There appears to have been insufficient effort made at NSF to understand why these rates remain low. Concerted effort would be needed to determine what practical steps can be taken to improve participation and hence both meet congressional objectives of the program and expand the pool of qualified applicants and capabilities.[25]

---

[22]Calculated from data reported in the text of Chapter 2, just prior to Table 2-9.

[23]Calculated from data reported in Table 2-7 and accompanying text.

[24]See Table 2-10.

[25]A discussion of conclusion about woman and minority participation and NSF's limited efforts to address the issue is provided in more detail in Chapter 2, "Conclusions: Woman and Minority Participation in the NSF SBIR Program."

III. Stimulating Technological Innovation and Meeting Agency Mission Needs

NSF's agency mission is focused on the advancement of science.[26] Thus for the NSF SBIR program, the twin objectives to use small business to meet federal R/R&D needs and to stimulate technological innovation are close complements and are therefore discussed together in this section.

A. **Meeting federal R/R&D needs: The SBIR program at NSF supports the development and adoption of technological innovations. Selection of both topics and individual projects for funding reflects a strong focus on technological innovation.**[27]
   - Focused Topic Selection: Program Directors strongly assert that topic selection is driven by efforts to enhance and support innovation, and topics undergo a review process within the agency toward that end.
   - Solicitations are not Exclusive: NSF does not formally require that projects specifically match the topics described in the solicitation. Under the terms of the solicitation, non-matching applications can also be accepted, in an effort to ensure that potentially important innovations are not excluded by the topic structure. However, NSF does not track whether applications match the solicitation topics, there is no evidence to confirm or refute if solicitations are nonexclusive.

B. **Emphasis on Technological Innovation: Scoring for individual projects is weighted toward technological innovation: NSF merit criteria emphasize the importance of technical innovation, and selection decisions are—according to Program Directors who make recommendations for those decisions—also heavily weighted toward the support of potentially transformational technologies. SBIR continues to connect companies and universities.** Survey data indicate that NSF SBIR projects continue to utilize universities in a variety of ways (excluding the even closer connection through Small Business Technology Transfer [STTR]):
   - Faculty and Student Participation: Just under 60 percent of Phase II survey respondents reported a link to a university. In about 35 per-

[26]"The National Science Foundation (NSF) is an independent federal agency created by Congress in 1950 "to promote the progress of science; to advance the national health, prosperity, and welfare; to secure the national defense…" See National Science Foundation, "NSF—At A Glance," accessed on June 2, 2014. A more extended description from the same source states that goals are "to advance the frontiers of knowledge, cultivate a world-class, broadly inclusive science and engineering workforce and expand the scientific literacy of all citizens, build the nation's research capability through investments in advanced instrumentation and facilities, and support excellence in science and engineering research and education through a capable and responsive organization."

[27]See Chapter 2 for a discussion of NSF's topic selection and funding decisions.

cent of responses, university faculty worked on the project (not as PI); 30 percent employed graduate students; and in almost a quarter of Phase II projects reporting, universities were subcontractors.[28] These figures are broadly similar to those reported from the Academies' 2005 survey.

- Project Partners: 114 different universities were identified by survey respondents as project partners; 13 were mentioned by more than 3 respondents.[29]
- Academic Founders: More than 80 percent of NSF Phase II SBIR companies responding to the survey reported at least one academic founder, and about a quarter reported that the most recent prior employment of the founder was at a university.[30]

C. Academies survey data show that SBIR projects generate substantial knowledge-based outputs such as patents and peer reviewed publications.

- Patents: Patenting remains an important component of knowledge diffusion (and protection).
  - More than 70 percent of the Academies' Phase II survey respondents reported filing at least one patent related to the surveyed project.
  - 23 percent reported filing five or more related patents.
- Publications: Publication of peer reviewed articles remains the primary currency of scientific discourse, and despite the need to protect ideas in the commercial environment of small businesses, the Academies' Phase II Survey shows that SBIR firms continue to participate deeply in scientific publication.
  - About 80 percent of surveyed Phase II projects reported at least one resulting peer reviewed publication.[31]
  - 32 percent reported more than three publications resulting from the surveyed project.[32]
  - Many of the companies interviewed for case studies made a point of indicating that they take a great deal of pride in the number of peer reviewed publications developed by their scientists and engineers, both within and outside of the SBIR program.

D. SBIR funds projects with social benefits that may not be attractive to commercial sources of funding.[33]

---

[28] See Table 4-17.

[29] See Table 4-18.

[30] See Table 4-28 and Chapter 4 text, just before "Conclusions: Knowledge Effects."

[31] See Table 4-16.

[32] See Table 4-16.

[33] See Chapter 6, "Funding Otherwise Un-fundable Projects."

- Projects with Social Benefits: NSF SBIR funds some projects that are socially desirable and market oriented but that are unlikely to generate the high returns need to attract venture-type funding. For example, NSF SBIR-funded projects supporting assistive technologies for the blind are now in use worldwide (for a detailed analysis of socially valuable projects with limited markets, see Touch Graphics case study[34]).
- Projects with Long Timelines: NSF also funds some projects with timelines that are too long for venture funders, some of which may turn out to be "platform technologies" with multiple applications (for example, new materials or battery technologies.)

IV. Fostering Innovative Companies

A. Company formation and early funding: Many SBIR companies surveyed by the Academies report reported that SBIR funding was instrumental in the founding of the company. The formation of new innovative companies is a positive outcome for the program.

- 45 percent of Phase II respondents to the Academies survey said that the company was founded entirely or in part because of the SBIR program.[35]
- For some companies included among the case studies, SBIR funding permitted the shift from an exploratory to a professional operation.[36]

B. De-risking subsequent investment and new markets

- Mitigating Risks: One important impact of the NSF SBIR program is to provide funding that reduces the risk for subsequent investors. Risk mitigation often leads to the leveraging of SBIR funding while retaining the power of markets to make final decisions about funding.[37]
- Matching Funds: The NSF Phase IIB program lowers risks by providing matching funds to companies with external investors or other approved sources of funding.[38]
- Early Stage: More widely, many respondents to the Academies survey and a number of case study interviewees said that NSF SBIR funding was provided at a stage when the project was sim-

[34]See Appendix E: Touch Graphics Case Study.

[35]See Table 4-29.

[36]See "Validation Effects" in Chapter 6, and Box 6-7.

[37]See section on "Risk" in Chapter 6, and survey results for risk reduction in Box 6-4.

[38]See discussion in Chapter 5 of requirements and effects of matching funding for Phase IIB projects, and tables 5-2 and 5-3 that show a 3:1 ratio of matching investments to NSF funding over a dozen years.

ply too risky for commercial sources of funding. Once the project proceeded further, risk was lower and additional funding could be acquired.[39]

- Support for Technology Development: NSF SBIR funding supports technology development, which can be supported through commercial funding further downstream. SBIR is particularly important for funding proof of concept for new technologies.[40]
- Validation: NSF SBIR funding has itself been important validation for companies seeking further investments, according to case study interviews and survey responses. The strength of the selection process and growing understanding of SBIR among investors appears to be strengthening this effect.[41]
- NSF funding can support company efforts to enter new markets.[42]
- Exploit technology platforms: In some cases, companies use SBIR funding to build off existing platform technologies specifically to enter new markets. This platform-driven approach is used by a number of the interviewed companies. (See chapter 6.)
- Strategic corrections: Innovative companies must often make mid-course corrections. NSF funding has—according to Academies survey recipients—helped a number of surveyed companies successfully make what are often difficult changes that are hard to fund.

C. **Long term positive effect on SBIR awardee companies.** Commercialization in the long run requires sustainable companies, and SBIR has supported the development of an ecosystem of small innovative companies in the United States.[43] The Academies survey provided SBIR companies with the opportunity both to report the overall impact of SBIR on the company, and to identify specific kinds of impacts.

- 35 percent of Phase II winning recipients indicated that the NSF SBIR program had a "transformative" effect on their company. Another 54 percent said that it had a "substantial positive long-term effect."[44]
- Of the 179 detailed comments received in the Academies survey, none reported negative effects. Widely differing kinds of impacts were reported, as summarized in Box 7-1.[45]

---

[39] See "Company Formation and Very Early-stage Funding" in Chapter 6, and Box 6-2.

[40] See "Funding Core Technology Development" in Chapter 6, and Box 6-5.

[41] See "Validation Effects" in Chapter 6, and Box 6-7.

[42] See "Expansion into New Markets" in Chapter 6.

[43] See "Capacity Building—Human Capital" in Chapter 6, and Box 6-8.

[44] See Table 4-23.

[45] See "Key Aspects of SBIR-Driven Transformation" in Chapter 4, as well as Chapter 6 and cases in Appendix E.

BOX 7-1
Different Ways in Which SBIR Awards Helped to Transform Companies

- Provided first dollars.
- Funded areas where venture capital and other funders were not interested.
- Provided funding during downturns in the business cycle.
- Opened doors to many potential stakeholders in specific technologies, including agencies, prime contractors, investors, suppliers, subcontractors, and universities.
- Helped address niche markets too small for major players/funders.
- Supported adaptation of technologies to new uses, markets, and industry sectors.
- Funded technology development.
- Enabled projects with high levels of technical risk.
- Reduced technological risk.
- Diversified expertise and allowed hiring of specialists.
- Attracted and developed young researchers.
- Redirected company activities to new opportunities.
- Reduced costs.
- Helped address needs that require high tech at low volume and relatively low cost.
- Gave companies added credibility because SBIR research is seen as peer reviewed.
- Funded researchers to enter business full time.
- Transformed company culture to become more market driven.
- Drove researchers to focus on technology transition.
- Created new companies and kept companies in business (that would not exist without SBIR funding).
- Supported feasibility testing for high-risk/high-payoff projects.
- Supported projects with longer time horizons/long sales cycles.
- Provided the basis for spin-off companies.
- Funded proof of concept.
- Encouraged R&D companies to transition into manufacturing.
- Funded disruptive technologies.
- Provided significant mentoring especially for new businesses.
- Stimulated international collaboration.

SOURCE: Analysis of company responses to 2011 Survey. For each bullet multiple responses indicated its existence and importance for surveyed projects and firms.

D. Limiting Company dependence on SBIR

- Limits on Multiple Awards: The NSF program tightly limits the capacity of firms to garner multiple awards.[46]
    - o Firms are permitted to apply for only two awards per year.[47]
    - o The most prolific firms received fewer than 10 SBIR Phase II awards over a period of 10 years.[48]
    - o Interviews with Program Directors indicate a preference for spreading Phase II awards more widely.[49]
- NSF firms are not dependent on SBIR awards: Less than 20 percent of companies responding to the 2011 Survey report that SBIR accounts for more than half of current revenues;[50] however, a considerable number of surveyed firms reported in textual responses that SBIR is the most important source of funding prior to reaching the market.[51]
- SBIR Innovation is nonlinear: Most projects at most companies do not proceed directly from Phase I to Phase II to commercialization.[52]
    - o About 85 percent of Phase II survey respondents reported at least one additional SBIR Phase II award related to the surveyed project.[53]
    - o About one third reported at least two additional related Phase II awards.[54]
    - o As noted above, more than 60 percent of Phase II respondents report additional investment funding related to the project subsequent to the SBIR award.[55]

V. The Phase IIB Program

The NSF Phase IIB program supports the accelerated commercialization of SBIR-funded research through the provision of matching NSF funds provided companies can generate additional investment funding of their own. The evidence shows that Phase IIB projects tend to commercialize more than Phase II-only projects. In part, this reflects the rigorous selection process—NSF selects Phase IIB projects specifically for their commercial potential, so improved commercial

---

[46]Previous Academies studies discussed the question of multiple SBIR awards to companies. The 2009 report concluded that "mills" are not a significant problem.

[47]National Science Foundation, June 2014 SBIR program Solicitation, NSF-13-599, p. 1.

[48]See Chapter 3, Annex Table 3-6.

[49]See text in Chapter 3 at the end of the section, "New Companies in the SBIR Program."

[50]See Table 4-36.

[51]See Chapter 6 and case studies in Appendix E.

[52]See Chapter 4's section, "Prior Use of the SBIR Program."

[53]See Table 4-38.

[54]See Table 4-38.

[55]See Table 4-6.

outcomes should be expected. Moreover, evidence from surveyed recipients and from case studies strongly suggests that the program does have a catalyzing effect, attracting additional funding, and that it does make a positive difference to company activities and outcomes.[56]

A. The Academies survey of Phase IIB projects reports increased commercialization and larger revenues.

The Phase IIB program attracts third-party investors, thereby accelerating and enhancing the commercialization of SBIR-funded technologies:

- Phase IIB projects reach the market at a high rate—81 percent of surveyed Phase II projects reported that they had achieved some sales.[57]
- Products in use. Just under half of the respondents to the Phase IIB survey report that their products, processes, or services are in use today.[58]
- Projects with large revenues. Almost a quarter of Phase IIB projects surveyed reported project-based revenues of $3 million or more percent. Median reported sales were $650,000.[59]
- Important role of Phase IIB. Eighty percent of respondents to the Phase IIB survey indicated that Phase IIB has accelerated commercialization of their product or services.[60] This is an important point because the program by design focuses on firms with maximum commercial potential. Confirmation from firms about the role and impact of the program is therefore highly relevant.

B. The Academies survey of Phase IIB projects reports employment effects and opportunities for female PI's.

- Employment growth. Respondents to the Phase IIB survey reported employment growth from a median of 8 employees at the time of the award to 17 at the time of the survey.[61]
- More opportunities for female principal investigators. Phase IIB respondents reported more female principal investigators than Phase II projects—18 percent versus 7 percent.[62]

C. The Academies survey of Phase IIB projects reports that by requiring matching funds, the Phase IIB program provides incentives for firms to acquire additional investment.

---

[56]See Chapter 5 for a description of the Phase IIB program and its effects.

[57]See Table 5-8. See also Table A-5 for information on the Phase IIB survey responses. A total of 281 survey responses were received, reflecting an effective response rate of 48.3 percent.

[58]See Table 5-7.

[59]See Table 5-9.

[60]See Table 5-27.

[61]See Tables 5-11 and 5-12.

[62]See Table 5-6.

- Phase IIB projects leverage NSF funding.
  - Respondents to the Phase IIB survey reported a mean of $651,000 in additional matching funds.[63]
  - Sixty-two percent of Phase IIB respondents reported at least $500,000 in additional funding, and 36 percent reported receiving at least $1 million.[64]
  - Phase IIB respondents reported funding from a range of sources: 23 percent of their matching funds from sales, 21 percent from another U.S. company, and 16 percent from U.S. angel investment. Nine percent came from venture capital investors.[65]
  - The requirement for matching funds works to screen out projects that are less able to attract additional investment funding. About two thirds of the Phase II projects that did not apply for Phase IIB funding reported that they could not raise the required matching funds.[66]
- Attracting New Investors: Some Phase IIB projects successfully attracted new investors, in some cases attracting substantial additional funding.
  - Thirty percent of Phase IIB projects reported that that their Phase IIB matching funding came from a new financial partner or investor.[67]
  - While sales were the single largest source of matching funds, three quarters of responses indicated that funds were acquired from financial, technical, or commercial partners with whom they had a long-term relationship.[68]
  - Sixty-one percent of respondents indicated that the new funding would not have been acquired without Phase IIB.[69]
- Providing Direct Incentives for Firms: Phase IIB provides direct incentives for firms, and imposes reasonable administrative burdens.
  - More than half of Phase IIB respondents indicated that the Phase IIB program provided a substantial incentive for or drove the process for seeking external funding.[70]
  - Forty-one percent of projects reported the total effort required to apply for Phase IIB as "no additional effort needed except

---

[63]There were no statistically significant differences between survey responses from Phase IIB respondents and agency data from the NSF awards database. See Table 5-23.

[64]See Table 5-23.

[65]See Table 5-24.

[66]See Table 5-17.

[67]See Table 5-26.

[68]See Table 5-26.

[69]See Table 5-29.

[70]See Table 5-30.

paperwork" or "less than 2 weeks FTE for company staff."[71] About one third of respondents believed the application process was easier than for comparable Federal R&D applications, while nineteen percent thought it more difficult.[72]

D. Companies report positive views of Phase IIB impacts.

Companies had remarkably positive views of the impact of Phase IIB funding on both their projects and their companies as a whole.

- Worth Applying: Ninety-eight percent of Phase IIB respondents believed that the funding they received was worth the effort involved in applying—even though more than a fifth had spent at least two months of senior management time on the application.[73]
- Long-term Impact: Ninety-six percent of Phase IIB respondents indicated that Phase IIB had a positive long-term impact on the company. Nearly two-thirds stated that the program had a substantial positive impact, and nearly one in ten responded that the program had had a "transformative effect" on the company.[74]
- Broad Support: Comments from respondents indicate that Phase IIB supports a wide range of activities tied to a considerable variety of commercialization strategies and approaches.

E. Other Aspects of the Phase IIB program.

- Strong support for companies from NSF program officers. Survey respondents and interviewees both had generally strongly positive views on the role of NSF program officers in the application process (and indeed during the course of the award—see Chapter 2 and Appendix E).

F. Improving Phase IIB. The study draws attention to potential areas of improvement for the Phase IIB program. These include the following:

- Confusion about matching funds criteria. There was considerable confusion even among recipients about the criteria for identifying appropriate matching funds. Many respondents identified funding sources that were in fact not acceptable as matching funds; and many failed to identify equity funding and sales, both of which are permitted.[75]
- Matching funds criteria. Aside from the confusion about criteria noted above, almost half of Phase IIB awardees indicated that NSF matching fund criteria were too stringent in excluding in kind con-

[71] See Table 5-19.
[72] See Table 5-20.
[73] See Table 5-34.
[74] See Table 5-28.
[75] See Table 5-21.

tributions. However, stringent criteria are a strength of the program, in the view of the Committee.[76]

- Commercialization focus. While the commercial focus of the NSF SBIR program is widely appreciated by survey respondents and interviewees (as reflected in comments in Chapter 6 and Appendix E) a number of respondents to the Phase IIB survey suggested that NSF should review the balance between rapid commercialization and powerful innovation, as the current balance might end up excluding potentially important projects.

VI. Program Management

A. The NSF Management Model

- Key Role of Program Directors: The NSF model relies on a small cadre of highly qualified Program Directors who have strong technical and commercial backgrounds to provide guidance and support to companies that apply for SBIR grants and that seek to commercialize SBIR technologies. Evidence from the Phase II survey and from case studies suggests that in many cases the relationship between the Program Director and the company is strong and that the model works as anticipated.[77]
- There is evidence from several sources that the Program Directors face a heavy workload.[78]
    - Program Directors interviewed for this study indicated that they are responsible for about 30 Phase II awards at any one time. Given this workload, it is not clear how much time they are able to devote to any one company.
    - Phase II Recipient survey responses indicate that in some cases companies believe that the workload for Program Directors is too great for them to provide appropriate support.

B. Best Practices

- Effective Program Directors: Companies strongly appreciate the guidance offered by program officers in most cases, and in interviews suggest that they benefit considerably from the high quality of the NSF program directors.[79]
- Limits on submissions per company: These limits are an appropriate mechanism for ensuring that the number of submissions is limited while at the same time not excluding any specific project.

---

[76]See Table 5-37.

[77]See "Role and Effectiveness of Program Directors" and "Qualifications of PDs" in Chapter 2 and Box 2-6.

[78]See Tables 2-11, 2-12, 2-13, 2-14, 2-15, 2-16, 2-17, and 2-18, and related text in Chapter 2.

[79]See Box 2-6 in Chapter 2.

Companies decide which projects to submit, reducing the number of applications that program staff must review and ensuring that if a project is funded, it will be one that was at or near the top of the company's priority list.[80]

- Inclusive Topic Deployment: The broad topics offered by NSF—in conjunction with the increase in the number of solicitations by topic to two per year—indicate that the agency is working hard to ensure that promising projects are not excluded unnecessarily. NSF program officers indicate that they believe any technically and commercially worthwhile project could be funded under the published solicitation guidelines.[81] Some companies, however, disagree.
- The Phase IIB Program: This program has been an important initiative at NSF, and has since been adopted in different forms at other agencies.[82]
- Partnerships: NSF is to be commended for exploring a range of partnerships and mechanisms through which to support the commercialization of funded projects.
- i-Corps, an NSF program that offers entrepreneurship training (primarily for academics), seeks to ensure that some key steps in the development of a sustainable company are taken at an early stage in the product development cycle.[83]
- The Innovation Accelerator, a private organization partnering with NSF to facilitate the commercialization efforts of high-technology small businesses funded by grants from the NSF SBIR program, is also potentially valuable, although its reach may have to be expanded beyond the information technology sector.

C. Key Concerns

- Data Tracking, Management, and Use: There are broad challenges in tracking commercialization, at both the company and project levels. Companies move in and out of the program, and tracking is harder once they have left. More generally, commercialization may come years after an award, and may involve multiple awards plus considerable additional funding. All this makes it difficult to assert that any specific outcome "results from" an SBIR award. But there are also specific challenges with existing tracking tools.
- Recognition of NSF efforts to track outcomes. NSF's existing effort to track outcomes is commendable. It predates similar efforts at other agencies by a number of years, and by seeking information

[80] See, for example, Chapter 3, last paragraph before "Demographics of Phase I Winning Companies."

[81] See "Topic Selection" in Chapter 2, and Box 2-9.

[82] For a description of the Phase IIB Program at NSF, see Chapter 5.

[83] See I Corps program highlights in Chapter 2, Box 2-3.

at set times after the end of each award provides the agency with longitudinal data.

- Limits to existing approach: The current effort is limited, as the collection mechanism may be insufficiently standardized (ad hoc interviews) and also relies entirely on self-reported data with no cross checks involved. It is also focused almost entirely on only one of the congressional objectives—commercialization.
- Need for Additional Tools: Additional data collection tools that reflect the changing information environment have not been deployed—NSF does not use additional methods of data collection beyond the flow of awards and applications data and the annual telephone survey.[84] Provisions of a web-based survey tool for use by companies periodically to update their progress metrics may facilitate company reporting with reduced burden.
- Need for a Broader Focus: The focus of outcomes data collection is almost entirely on commercialization that, as a result, misses opportunities to provide detailed feedback about program operations as well as other congressionally-mandated outcomes.
- Need for an Information Management System: There is no single information management system through which agency staff can visualize the complete range of relevant program metrics and data.
- Uses of the data are limited. Interviews with program officials suggest that the data collected through the internal exercise are not systematically used to improve program management. Given that the agency has collected commercialization data for a number of years, it would seem worthwhile for the agency to explore more deeply potential causal variables related to commercial success and to use data in program management.
- Fixed Award Sizes: Like most other agencies NSF provides award funding that is largely fixed in size: prior to reauthorization, Phase II was $500,000 plus additional Phase IIB funding for some companies. Phase II funding is now a maximum of $1 million, and most awards will likely be made at that level. This fixed approach does not seem best designed to ensure that each project receives the appropriate amount of funding.
- High Rate of Failed Applications: Some see the rate of failed applications to be burdensome for both applicants and agency staff. (At best 1 in 6 Phase I applications and less than 50 percent of Phase II

---

[84]It appears that this may be in part a result of compliance with the Paperwork Reduction Act, which limits agency tracking capabilities.

applications are successful for any given solicitation.)[85] These success rates impose costs on both companies and the agency: 5 out of 6 Phase I applicants are not funded so their application effort is largely wasted; and agency reviewers must spend considerable time and effort on projects that are not funded. On the other hand, public R&D programs have traditionally encouraged companies to participate and have developed review strategies for dealing with the larger numbers and varied quality of proposals that invariably result.

RECOMMENDATIONS

While the NSF SBIR program generates substantially positive outcomes, the committee has identified a series of recommendations to improve its processes and outcomes.

I. Addressing Underserved Populations

NSF should immediately examine past and current efforts to address the clear congressional mandate to foster the participation of underserved populations in the SBIR program, examine and report on best practices and create benchmarks to relate the impact of such activities.

A. **Quotas are not necessary.** NSF should not develop quotas for the inclusion of selected populations into the SBIR program. Such an approach is not necessary to meet congressional intent, and is likely to reduce program effectiveness.

B. **NSF should develop new benchmarks and metrics.**[86]
   - Improve participation metrics: The SBIR/STTR program office should work with the NSF indicators group to develop much improved metrics for benchmarking the participation of underserved populations, developing and publishing clear benchmarks based on a defensible analysis of existing data.
   - Disaggregate Benchmarks: Measures of the participation of socially disadvantaged groups must be disaggregated by race or ethnicity, and attention focused on the clear congressional intent to support "minority" participation. We do not believe a focus on the current SBA definition of "socially and economically disadvantaged" is sufficient to meet this objective.

---

[85]But others see the NSF Phase I and Phase II success rates as not low relative to the acceptance rates of other competitive R&D funding programs. Similarly, some see failed applications as a waste of company and reviewer time, while others see knowledge benefits and capacity building in company efforts to develop proposals for the SBIR program even if they are not successful in obtaining an SBIR grant. Formulated plans may go on to generate benefits in other ways.

[86]See Finding II-A and II-C.

- Customize Benchmarks: Points of reference should be developed separately (though perhaps drawing on a shared methodology) for women and minorities. These benchmarks should be shared with other SBIR agencies. Benchmarks should address key questions that would include the following metrics, all of which should include both absolute levels and trends over time:
    - Shares of applications from companies owned by women and minorities
    - Shares of applications with woman and minority principal investigators
    - Shares of Phase I awards to companies owned by women and minorities
    - Shares of Phase I awards with woman and minority principal investigators
    - Shares of Phase II awards to companies owned by women and minorities
    - Shares of Phase II awards with woman and minority principal investigators
    - Shares of Phase IIB awards to woman and minority owned firms
    - Shares of Phase IIB awards with woman and minority principal investigators
- Track Related Program Operations: Metrics should also track related program operations including outreach efforts. (See below.)

C. **NSF should undertake an investigation to understand better why its efforts to date to expand the participation of under-served populations have been largely unsuccessful.**[87]

- Develop Outreach Strategy: Based on this investigation, NSF should develop a coherent and systematic outreach strategy that provides for cost-effective approaches to enhance recruitment of both woman- and minority-owned companies and female and minority PI's, developed in conjunction with other stakeholders and with experts in the field, including relevant programs at the state level. Outreach efforts should aim to expand SBIR awareness among potential applicants from underserved demographics.
- Integrate Outreach Effort: NSF should ensure that outreach to selected populations is an integral part of its overall outreach.
- Provide Management Resources: NSF should provide significant management resources given that improving participation is likely to be both difficult and a long term effort.

[87] See Finding II-B.

- Designate Staff: NSF should consider designating a senior staffer to work exclusively on improving women and minority participation in order to identify strategic initiatives, implement them, and provide consistent and timely reporting on their impacts.

D. NSF should review internal award and selection data and processes to identify and address disparities between Phase I and Phase II awards to selected populations. The goal is to ensure that there are no biases in the selection process that are adversely affecting the selection of women and minorities.[88]

- NSF should carefully review patterns of success rates for WOSB and MOSB to identify potential disparities.
- As part of the review, NSF should investigate the possibility of selection-bias in the review process and take actions to make sure that it is avoided—including ensuring that review panels are reasonably diverse.

II. Commercialization

The NSF SBIR program is tightly focused on commercialization that is based on a venture capital model. However, it is worth considering some possible adjustments to the current approach.

- **Broaden Perspective to include strategic partners:**[89] The limited funding for seed and startup projects from U.S. venture capital in general, and the low numbers of NSF firms that report venture capital funding, suggests that NSF should not focus too tightly on commercialization models that rest on venture capital funding. Many alternatives exist—for example, a deeper focus on finding strategic partners. The VC-focused commercialization model narrows the program in several ways—by limiting the timeframe viewed as appropriate for commercialization, and also by anticipating certain levels of commercial scale needed to attract VC-type funding. We recommend that NSF review its conceptual approach to commercialization with a view to ensuring that different paths to commercial success are fully included, such as angel funding and strategic investments by other companies. In addition, NSF should explore newer alternatives—for example, open source models or use of equity crowdfunding. This is to some degree implicit in the I-Corps model, and NSF should ensure that its grantees are aware of alternative funding paths.

[88] See Finding II-A, points 2 and 3.

[89] See Finding I-F.

III. Phase IIB

A. NSF should continue to operate the Phase IIB program.[90]

- The program has had a positive impact on the commercialization of SBIR funded research.
- Respondents strongly support retention of the program.

B. NSF could consider expanding the size of the Phase IIB program.[91]

- The strongly positive impact suggests that there may be additional projects that could benefit from Phase IIB funding.
- But NSF should be cautious about the impact of expansion on other aspects of the program. A majority of Phase IIB respondents would not support the addition of Phase IIB awards at the cost of fewer Phase I or Phase II awards.

C. NSF should not increase the dollars per Phase IIB maximum award if that reduces the number of Phase IIB awards.[92]

- The program appears to successfully meet its mission at current award levels. Two-thirds of respondents agreed that the amount of funding was about right.
- NSF should be cautious about reducing the number of Phase IIB awards in order to increase the size of each award. The analysis identified no evidence to suggest that this would be an improvement to the program or that recipients would support such an adjustment.
- Survey responses indicated that while some supported such an adjustment, more than twice as many did not favor a trade-off of larger awards for fewer awards.

D. NSF should clarify the criteria under which funding qualifies as an acceptable match for Phase IIB purposes.[93]

- Survey and interview data indicated confusion among recipients (and presumably potential applicants) about acceptable criteria, and there are inconsistencies between different statements on the NSF Phase IIB web page.

E. Matching funds requirements.[94]

- <u>Explore use of In-Kind Contributions</u>: NSF may wish to explore allowing the limited use of some specified in-kind contributions as part or all of the matching funds. Nearly half of survey respondents agreed. Perhaps NSF might opt to allow in-kind contributions in "exceptional circumstances," offering a limited but potentially important degree of flexibility.

---

[90] See Finding V-A.
[91] See Finding V-A.
[92] See Finding V-C.
[93] See Finding V-F
[94] See Finding V-F.

- Review Matching Requirement: Similarly, NSF may wish to review the requirement that matching funds be developed during the timeline of the Phase II. Some respondents indicated that this excludes funding developed during Phase I, for example. It also implies a simple linear progression toward market, which may also tend to exclude projects otherwise appropriate for funding.

IV. Improving Monitoring, Evaluation, and Assessment

The development of more careful monitoring and more sophisticated analysis of key variables are necessary to improve program outcomes. The NSF deserves considerable credit for its early efforts to track outcomes, but more remains to be done in this area. Newly available administrative funding can improve this area of program operations.

A. Improving current data collection approaches and methodologies.[95]

Data collected through the current process are a good start but are far from sufficient to underpin a data driven program.

- Improve Data Collection and Organization:
  - NSF should more systematically collect a range of quantitative data and standardize key questions to improve program evaluation, management, and outcomes.
  - This data collection effort should address the entire range of congressionally mandated outcomes, not just commercialization.
  - NSF should develop a dataset that can provide a basis for longitudinal analysis.
- Track Commercialization Outcomes:
  - NSF should track commercialization outcomes in ways similar to the now widely-accepted methodology developed for the Academies studies. This approach focuses on multiple metrics in order to provide a deeper and more nuanced basis for analysis. (See methodology discussion in Chapter 1 and Appendix A).
  - Although NSF already tracks outcomes for several years after the end of an award, this period should be extended to more fully assess long term impacts as these may not become fully discernible during a more limited period of review.
- Manage Information:
  - NSF should explore the development of an integrated information management system to improve the management of its SBIR program.
  - NSF is now developing a database of Phase IIB projects for which it is to be commended. However, program management

[95]See Findings II-A and VI-C.

would be on a firmer basis if managers had readily at hand information about company participation in all of the program's activities, including Phase IIB, various supplements, I-Corps, and the Innovation Accelerator.

- Collect Demographic Data:
  - NSF should take immediate steps to improve its collection of demographic data, using the Fastlane electronic submission program.
  - NSF should extend its collection of the demographics of company ownership to show which of the SBA's socially and economically disadvantaged categories an applicant belongs to. In addition, applicants should be asked the same demographic questions about the principal investigator.
- Collect qualitative outcomes data:
  - NSF should also develop and adopt a more systematic and critical approach to the use of detailed case studies and success stories. Success stories can provide inspiration lessons learned, and important information not available elsewhere about program impacts. Case studies can inform the roles played by SBIR awards, the challenges faced by small businesses, insights into needed improvements in process, lessons learned, and other important information not available elsewhere about program impacts.

B. Use available commercial tools to gather more current data on a long-term basis. While we recommend utilizing modern information management and data visualization tools, we stress that effective products are widely available on the commercial market.[96]

- Use New Technologies: NSF should explore ways to use new technology such as social media to collect more current data. SBIR companies—like "customers" in other markets—are an important source of information about program strengths and weaknesses. This knowledge is currently not systematically included in internal program evaluation by NSF's SBIR program.
- Develop Feedback Tools: NSF should develop pathways to provide ongoing feedback from companies about program activities and operations. These could include various electronic communication tools.
- Improve Networking: Similarly, NSF should consider developing mechanisms—like electronic tools—through which recipients can share information about their SBIR projects, helping them both to find technical or marketing or investment partners and to navigate

---

[96]See Finding VI-C.

the often-complex regulatory and technical environment of NSF programs.

C. **Improved Data Utilization.** NSF has collected data on outcomes for some years, but it is not clear whether these data are systematically employed to guide program management.[97]

- Develop a Plan for Data Analysis: NSF should seek to develop a more sophisticated approach to analyzing and applying the data that are already collected. For example, there appear to be no plans in place within NSF to use the data collected over a number of years to evaluate factors that tend to encourage successful transitions between Phases, into Phase IIB, and then into full scale commercialization. Such an approach could identify key issues for program management.
- Undertake Regular Analysis of Data: By collecting more and better data on outcomes and participation, NSF will be positioned to undertake regular analysis—either internally or with third-party help—on key program management issues, such as
  - What is the long term impact of partnership programs and other commercialization supports?
  - Is Phase IIB simply picking successful companies or is it at least, in part, causing companies to be successful?
  - How well do NSF selection processes predict eventual successful projects?
- Recognize Impacts of Data Collection and Analysis: In some cases, simply measuring something more closely will likely provoke action. Closely tracking the participation of women and minorities could help assure a fair process and surface problem issues early, when they can be most easily corrected. For example, systematically tracking woman and minority participation would have surfaced the issue of differential Phase I and Phase II success rates identified earlier.

D. **Annual Report to the National Science Board and Congress:** The SBIR/STTR program should provide a comprehensive annual report to the National Science Board and Congress and the public on its operations. While the precise details should be left to the agency we would recommend that it consider including six areas of program operations:[98]

- Inputs, including aggregate current and longitudinal data on numbers of applications and awards, broken out by relevant subgroups (such as demographics, state)

[97] See Finding VI-C.

[98] See Findings II-B and VI-C.

- Program initiatives, which would provide NSF with an opportunity to describe the numerous and impressive initiatives undertaken by the program
- Aggregated outputs drawn from an improved tracking, outcomes collections, and analysis process
- Qualitative review, based on improved use of case studies, as well as success stories and social media
- Impact assessment, focused on the extent to which NSF meets congressional objectives for the program. Impact assessment should be contracted out to ensure objectivity.
- Summary conclusions, including prospective views on program activities and improvements for the coming year.

While caution should be employed when imposing new reporting burdens on the NSF SBIR program, implementation of an improved information management system would provide a cost- and time-effective basis on which to provide better reporting on the program. An annual report would provide much improved transparency and would provide a coherent point of discussion for other stakeholders. It would effectively replace the existing report to SBA which is of limited utility for NSF or other stakeholders, and could be organized to meet some of the new reporting requirements imposed under reauthorization.

E. **Additional and better data collection tools and analysis consume staff time and resources. NSF should ensure that its staff is incentivized, and that sufficient funding is available for this purpose.**[99]

V. Improving Program Management

Recommendations in this section are designed to improve program operations in ways that should enhance the program's ability to address some or all of its objectives.

A. **NSF should seek to reduce the burden on program directors.**[100]
   - Review Program Director Workloads: NSF should consider whether the current workload is appropriate. The NSF model calls for a mentoring relationship between Phase II winners and Program directors, and for a closer than normal relationship between Phase I winners and Program directors. This is hard to sustain given current levels of activity with each program director responsible for about 30 Phase II awards and perhaps twice that many Phase I awards.
   - Consider Ways to Make the NSF Model Work Better: While Providing Individual Attention to Projects and Not Overburdening PDs:

---

[99] See Finding VI-C.
[100] See Finding VI-A.

For example, consider adding part-time support staff or an additional PD to make the workloads more manageable, such as a PD to manage proposals in topics outside the announced solicitation.

- Consider Virtual Meetings: NSF should also consider whether the current approach to identifying candidates to be recommended for awards could be streamlined as suggested by the NSF SBIR/STTR Advisory Committee through the use of more virtual meetings—both for review panels and for review of Phase IIB candidates where currently all have to go through an in-person interview process.

B. NSF should consider ways to improve the average quality of applications.[101]

- Encourage firms to submit higher quality proposals: NSF's use of a pre-selection White Paper approach, which provides guidance and feedback to prospective applicants, appears to be partially effective in improving quality while reducing the number of applications, as is the 2 per company limit. NSF should look to find additional ways to attract fewer and higher quality applications.
- Eliminate use of applicant success rates to validate program quality. NSF should at a minimum not use low success rates to validate program quality, as this in part simply reflects the receipt of weak proposals. Screening out more weak proposals before application would raise success rates for the remaining applications, while maintaining or even improving the quality of the awardees.

C. NSF should consider the adoption of further mechanisms to reduce funding gaps between Phase I and Phase II.[102]

Some of the mechanisms used at other agencies include a Fast Track process for merging Phase I/Phase II applications; use of agency options as bridge funding; early identification of projects that are likely to be funded. NSF could explore these and other options.

D. NSF should investigate how open in its solicitations are in practice.

It should track the number of proposals by topic and determine the percentage that fit the announced solicitation topic(s). It should explore whether NSF approach to topics is sufficient to keep up with evolving opportunities. The relationship between topics and the number of applications should also be subject to ongoing review.

[101] See Finding VI-C.

[102] See Finding VI-C.

Appendixes

Appendix A

Overview of Methodological Approaches, Data Sources, and Survey Tools

This series of reports on the Small Business Innovation Research (SBIR) programs at the Department of Defense (DoD), National Institutes of Health (NIH), National Aeronautics and Space Administration (NASA), Department of Energy (DoE), and National Science Foundation (NSF) represents a second-round assessment of the program undertaken by the National Academies of Sciences, Engineering, and Medicine.[1] The first-round assessment, conducted under a separate ad hoc committee, resulted in a series of reports released from 2004 to 2009, including a framework methodology for that study and on which the current methodology builds.[2]

The current study is focused on the twin objectives of assessing outcomes from the programs and of providing recommendations for improvement.[3] The

[1]Effective July 1, 2015, the institution is called the National Academies of Sciences, Engineering, and Medicine. References in this report to the National Research Council, or NRC, are used in an historic context identifying programs prior to July 1.

[2]National Research Council, *An Assessment of the Small Business Innovation Research Program: Project Methodology* (Washington, DC: The National Academies Press, 2004).

[3]The methodology developed as part of the Academies' first-round assessment of the SBIR program also identifies two areas that are excluded from the purview of the study: "The objective of the study is *not* to consider if SBIR should exist or not—Congress has already decided affirmatively on this question. Rather, the NRC committee conducting this study is charged with providing assessment-based findings of the benefits and costs of SBIR . . . to improve public understanding of the program, as well as recommendations to improve the program's effectiveness. It is also important to note that, in accordance with the Memorandum of Understanding and the Congressional mandate, the study will *not* seek to compare the value of one area with other areas; this task is the prerogative of the Congress and the Administration acting through the agencies. Instead, the study is concerned with the effective review of each area." National Research Council, *Assessment of the SBIR Program: Project*

SBIR programs are unique efforts designed by Congress to provide funding via government agencies in pursuit of four objectives[4]:

1. Stimulate technological innovation;
2. Use small business to meet Federal research and development needs;
3. Foster and encourage participation by minority and disadvantaged persons (including woman owners of small business) in technological innovation; and
4. Increase private-sector commercialization innovation derived from Federal research and development.

The SBIR programs, on the basis of highly competitive solicitations, provide modest initial funding for selected Phase I projects (up to $150,000), and for feasibility testing and further Phase II funding (up to $1 million) for qualifying Phase I projects.

From a methodology perspective, assessing this program presents formidable challenges. Among the more difficult are the following:

- **Lack of data.** NSF does track outcomes data, but these data were not made available to us. There are no published or publicly available outcomes data.
- **Intervening variables.** Analysis of small innovative businesses suggests that they are often very path dependent and, hence, can be deflected from a given development path by a wide range of positive and negative variables. A single breakthrough contract—or technical delay—can make or break a company.
- **Lags.** Not only do outcomes lag awards by a number of years, but also the lag itself is highly variable. Some companies commercialize within 6 months of award conclusion; others take decades. And often, revenues from commercialization peak many years after products have reached markets.

ESTABLISHING A METHODOLOGY

The methodology utilized in this second-round study of the SBIR program builds on the methodology established by the committee that completed the first-round study.

Methodology. In implementing this approach in the context of the current round of SBIR assessments, the Committee has opted to focus more deeply on operational questions.

[4]Public Law 97–219, § 2, July 22, 1982, 96 Stat. 217.

Publication of the 2004 Methodology

The committee that undertook the first-round study and the agencies under study acknowledged the difficulties involved in assessing SBIR programs. Accordingly, that study began with development of the formal volume on methodology, which was published in 2004 after completing the standard Academies peer review process.[5]

The established methodology stressed the importance of adopting a varied range of tools based on prior work in this area, which meshes with the methodology originally defined by the first study committee. The first committee concluded that appropriate methodological approaches

> "build from the precedents established in several key studies already undertaken to evaluate various aspects of the SBIR. These studies have been successful because they identified the need for utilizing not just a single methodological approach, but rather a broad spectrum of approaches, in order to evaluate the SBIR from a number of different perspectives and criteria.
>
> This diversity and flexibility in methodological approach are particularly appropriate given the heterogeneity of goals and procedures across the five agencies involved in the evaluation. Consequently, this document suggests a broad framework for methodological approaches that can serve to guide the research team when evaluating each particular agency in terms of the four criteria stated above. Table A-1 illustrates some key assessment parameters and related measures to be considered in this study."[6]

The tools identified in the illustration above include many of those used by the committee that conducted the first-round study of the SBIR program. Other tools emerged since the initial methodology review.

Tools Utilized in the Current SBIR Study

Quantitative and qualitative tools being utilized in the current study of the SBIR program include the following:

- **Surveys.** The Committee commissioned two extensive surveys of NSF SBIR award recipients as part of the analysis. These are described in depth below.[7]
- **Case studies.** The Committee commissioned in-depth case studies of 10 SBIR recipients at NSF. These companies were geographically and demographically diverse and were at different stages of the company lifecycle.

---

[5]National Research Council, *Assessment of the SBIR Program: Project Methodology*, p. 2.

[6]Ibid, p. 2.

[7]The surveys conducted as part of the current, second-round assessment of the SBIR program are referred to below as the "2011 Survey" and the "2010 Phase IIB Survey."

TABLE A-1 Overview of Approach to SBIR Program Assessment

SBIR Assessment Parameters →	Quality of Research	Commercialization of SBIR Funded Research/Economic and Non-economic Benefits	Small Business Innovation/ Growth	Use of Small Businesses to Advance Agency Missions
Questions	How does the quality of SBIR funded research compare with that of other government funded R&D?	What is the overall economic impact of SBIR funded research? What fraction of that impact is attributable to SBIR funding?	How to broaden participation and replenish contractors? What is the link between SBIR and state/regional programs?	How to increase agency uptake while continuing to support high risk research
Measures	Peer review scores, Publication counts, Citation analysis	Sales; follow up funding; progress; IPO	Patent counts and other IP/ employment growth, number of new technology firms	Agency procurement of products resulting from SBIR work
Tools	Case studies, agency program studies, study of repeat winners, bibliometric analysis	Phase II surveys, program manager surveys, case studies, study of repeat winners	Phase I and Phase II surveys, case studies, study of repeat winners, bibliometric analysis	Program manager surveys, case studies, agency program studies, study of repeat winners
Key Research Challenges	Difficulty of measuring quality and of identifying proper reference group	Skew of returns; significant interagency and inter-industry differences	Measures of actual success and failure at the project and firm level; relationship of federal and state programs in this context	Major interagency differences in use of SBIR to meet agency missions

NOTE: Supplementary tools may be developed and used as needed. In addition, since publication of the methodology report, this Committee has determined that data on outcomes from Phase I awards are of limited relevance.
SOURCE: National Research Council, *An Assessment of the Small Business Innovation Research Program: Project Methodology* (Washington, DC: The National Academies Press, 2004), Table 1, p. 3.

- **Workshops.** The Committee convened several workshops to allow stakeholders, agency staff, and academic experts to provide insights into the program's operations, as well as to identify questions that should be addressed.
- **Analysis of agency data.** NSF provided the Committee with a range of datasets covering various aspects of agency SBIR activities. The Committee analyzed and included these data as appropriate.
- **Open-ended responses from SBIR recipients.** For the first time, the Committee solicited textual survey responses in the context of the 2011 Survey. More than 400 recipients provided narrative comments.
- **Agency interviews.** The Committee discussed program operations with NSF staff. Most were helpful in providing information both about the program and the challenges that they faced.
- **Literature review.** Since the start of our research in this area, a number of papers have been published addressing various aspects of the SBIR program. In addition, other organizations— such as the Government Accountability Office (GAO)—have reviewed particular parts of the SBIR program. The Committee where useful has included references to these works in the course of this analysis.

Taken together with our deliberations and the expertise brought to bear by our individual members, these tools provide the primary inputs into the analysis.

We would stress that, for the first-round study and for the current study, multiple research methodologies feed into every finding and recommendation. No findings or recommendations rested solely on data and analysis from Academies surveys; conversely, survey data was used to support analysis throughout the report.

COMMERCIALIZATION METRICS AND DATA COLLECTION

Recent congressional interest in the SBIR program has to a considerable extent focused on bringing innovative technologies to market. This enhanced attention to the economic return from public investments made in small business innovation is understandable.

However, in contrast to the Department of Defense, which may procure selected SBIR technologies, commercialization at NSF takes place solely in private-sector markets; to a considerable extent, NSF has defined successful commercialization in the past as projects that reach the market—where at least $1 of sales or revenues has been generated.

In its 2008 report on the NSF SBIR program,[8] the Academies committee charged with the first-round assessment held that a binary metric of commer-

[8]National Research Council, *An Assessment of the SBIR Program at the National Science Foundation*, (Washington, DC: The National Academies Press, 2008).

cialization was insufficient. It noted that the scale of commercialization is also important and that there are also other important milestones both before and after the first dollar in sales that should be included in an appropriate approach to measuring commercialization.

Challenges in Tracking Commercialization

Despite substantial efforts at NSF, described below, significant challenges remain in tracking commercialization outcomes for the NSF SBIR program. These include the following:

- **Data limitations.** NSF, like most other agencies, does not maintain an electronic reporting system for post-award data, nor are companies penalized for failing to report outcomes. In addition, companies face few incentives to report their successes and failures in commercialization.
- **Linear linkages.** Tracking efforts usually seek to link a specific project to a specific outcome. Separating the contributions of one project is difficult for many companies, given that multiple projects typically contribute to both anticipated and unanticipated outcomes.
- **Lags in commercialization.** Data from the extensive DoD commercialization database suggests that most projects take at least 2 years to reach the market *after the end of the Phase II award.* They do not generate peak revenue for several years after this. Therefore, efforts to measure program productivity must take these significant lags into account.
- **Attribution problems.** Commercialization is often the result of several awards, not just one, as well as other factors, such that attributing company-level success to specific awards is challenging at best.

Why New Data Sources Are Needed

Congress often seeks evidence about the effectiveness of programs or indeed about whether they work at all. This interest has in the past helped to drive the development of tools such as the Company Commercialization Report (CCR) at DoD, which captures the quantitative commercialization results of companies' prior Phase II projects. However, in the long-term the importance of tracking may rest more in its use to support program management. By carefully analyzing outcomes and CCR's associated program variables, program managers will be able to manage their SBIR portfolios more successfully.

In this regard, the NSF SBIR program can benefit from access to the Academies survey data. The survey work provides quantitative data necessary to provide an evidence-driven assessment and, at the same time, allows management to focus on specific questions of interest. For example, it focused in particular on the impact of Phase IIB awards—an important NSF initiative, for which the survey's outcomes provided the first hard evidence of effectiveness. (The Phase IIB program is discussed in detail in Chapter 5.)

THE ACADEMIES SURVEY ANALYSIS

Traditional modes of assessing the NSF SBIR program include case studies, interviews, and other qualitative methods of assessment. These remain important components of the Academies' overall methodology, and a chapter in the current report is devoted to lessons drawn from case studies. But qualitative assessment alone is insufficient. This study thus also draws on the results of two surveys conducted on our behalf by Grunwald Associates LLC—the 2010 Phase IIB Survey and the 2011 Survey.

The 2011 Survey

The 2011 Survey offers some significant advantages over other data sources. Specifically, it

- provides a rich source of textual information in response to open-ended questions;
- probes more deeply into company demographics and agency processes;
- for the first time addresses principal investigators (PIs), not just company business officials;
- allows comparisons with previous data-collection exercises;
- generates the first comparative data on Phase IIB[9]; and
- addresses other congressional objectives for the program beyond commercialization.

For these and other reasons, we determined that a survey would be the most appropriate mechanism for developing quantitative approaches to the analysis of the SBIR programs. At the same time, however, we are fully cognizant of the limitations of survey research in this case. Box A-1 describes a number of areas where caution is required when reviewing results.

To take account of these limits, while retaining the utility and indeed explanatory power of survey-based methodology, the current report contextualizes the data by comparing results to those from the Academies survey conducted as part of the first-round assessment of the SBIR program (referred to below as the "2005 Survey"). This report also adds transparency by publishing the number of responses for each question and indeed each subgroup. As noted later in the discussion, the use of a control group was found infeasible for comparing

[9]This comparison is made possible through the incorporation of data from the 2010 Phase IIB Survey, discussed later in this appendix.

BOX A-1
Multiple Sources of Bias in Survey Response[a]

Large innovation surveys involve multiple sources of potential bias that can skew the results in different directions. Some potential survey biases are noted below.

- Successful and more recently funded companies are more likely to respond. Research by Link and Scott demonstrates that the probability of obtaining research project information by survey decreases for less recently funded projects and increases the greater the award amount.[b] Nearly 75 percent of Phase II respondents to the 2011 Survey received awards after 2003, largely because winners from more distant years are difficult to reach: small businesses regularly cease operations, are acquired, merge, or lose staff with knowledge of SBIR awards. This may skew commercialization results downward, because more recent awards will be less likely to have completed the commercialization phase.
- Non-respondent bias. Very limited information is available about SBIR award recipients: company name, location, and contact information for the PI and the company point of contact, agency name, and date of award (data on woman and minority ownership are not considered reliable). No detailed data are available on applicants who did not win awards. It is therefore not feasible to undertake detailed analysis of non-respondents, but the possibility exists that they would present a different profile than would respondents.
- Success is self-reported. Self-reporting can be a source of bias, although the dimensions and direction of that bias are not necessarily clear. In any case, policy analysis has a long history of relying on self-reported performance measures to represent market-based performance measures. Participants in such retrospective analyses are believed to be able to consider a broader set of allocation options, thus making the evaluation more realistic than data based on third-party observation.[c] In short, company founders and/or PIs are in many cases simply the best source of information available.
- Survey sampled projects from PIs with multiple awards. Projects from PIs with large numbers of awards were under-represented in the sample, because PIs could not be expected to complete a questionnaire for each of numerous awards over a 10-year time frame.
- Failed companies are difficult to contact. Survey experts point to an "asymmetry" in the survey's ability to include failed companies for follow-up surveys in cases where the companies no longer exist.[d] It is worth noting that one cannot necessarily infer that the SBIR project failed; what is known is only that the company no longer exists.
- Not all successful projects are captured. For similar reasons, the survey could not include ongoing results from successful projects in companies that merged or were acquired before and/or after commercialization of the project's technology.
- Some companies are unwilling to fully acknowledge SBIR contribution to project success. Some companies may be unwilling to acknowledge that they received important benefits from participating in public programs for a variety

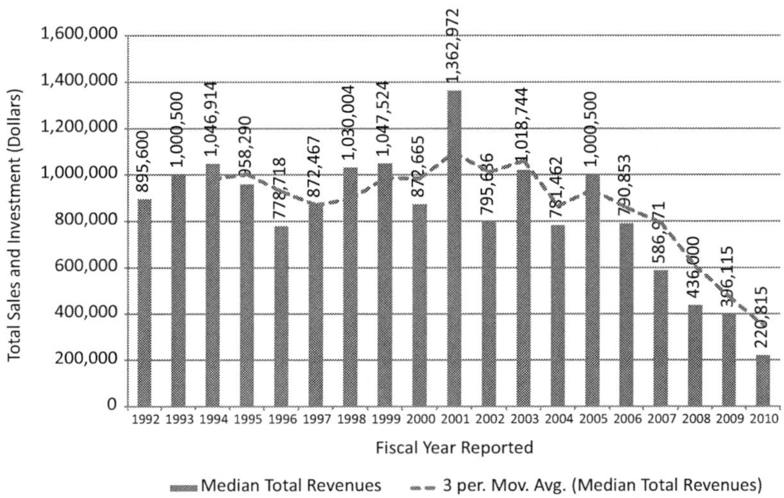

FIGURE Box A-1 The impact of commercialization lag.
SOURCE: DoD Company Commercialization Database.

of reasons. For example, some may understandably attribute success exclusively to their own efforts.

- Commercialization lag. Although the 2005 Survey broke new ground in data collection, the amount of sales made—and indeed the number of projects that generate sales—are inevitably undercounted in a snapshot survey taken at a single point in time. On the basis of successive data sets collected from National Institutes of Health (NIH) SBIR award recipients, it is estimated that total sales from all responding projects will be considerably greater than can be captured in a single survey.[e] This underscores the importance of follow-on research based on the now-established survey methodology. Figure Box A-1 illustrates this impact in practice: projects from 2006 onward have not yet completed commercialization as of August 2013.

---

[a] The limitations described here are drawn from the methodology outlined for the previous Academies survey in National Research Council, *An Assessment of the SBIR Program at the Department of Defense*, (Washington, DC: The National Academies Press, 2009).

[b] Albert N. Link and John T. Scott, *Evaluating Public Research Institutions: The U.S. Advanced Technology Program's Intramural Research Initiative* (London: Routledge, 2005).

[c] While economic theory is formulated on what is called "revealed preferences," meaning that individuals and companies reveal how they value scarce resources by how they allocate those resources within a market framework, quite often expressed preferences are a better source of information, especially from an evaluation perspective. Strict adherence to a revealed preference paradigm could lead to misguided policy conclusions because the

continued

BOX A-1 Continued

paradigm assumes that all policy choices are known and understood at the time that an individual or company reveals its preferences and that all relevant markets for such preferences are operational. See Gregory G. Dess and Donald W. Beard, "Dimensions of Organizational Task Environments," *Administrative Science Quarterly* 29(1984):52-73; (2) Albert N. Link and John T. Scott, *Public Accountability: Evaluating Technology-Based Institutions* (Norwell, MA: Kluwer Academic Publishers, 1998).

[d] Albert N. Link and John T. Scott, *Evaluating Public Research Institutions: The US Advanced Technology Program's Intramural Research Initiative* (London: Routledge, 2005.

[e] Data from the Academies assessment of the SBIR program at NIH indicate that a subsequent survey taken 2 years later would reveal substantial increases in both the percentage of companies reaching the market and the amount of sales per project. See National Research Council, *An Assessment of the SBIR Program at the National Institutes of Health*, (Washington, DC: The National Academies Press, 2009).

Phase II and Phase I recipients, but feasible for comparing Phase IIB and Phase II recipients.

We contracted with Grunwald Associates LLC to administer a survey to award recipients. The 2011 Survey is built closely on the 2005 Survey, but is also adapted to draw on lessons learned and includes some important changes discussed in detail below. A subgroup of this Committee with particular expertise in survey methodology also reviewed the survey and drew in current best practices. The 2010 Phase II/IIB survey covered only NSF, while the 2011 survey covered NSF, DoD, and NASA simultaneously.[10]

The primary objectives of the 2011 Survey (in combination with the 2010 Phase IIB Survey) were as follows:

- Provide an update of the program "snapshot" taken in 2005, maximizing the opportunity to identify trends within the program.
- Probe more deeply into program processes, with the help of expanded feedback from participants and better understanding of program demographics.
- Reduce costs and shrink the time required by combining three 2005 questionnaires—for the company, Phase I, and Phase II awards, respectively—into a single 2011 Survey questionnaire.

[10]Delays at NIH and DoE in contracting with the Academies combined with the need to complete work contracted with DoD NSF and NASA led the Committee to proceed with the survey at three agencies only.

The survey was therefore designed to collect the maximum amount of data, consistent with our commitment to minimizing the burden on individual respondents.

In light of these competing considerations, we decided that it would be more useful and effective to administer the survey to PIs—the lead researcher on each project—rather than to the registered company point of contact (POC), who in many cases would be an administrator rather than a researcher. This decision was reinforced by difficulties in accessing current POC information. Key areas of overlap between the 2005 and 2011 surveys are captured in Table A-2.

Starting Date and Coverage

The 2011 Survey included awards made from fiscal year FY1998 to FY2007 inclusive. This end date allowed completion of Phase II-awarded projects (which

TABLE A-2 Similarities and Differences: 2005 and 2011 Surveys

Item	2005 Survey	2011 Survey
Respondent selection		
Focus on Phase II winners	✓	✓
All qualifying awards		✓
PIs		✓
POCs	✓	
Max number of questionnaires per respondent	< 20	2
Distribution		
Mail	✓	No
Email	✓	✓
Telephone follow-up	✓	✓
Questionnaire		
Company demographics	Identical	Identical
Commercialization outcomes	Identical	Identical
IP outcomes	Identical	Identical
Woman and minority participation	✓	✓
Additional detail on minorities		✓
Additional detail on PIs		✓
New section on agency staff		✓
New section on company recommendations for SBIR		✓
New section capturing open-ended responses		✓

nominally fund 2 years of research) and provided a further 2 years for commercialization. This time frame was consistent with the previous Academies survey, administered in 2005, which surveyed awards from FY1992 to FY2001. It was also consistent with a previous GAO study, which in 1991 surveyed awards made through 1987.

The aim of setting the overall time frame at 10 years was to reduce the impact of difficulties generating information about older awards, because some companies and PIs may no longer be in place and memories fade over time. Reaching back to awards made in 1998 while ensuring comparability generated few results from older awards.

Determining the Survey Population

Following the precedent set by both the original GAO study and the first round of Academies analysis, this Committee differentiates between the total population of SBIR recipients, the preliminary survey target population, and the effective population for this study, which is the population of respondents that were reachable.

Initial Filters for Potential Recipients

Determining the effective study population required the following steps:

- acquisition of data from the sponsoring agencies—DoD, NSF, and NASA—covering record-level lists of award recipients;
- elimination of records that did not fit the protocol agreed upon by the Committee—namely, a maximum of two questionnaires per PI (in cases where PIs received more than two awards), awards were selected first by agency (NASA, NSF, DoD, in that order), then by year (oldest first), and finally by random number; and
- elimination of records for which there were significant missing data—in particular, where emails and/or contact telephone numbers were absent.

This process of excluding awards either because they did not fit the selection profile approved by the Committee or because the agencies did not provide sufficient or current contact information reduced the total award list provided by the agencies to a preliminary survey population of approximately 15,000 awards.

Secondary Filters to Identify Recipients with Active Contact Information

This nominal population still included many potential respondents whose contact information was complete but who were no longer associated with the contact information provided and hence effectively unreachable. This is unsurprising given that there is considerable turnover in both the existence of and the personnel working at small businesses, and the survey reaches back 13 years to awards made in FY1998. Recipients may have switched companies, the company may have ceased to exist or been acquired, or telephone and email contacts may have changed, for example. Consequently, we utilized two further filters to help identify the effective survey population.

- First, contacts for which the email address bounced twice were eliminated. Because the survey was delivered via email, the absence of a working email address disqualified the recipient. This eliminated approximately 30 percent of the preliminary population.
- Second, email addresses that did not officially "bounce" (i.e., return to sender) may still in fact not be active. Some email systems are configured to delete unrecognized email without sending a reply; in other cases, email addresses are inactive but not deleted. So a non-bouncing email address did not equal a contactable PI.

In order to identify non-reachable PIs, the Committee undertook an extensive telephone survey. For NSF, telephone calls were made to every targeted award recipient in the preliminary survey population that did not respond to the initial email invitation to participate.

Deployment

The 2011 Survey opened on October 4, 2011, and was deployed by email, with voice follow-up support. Up to four emails were sent to the effective population (emails discontinued once responses were received). In addition, two voice mails were delivered to non-respondents between the second and third and between the third and fourth rounds of email. In total, up to six efforts were made to reach each questionnaire recipient. After the members of the data subgroup of our committee concluded that sufficient data for the purposes had been collected, the survey closed on December 19, 2011. It was open for a total of 11 weeks.

Response Rates

Standard procedures were followed to conduct the survey. These data collection procedures were designed to increase response to the extent possible within the constraints of a voluntary survey and the survey budget. The population surveyed is a difficult one to contact and obtain responses from as evidence from the literature shows. Under these circumstances, the inability to contact and obtain responses always raises questions about potential bias of the estimates that cannot be quantified without substantial extra efforts that would require resources beyond those available for this work.

The lack of detailed applications data from the agency makes it impossible to estimate the possible impact of non-response bias. We, therefore, have no evidence either that non-response bias exists or that it does not.

Table A-3 shows the response rates at NSF by phase, based on both the preliminary study population prior to adjustment and the effective study population after all adjustments. The results are only for the 2011 Survey, and do not include data from the 2010 Phase IIB Survey.

All subsequent references to the 2011 Survey in this report address only responses for awards made by NSF.

Effort at Comparison Group Analysis

Several readers of the reports in the first round analysis of SBIR suggested the inclusion of comparison groups in the analysis. The Committee concurred that this should be attempted. There is no simple and easy way to acquire a comparison group for Phase II SBIR awardees. These are technology-based companies at an early stage of company development, which have the demonstrated capacity to undertake challenging technical research *and* to provide evidence that they are potentially successful commercializers. Given that the operations of the SBIR program are defined in legislation and limited by the Policy Guidance provided by SBA, randomly assigned control groups were not a possible alternative. Efforts to identify a pool of SBIR-like companies were made by contacting the most likely sources—Dunn and Bradstreet and Hoovers—but these efforts were not successful, as insufficiently detailed and structured information about companies was available.

In response, this Committee sought to develop a comparison group from among Phase I awardees that had not received a Phase II award from the three surveyed agencies (DoD, NSF, and NASA) during the award period covered by

TABLE A-3 2011 Survey Response Rates at NSF

Preliminary population	996
Missing contact information	–212
Contact moved/uncontactable	–376
Effective population	408
Responses	186
Surveys as Percentage of Awards Contacted	45.6
Surveys as Percentage of Sample	18.7

SOURCE: 2011 Survey, without inclusion of data from Phase IIB.

the survey (1999-2008). After considerable review, however, we concluded that the Phase I-only group was not appropriate for use as a statistical comparison group.

Including Results from the 2010 Phase IIB Survey

The 2010 Phase IIB Survey of NSF Phase II awards was carried out with the objective of comparing outcomes between award recipients receiving standard Phase II awards and those receiving Phase IIB enhancements in addition. This survey generated interesting responses and highlighted some important differences between the groups, and these are addressed in Chapter 5. A detailed discussion of the methodology for the 2010 Phase II Survey appears later in this appendix.

In addition, for purposes of analyzing the NSF program as a whole, Phase II responses from the 2010 and 2011 surveys were combined in cases where the questions asked were identical. This decision was reinforced by the fact that all of the respondents to the 2010 survey were excluded from the population of potential recipients for the 2011 Survey, on the grounds that the responses would to a considerable degree involve answering identical questions.

To conclude, we aggregated into the combined dataset the 2011 Survey analysis as well as all responses from the Phase IIB Survey where the wording of questions and responses was identical to that for the 2011 survey. In these cases, the inclusion of Phase IIB Survey responses is reflected in the number of responses reported for each question.

NSF Responses and Respondents

Table A-4 shows NSF SBIR responses by year of award. The survey primarily reached companies that were still in business: overall, 94 percent of respondents indicated that the companies were still in business.[11]

2010 Phase IIB Survey

The initial challenge in examining the NSF Phase IIB program was to develop data on which to base the assessment and hence eventual recommendations. Like most research and development (R&D) programs, there are only limited data available on both the program itself and its outcomes, and many of these—including the survey data developed by the Academies during the first-round assessment of the SBIR program—were in need of updating. Accordingly, the Committee determined that along with case studies and interviews with agency staff, it would be appropriate to survey participants in Phase IIB projects.

[11] 2011 Survey, Question 4A.

TABLE A-4 SBIR Responses by Year of Award (Percent Distribution)

	Phase II (Percent)
1998	2.2
1999	5.9
2000	3.7
2001	2.4
2002	8.6
2003	6.8
2004	11.0
2005	20.5
2006	12.7
2007	11.2
2008	6.8
2009	8.1
	100.0

NOTE: Phase II N=186
SOURCE: 2011 Survey; 2010 Phase IIB Survey.

The Phase IIB Survey Population

NSF provided the Academies with a data set containing records on 285 Phase IIB awards; data included the PI's name and email address. Four of the email addresses were invalid. No companies received more than six awards, and fewer than 20 companies received three or more.

Accordingly, it was determined that all Phase IIB award recipients with active email addresses that received their awards in FY1999-2009 should be surveyed. Awards made in 2010 would be too recent to address questions of commercialization impacts, an important aspect of the assessment.

Effort at Comparison Group Analysis

As with the later 2011 Survey, the Committee determined that if feasible, a control group should be developed against which to compare outcomes for Phase IIB projects. We ultimately concluded that the group, selected from the population of Phase II award recipients that did not receive Phase IIB funding, was not appropriate for use as a statistical comparison group.

Questionnaire Development

In order to ensure maximum comparability with previous Academies surveys, the survey questionnaire was based on the 2005 Survey instrument.[12]

Three changes were made to the previous instrument:

- Some questions were eliminated on the grounds that they had not generated knowledge used in previous assessments.
- Questions from both the project survey and the company survey were integrated into a single instrument.
- A section was added, focused on company experience with—and impacts from—Phase IIB.

Survey Deployment

Deployment faces twin challenges: it must generate an adequate number of responses, and it must be cost effective. An approach based on multiple iterations via different deployment vectors was adopted, as follows:

1. The survey was deployed on the web, and individualized links were created for each recipient, so as to track individual responses.
2. Four rounds of emails were sent to each recipient, separated by approximately 1-2 weeks, seeking response via an embedded link.
3. One further round was sent by the NSF's Dr. Joseph Hennessey.
4. Two additional rounds of voicemails were delivered (directing respondents to emails sent within the previous 24 hours).

Survey Responses

Based on the provision of 281 apparently valid email addresses by NSF and 281 control group awards, the preliminary survey deployment list included 562 email addresses, one per project. Based on two rounds of emails, it was determined that a number of these email addresses were nonfunctional and, after research, that 104 could not be replaced with functional equivalents. This had the effect of reducing the sample size of the project to 458 projects. Response rates for both groups are summarized in Table A-5.

As a result of these coordinated efforts, effective survey response rates were 51 percent for Phase II and 48 percent for Phase IIB only, after elimination of bad and non-replaceable email addresses (in some but not all cases alternative email addresses were identified).

---

[12] See National Research Council, *An Assessment of the SBIR Program at the National Science Foundation*, Appendix B.

TABLE A-5 2010 Phase IIB Survey Responses

	Phase IIB	Phase II
Excluded		
Bounced	36	62
No Email	2	0
Deleted Duplicate	1	3
All excluded	39	65
Included		
Complete	117	110
Partial	3	2
No Response	122	104
All included	242	216
Total	281	281
Gross response rate	41.6%	39.1%
Effective response rate	48.3%	50.9%

SOURCE: 2010 Phase IIB Survey

The slightly lower Phase IIB net response rate is explained by the preference of some projects with more than one award to provide only a single response. The Phase IIB response rate by company was essentially identical to that for Phase II respondents, at 51.2 percent.

A more detailed breakout of survey responses by year indicates that with two exceptions, the two groups are quite similar (see Table A-6). Phase IIB responses tended to cluster more in the 2002 and 2006 award years.

TABLE A-6 2010 Phase IIB Survey Response Distribution by Year and Group

Year of Award	Phase II (Percent)	Phase IIB (Percent)
1998	2.7	2.6
1999	8.2	3.4
2000	3.6	2.6
2001	1.8	0.0
2002	2.7	11.1
2003	2.7	5.1
2004	10.9	8.5
2005	21.8	16.2
2006	8.2	20.5
2007	7.3	6.0
2008	12.7	12.0
2009	17.3	12.0
Total	100.0	100.0

NOTE: N=120 (Phase II) and 112 (Phase IIB)
SOURCE: 2010 Phase IIB Survey.

Appendix B

Major Changes to the SBIR Program Resulting from the 2011 SBIR Reauthorization Act, Public Law 112-81, December 2011

1. **The SBIR program received an increased share of federal agencies' extramural budget:**[1]
   a. Congress increased the SBIR/STTR share from 2.5 percent to 2.6 percent in FY2012 and by 0.1 percent per year thereafter through FY2017, when the share would be 3.2 percent.
2. **STTR's share of the overall combined program was increased:**[2]
   a. It is to grow from 0.25 percent to 0.3 percent in FY2011, 0.35 percent in FY2012, 0.4 percent in 2013, and 0.45 percent thereafter.
3. **Award levels were increased:**[3]
   a. The existing limit of $100,000 for Phase I SBIR and STTR awards was increased to $150,000.
   b. The existing limit of $750,000 for Phase II SBIR and STTR awards was increased to $1,000,000.
   c. These limits were also for the first time indexed to inflation.
4. **Agency flexibility to issue larger awards was curtailed:**[4]
   a. Awards may no longer exceed 150 percent of guidelines (i.e., $1.5 million for Phase II) without a specific waiver from the SBA Administrator.
   b. The waiver can apply only to a specific topic, not to the agency as a whole. The agency must meet specific criteria and must show in its application that these criteria have been met before a waiver can be issued.

---

[1]U.S. Congress, Public Law 112-81, Sec. 5102 (a)(1)(a).
[2]Sec. 5102(b).
[3]Sec. 5103.
[4]Sec. 5103.

c. For every award under a waiver, agencies must maintain additional information about the recipient, including the extent to which they are owned or funded by venture capital or hedge fund investors.

5. **Agencies are permitted to utilize awards from other agencies:**[5]
   a. Agencies gained the ability to adopt Phase I awards from other agencies for Phase II funding; however, senior agency staff must certify that this is appropriate.
   b. Similarly, the legislation now permits between-phase crossovers between SBIR and STTR.
6. **Phase II invitations were eliminated:**[6]
   a. Previously some agencies—especially DoD—required that a company be invited by the agency before it could propose work for Phase II. This requirement is now prohibited.
7. **Pilot programs to skip Phase I were established:**[7]
   a. The legislation allows NIH, DoD, and the Department of Education to undertake pilot programs in this area. Discussions with agency staff indicate that for now DoD does not expect to utilize this new flexibility.
8. **Limited participation by previously excluded firms with majority venture capital or hedge fund ownership is now permitted (although subsidiaries of large operational companies are still excluded):**[8]
   a. NIH, NSF, and DoE are permitted to award up to 25 percent of their program funding to such companies.
   b. Other agencies are limited to 15 percent.
   c. For each award to such an entity, the Agency or component head must certify that this award is in the public interest based on criteria laid out in Sec. 5107(A)(dd)(2).
   d. Access to venture capital or hedge fund support may not be used as an award selection criterion by agencies.
   e. Special "affiliation" rules are provided for venture capital- and hedge fund-owned companies:
      i. Portfolio companies partially owned by venture firms or hedge funds are not deemed to be "affiliated" for purposes of determining whether an applicant meets size limitations, unless they are wholly owned or the owning company has a majority of board seats on the portfolio company.
9. **Explicit procurement preference were given for SBIR and STTR projects:**[9]
   a. The legislation states that agencies *and prime contractors* (emphasis added) must give preference to SBIR and STTR projects where practicable. However, there are no explicit targets included in the legislation.

[5]Sec. 5104.
[6]Sec. 5105.
[7]Sec. 5106.
[8]Sec. 5107.
[9]Sec. 5108.

10. **Sequential Phase II awards were permitted:**[10]
    a. The legislation now explicitly permits agencies to award one additional Phase II award after the first Phase II has been completed.
    b. The language implies that the provision of more than one sequential Phase II is prohibited.
11. **Commercialization support was expanded:**[11]
    a. Agencies are permitted to spend up to $5,000 per year per award on support for commercialization activities.
    b. Individual firms can now request up to $5,000 per year *in addition to their SBIR or STTR award* (emphasis added) to pay for commercialization activities from agency-approved vendors.
12. **The commercialization readiness pilot at DoD was converted to a permanent program—the Commercialization Readiness Program (CRP).** Details include in particular the following:[12]
    a. An SBIR Phase III insertion plan is now required for all DoD acquisition programs with a value of $100 million or more.
    b. SBIR/STTR Phase III reporting is now required from the prime contractor for all such contracts.
    c. The Secretary of Defense (SecDef) is now required to set goals for the inclusion of SBIR/STTR Phase II projects in programs of record and fielded systems and must report on related plans and outcomes to the SBA Administrator.
    d. The legislation explicitly requires the SecDef to develop incentives toward this purpose and to report on the incentives and their implementation.
13. **CRP may be expanded to other agencies:**[13]
    a. Other agencies may spend up to 10 percent of their SBIR/STTR program funds on commercialization programs.
    b. CRP awards may be up to 3 times the maximum size of Phase II awards.
    c. CRP authority expires after FY2017.
14. **Phase 0 pilot partnership program at NIH was enabled:**[14]
    a. NIH is permitted to use $5 million to establish a Phase 0 pilot program.
    b. The funding must go to universities or other research institutions that participate in the NIH STTR program.
    c. These institutions must then use the funding for Phase 0 projects for individual researchers.

[10]Sec. 5111.
[11]Sec. 5121.
[12]Sec. 5122.
[13]Sec. 5123.
[14]Sec. 5127.

15. Data collection and reporting were enhanced.[15]

a. Overall, the legislation calls for substantially increased data collection for individual recipients and for much more detailed reporting from agencies to SBA and to Congress.
b. Specific areas for improved reporting include
   i. Participation of (and outreach toward) woman- and minority-owned firms and the participation of woman and minority principal investigators;
   ii. Phase III take-up (from both agencies and prime contractors);
   iii. Participation of venture capital- and hedge fund-owned firms;
   iv. Appeals and noncompliance actions taken by SBA;
   v. Sharing of data between agencies electronically;
   vi. Extra-large awards;
   vii. SBIR and STTR project outcomes (from participants);
   viii. University connections (especially for STTR projects);
   ix. Relations with the FAST state-level programs;
   x. Use of administrative funding;
   xi. Development of program effectiveness metrics at each agency; and
   xii. SBIR activities related to Executive Order 1339 in support of manufacturing.
c. SBA is charged with developing a unified database to cover all SBIR and STTR awards at all agencies, as well as company information and certifications.[16]

16. Funding was provided for a pilot program to cover administrative, oversight, and contract processing costs:[17]

a. Agencies are limited to spending 3 percent of their SBIR/STTR funding on this pilot.
b. The pilot is initially designated to last for 3 fiscal years following enactment.
c. Part of the funding must be spent on outreach in low-award states.

17. Minimum commercialization rates for participating companies are required:[18]

a. Agencies must establish appropriate commercialization metrics and benchmarks for participating companies, for both Phase I and Phase II (subject to SBA Administrator approval).
b. Failure to meet those benchmarks must result in 1-year exclusion for that company from the agency's SBIR and STTR programs.

---

[15]Especially Sec. 5132, Sec. 5133, Sec. 5138, and Sec. 5161, but specific requirements are found throughout the legislation.

[16]Sec. 5135.

[17]Sec. 5141.

[18]Sec. 5165.

Appendix C

2011 Survey Instrument

INTRODUCTION

Welcome to the National Academy SBIR Survey. Thank you for participating. This survey seeks responses related to the [Phase 1 or Phase II] project entitled [insert project title], funded by [insert agency name], at the following company [insert company name]. Funding was awarded in [insert FY].

Note: If you need to revisit the survey before finally completing it, you can return at the point you left off by clicking on the survey link in your email.

[Project title will be piped into the survey header throughout the survey]

PART 1. INFORMATION ABOUT YOU

This information is required only to determine your current status, and to ensure that we have accurate contact information. This information will be strictly private and will not be shared with any private entity or government agency.

1. For the project referenced above, were you (during the time period covered by this award) (select all that apply)
   a. Principal Investigator (PI) on this project
   b. Senior researcher (other than PI)
   c. the CEO
   d. not CEO but a senior executive with the company identified above
   e. None of the above (exit questionnaire)

PART 2. COMPANY INFORMATION SECTION

2. Have you already completed a questionnaire about another SBIR project for this National Academy survey related to [insert company name].
   [Yes/No. If yes, skip to Part 3]

3. Is [insert company name] still in business?
   [Yes/No]

4. Thinking about the number of founders of the company, what was…?
   a. The total number of founders [number box]
   b. The number of other companies started by one or more of the founders (before starting this one) [0,1,2,3,4,5 or more]
   c. The number of founders who have a business background [number box]
   d. The number of founders who have an academic background [number box]
   e. The number of founders with previous experience as company founders

5. What was the most recent employment of the company founders prior to founding the company? Select all that apply.
   a. Other private company
   b. Government
   c. College or University
   d. Other

6. Was the company founded because of the SBIR program?
   Yes
   No
   In part

7. What percentage of the company's total R&D effort (man-hours of scientists and engineers) was for SBIR activities during the most recent fiscal year ___%
   0%
   1-10%
   11-25%
   26-50%
   51-75%
   76-100%

8. What was the company's total revenue for the most recent fiscal year
   - <100,000
   - 100,000-499,999
   - 500,000-999,999
   - 1,000,000-4,999,999
   - 5,000,000-19,999,999
   - 20,000,000-99,999,999
   - 100,000,000+

9. What percentage of the company's revenues during its most recent completed fiscal year was Federal SBIR funding (Phase I and/or Phase II)
   - 0%
   - 1-10%
   - 11-25%
   - 26-50%
   - 51-75%
   - 76-99%
   - 100%

10. Which if any of the following has the firm experienced as a result of the SBIR program? Select all that apply.
    - Made an initial public offering
    - Planning to make an initial public offering in 2011-2012
    - Established one or more spin off companies
    - Been acquired by/merged with another firm
    - None of the above

11. How many patents have resulted, at least in part, from the company's SBIR awards [number box]

12. Does the company have one or more full time staff for marketing?
    [Yes/No]

PART 3. PI/SENIOR EXECUTIVE INFORMATION

13. Please verify or correct the following information about yourself. Please indicate any corrections in the boxes provided. If all this information is accurate, click "Next" to continue. [Information will be piped in from respondent database to pre-populate editable text fields]
    a. Last name
    b. First name
    c. Current email address
    d. Current work telephone number (for follow up questions if necessary)

14. The Principal Investigator for this [SBIR] Award was a (check all that apply) (3 part question—14a, 14b, 14c)
    a. Woman
    b. Minority
    c. For those checking minority PI, add drop down list from SBA
        Asian-Indian
        Asian-Pacific
        Black
        Hispanic
        Native American
        Other

15. At the time of the award, the age of the leading PI was
    [20-24, 25-29, 30-34, 35-39, 40-44, 45-49, 50-54, 55-59, 60-64, 65+]

16. What was the immigration status of the PI at the time of the award?
    American-born US citizen
    Naturalized US citizen
    US Green card
    H1 visa
    Other (please specify—box)

PART 4. POST-AWARD INFORMATION

17. Many agencies offer commercialization training in connection with SBIR awards. Did you (or another company staff member) participate in training related to this award?
    [Yes/No]

18. Number of company employees (including all affiliates)
    a. at the time of the award [pipe in award year] [Number box]
    b. Currently [Number box]

19. What was the ownership status of the company at the time of the award? (3 part question—19a, 19b, 19c)
    a. Woman-owned
    b. Minority-owned
    c. For those checking minority-owned, add drop down list from SBA
        Asian-Indian
        Asian-Pacific
        Black
        Hispanic
        Native American
        Other

PART 5. PROJECT STATUS INFORMATION

20. Please select the technology sector or sectors that most closely fit(s) the work of the SBIR project. Select all that apply.
    - Aerospace
    - Defense-specific products and services
    - Energy and the environment
        - Sustainable energy production (solar, wind, geothermal, bio-energy, wave)
        - Energy storage and distribution
        - Energy saving
        - Other energy or environmental products and services
    - Engineering
        - Engineering services
        - Scientific instruments and measuring equipment
        - Robotics
        - Sensors
        - Other engineering
    - Information technology
        - Computers and peripheral equipment
        - Telecommunications equipment and services
        - Business and productivity software
        - Data processing and database software and services
        - Media products (including web-, print- and wireless-delivered content)
        - Other IT
    - Materials (including nanotechnology for materials)
    - Medical technologies
        - Pharmaceuticals
        - Medical devices
        - Other biotechnology products
        - Other medical products and services
    - Other (please specify—box)

21. Prior to this SBIR [Phase I/Phase II] award, did the company receive funds for research or development of the technology in this project from any of the following sources?
    a. Prior SBIR (Excluding the Phase I which preceded this Phase II.) [this parenthetical not shown to Phase Is]
    b. Prior non-SBIR federal R&D
    c. Venture capital
    d. Other private company
    e. Private investor (including angel funding)
    f. Internal company investment (including borrowed money)

g. State or local government
h. College or university
i. Other Specify _________

[Phase 1s continue/skip to question 30]

22. Did you experience a gap between the end of Phase I and the start of Phase II for this award? [P2 only]
    a. Yes Continue.
    a. No Skip to question 24

23. During the funding gap between Phase I and Phase II for this award, which of the following occurred? Select all answers that apply [P2 only]
    a. Stopped work on this project during funding gap.
    b. Continued work at reduced pace during funding gap.
    c. Continued work at pace equal to or greater than Phase I pace during funding gap.
    d. Received bridge funding between Phase I and II.
    e. Company ceased all operations during funding gap.
    f. Other [specify]

24. In your opinion, in the absence of this SBIR award, would the company have undertaken this project? [P2 only] Select one.
    a. Definitely yes
    b. Probably yes [If selected a or b, go to question 25]
    c. Uncertain
    d. Probably not
    e. Definitely not [If c, d or e, skip to question 27]

25. If you had undertaken this project in the absence of SBIR, this project would have been [P2 only]
    a. Broader in scope
    b. Similar in scope
    c. Narrower in scope

26. In the absence of SBIR funding… (please provide your best estimate of the impact) [P2 only]
    a. how long would the start of this project have been delayed? [text box - months]
    b. the expected duration/time to completion would have been…
        1) longer
        2) the same
        3) shorter

c. in achieving similar goals and milestones, the project would be…
   1) ahead
   2) the same place
   3) behind

27. Did this award identify matching funds or other types of cost sharing in the Phase II Proposal? [P2 only]
   a. Yes.
   b. No. [If b, skip to question 30]

28. Matching or co-investment funding proposed for Phase II was received from (check all that apply). [P2 only]
   a. Our own company (includes borrowed funds).
   b. Federal non-SBIR funding.
   c. Another company.
   d. An angel or other private investment source.
   e. Venture capital.
   f. Other [specify]

29. How difficult was it for the company to acquire the funding needed to meet the matching funds requirements? [P2 only]
   a. No additional effort needed except paperwork
   b. Less than 2 weeks Full Time Equivalent (FTE) for senior company staff
   c. 2-8 weeks effort FTE for senior company staff
   d. 2-6 months of effort FTE for senior company staff
   e. More than 6 months of effort FTE for senior company staff

30. What is the current status of the project funded by the referenced award? Select the one best answer.
   a. Project has not yet completed SBIR funded research. Go to question 33.
   b. Efforts at this company have been discontinued. No sales or additional funding resulted from this project. Go to question 31.
   c. Efforts at this company have been discontinued. The project did result in sales, licensing of technology, or additional funding. Go to question 31.
   d. Project is continuing post-award technology development. Go to question 33.
   e. Commercialization is underway. Go to question 33.
   f. Products/Processes/ Services are in use by target population/customer/consumers. Go to question 33.
   g. Products/Processes/ Services are in use by population/customer/consumers not anticipated at the time of the award (for example, in a different industry). Go to question 33.

31. Did the reasons for discontinuing this project include any of the following?

		Yes
a.	Technical failure or difficulties	____
b.	Market demand too small	____
c.	Level of technical risk too high	____
d.	Not enough funding	____
e.	Company shifted priorities	____
f.	Principal investigator left	____
g.	Project goal was achieved (e.g. prototype delivered for federal agency use)	____
h.	Licensed to another company	____
i.	Product, process, or service not competitive	____
j.	Inadequate sales capability	____
k.	Another firm got to the market before us	____
l.	Failed to receive Phase II award funding	____
m.	Other (please specify):	____

32. Which of these was the primary reason for discontinuing the project? (pipe in reasons marked "yes" in question 31 for respondents to choose from)

PART 6. PROJECT OUTCOMES

33. Have you received or invested any additional developmental funding in this project since the SBIR award?
    a. Yes
    b. No [if no, skip to Q35]

34. To date, what has been the total additional developmental funding for the technology developed during this project? Enter dollars provided in drop down list provided for each of the listed sources below. [If none for a particular source, enter 0 (zero)]
    <100,000
    100,000-499,999
    500,000-999,999
    1,000,000-4,999,999
    5,000,000-9,999,999
    10,000,000-19,999,999
    20,000,000-49,999,999
    50,000,000+

Source of Developmental Funding Since Receiving SBIR Award

a. Non-SBIR federal funds
b. Private Investment
   (1) U.S. venture capital
   (2) Foreign investment
   (3) Other Private equity (including angel funding)
   (4) Other domestic private company
c. Other sources
   (1) State or local governments
   (2) College or Universities
d. Not previously reported
   (1) Your own company (including money you have borrowed)
   (2) Personal funds

35. Has the company and/or licensee had any actual sales of products, processes, services or other sales incorporating the technology developed during this project? Select all that apply.
   a. No sales to date nor are sales expected. Skip to question 38.
   b. No sales to date, but sales are expected. Skip to question 38.
   c. Sales of product(s)
   d. Sales of process(es)
   e. Sales of services(s)
   f. Other sales (e.g. rights to technology, licensing, etc.)

36a. For the company and/or the licensee(s), when did the first sale occur resulting from the technology developed during [name of project]?

If multiple SBIR Awards contributed to the ultimate commercial outcome, report only the share of total sales appropriate to this SBIR project.

For the company [Pulldown with choices from 1990-2011]
For any licensees [Pulldown with choices from 1990-2011]

36b. For the company and/or the licensee(s), what is the approximate amount of <u>total sales dollars of product(s), process(es) or services to date</u> resulting from the technology developed during the [name of project]?

For the company [Pulldown with choices: 0, < $100,000, $100,000-$499,999, $500,000-$999,999, $1,000,000-$4,999,999, $5,000,000-$9,999,999, $10,000,000-$19,999,999, $20,000,000-$49,999,999, $50,000,000+]

For any licensees [Pulldown with same choices]

36c. For the company and/or the licensee(s), what is the approximate amount of other total sales dollars (e.g. rights to technology, sale of spin-off company, etc.) to date resulting from the technology developed during the [name of project]?

For the company [Pulldown with choices: 0, < $100,000, $100,000-$499,999, $500,000-$999,999, $1,000,000-$4,999,999, $5,000,000-$9,999,999, $10,000,000-$19,999,999, $20,000,000-$49,999,999, $50,000,000+]

For any licensees [Pulldown with same choices]

37. To date, approximately what percent of total sales from the technology developed during this project have gone to the following customers? If none, enter 0 (zero). Round percentages. Answers required to add to 100%.
    a. Domestic private sector
    b. Department of Defense (DoD)
    c. NASA
    d. Prime contractors for DoD
    e. Prime contractor for NASA
    f. Agency that awarded the Phase II (if not NASA or DoD)
    g. Other federal agencies
    h. State or local governments
    i. Export Markets
    j. Other (Specify)____________

38. As a result of the technology developed during this project, which of the following describes the company's activities with other companies and investors? Select all that apply.

	U.S. Companies/Investors		Foreign Companies/Investors	
Activities	Finalized Agreements	Ongoing Negotiations	Finalized Agreements	Ongoing Negotiations
a. Licensing Agreement(s)				
b. Sale of Company				
c. Partial sale of Company				
d. Sale of technology rights				
e. Company merger				
f. Joint Venture agreement				
g. Marketing/distribution agreement(s)				
h. Manufacturing agreement(s)				
i. R&D agreement(s)				
j. Customer alliance(s)				
k. Other (specify) ____________				

39. Please give the number of patents, copyrights, trademarks and/or scientific publications for the technology developed as a result of [name of project]. Enter numbers. If none, enter 0 (zero).

Number Applied For/Submitted		Number Received/ Published
	Patents	
	Copyrights	
	Trademarks	
	Scientific Publications	

40. How many SBIR awards has the company received <u>that are related to the project/technology</u> supported by this award?
   a. Number of related Phase I awards
   b. Number of related Phase II awards

Phase I recipients skip to Q44

PART 7. SBIR PROCESS AND RECOMMENDATIONS

41. In comparison to other Federal awards or Federal funding, how would you rate the process of applying for Phase II funding? Applying for Phase II funding was..." [Phase 2 only]
   a. Much easier than applying for other Federal awards
   b. Easier
   c. About the same
   d. More difficult
   e. Much more difficult
   f. Not sure, not applicable, or not familiar with other Federal awards or funding

42. How adequate was the amount of money you received through Phase II funding for the purposes you applied for? Was it.. [P2 only]
   a. More than enough
   b. About the right amount
   c. Not enough

43. Should the size of Phase II awards be increased even if that means a proportionately lower number of Phase II awards are made? [P2 only]
   a. Yes
   b. No
   c. Not sure

44. Overall, would you recommend that the SBIR program be...?
    a. Expanded (with equivalent funding taken from other federal research programs that you benefit from and value)
    b. Kept at about the current level
    c. Reduced (with equivalent funding applied to other federal research programs you benefit from and value)
    d. Eliminated (with equivalent funding applied to other federal research programs you benefit from and value)

45. To what extent did the SBIR funding significantly affect long term outcomes for the company?
    a. Had a negative long term effect
    b. Had no long term effect
    c. Had a small positive effect
    d. Had a substantial positive long term effect
    e. Had a transformative effect

46. Can you explain these impacts in your own words? [memo field]

PART 8. WORKING WITH PROJECT MANAGERS

Project Managers take on different names at different agencies. At DoD they are called Technical Points of Contact (TPOCs); at NASA they are the Contracting Officer's Technical Representative (COTR); at NSF they are the Program Officer. We use Project Manager in the questions below to refer to all of these.

47. How often did you engage with your Project Manager in the course of your award?
    a. weekly
    b. monthly
    c. quarterly
    d. annually

48. How valuable was your Project Manager on a scale of 1-5, with 1 being no help and 5 being invaluable.

49. How knowledgeable was your Project Manager about the SBIR program. Were they able to guide you effectively through the SBIR process?
    a. Not at all knowledgeable
    b. Somewhat knowledgeable
    c. Quite knowledgeable
    d. Extremely knowledgeable

Phase I recipients skip to Q53

50. On a scale of 1-5, with one being least and 5 being most, how much did your project manager help during the Phase II award in the following areas: [1-5 scale for each row] [P2 only]
    a. The Phase II application process
    b. Providing direct technical help
    c. Introducing us to university personnel that could contribute to the project
    d. Introducing us to other firms that could provide technical expertise
    e. Finding markets for our technology or products/services

51. How closely did you work with your Project Manager as you pursued Phase III funding? [P2 only]
    a. Not at all
    b. Not much
    c. We discussed the application in detail
    d. The officer provided a lot of guidance during the application process
    e. We did not apply for Phase III funding

52. How effective was the Project Manager in connecting the company to sources of Phase III funding (such as acquisition programs or venture/angel funding)? [1-4 scale] [P2 only]
    Very helpful
    Somewhat helpful
    Not very helpful
    Not at all helpful

53. How easy was it to reach your Project Manager when you had questions or concerns? (New) [1-4 scale]
    Very hard
    Hard
    Easy
    Very easy

54. Was your Project Manager replaced during the course of your award?
    [Yes/No]

55. How do you see the time allocated for your Project Manager to work on your project? [1-3 scale]
    Insufficient
    Sufficient
    More than sufficient

56. Deleted during final instrument review

57. Additional comments on working with your TPOC or Program Officer
    [memo field]

58. Is a Federal System or Acquisition Program using the technology from this award?
    Yes (go to question 59)
    No (skip to question 60)

59. If yes, please provide the name of the Federal system or acquisition program that is using the technology. ___________________

60. This question addresses any relationships between your firm's efforts on this project and any University or College. Select all that apply.
    a. The PI for this project was at the time of the project a faculty member
    b. The PI for this project was at the time of the project an adjunct faculty member
    c. Faculty member(s) or adjunct faculty member(s) worked on this project in a role other than PI
    d. Graduate students worked on this project
    e. The technology for this project was licensed from a University or College
    f. The technology for this project was originally developed at a University of College by one of the participants in this project
    g. A University or College was a subcontractor on this project
    h. None of the above

If any of these are checked (other than "none of the above"), continue to 60a; else skip to Q61 [if you do not check a-g, you should skip 60a as well]

60a. Which university (or universities) worked with your firm on this project?

61. Other comments on your experience with SBIR [memo field]

Appendix D

2010 Phase IIB Survey Instrument

INTRODUCTION

Dear colleagues,
Thank you for participating in this survey. Please answer these questions from the perspective of the firm mentioned in the email inviting your participation, even if you are no longer with that firm. If you are unable to do this, please do not answer the survey.

With your help, we can gather information that will better inform policy makers' choices here in Washington thereby enabling us to maintain and strengthen the SBIR program at NSF.

Please note that the individual details of your survey response will be kept strictly confidential by the National Academies, and will not be shared with NSF, SBA, or any other government (or nongovernmental) agency.

Thank you for responding to our request. If you have any questions, please do not hesitate to contact me directly.

Dr. Charles Wessner
NRC Study Director
202-334-3801

This information is required only to determine your current status, and to ensure that we have accurate contact information. This information will be strictly private and will not be shared with any private entity or government agency—including NSF.

1. **Please provide the following information.**

   Last name: ________________________
   First name: ________________________
   Name of your current employer: ________________________
   Your current position: ________________________
   Current email address: ________________________
   Current work telephone number (for follow up questions if necessary):
   ____

2. **Which of the following describes your position on this project during the time period covered by this award?** *(If you were a Principal Investigator (PI), please select PI as your response even if you held other positions as well.)*
   [ ] Principal Investigator
   [ ] CEO
   [ ] Other senior executive
   [ ] None of the above

3. **Is the company still in business?**
   [ ] Yes
   [ ] No

4. **Thinking about the number of founders of your company, what was...**
   the total number of company founders: ________________________
   the number of founders with university backgrounds: ________________
   the number of founders with business background: ________________
   the number of founders with previous experience as company founders:

5. **What was total company revenue for the Awardee company during the most recent year (in dollars)?**
   [ ] Under $100,000
   [ ] $100,000 to < $500,000
   [ ] $500,000 to
   [ ] $1 million to
   [ ] $5 million to
   [ ] $20 million to
   [ ] $100 million or more

6. **At the time of the award, what was the ownership status of the company?** *(check all that apply)*
   The company was...
   [ ] Woman owned
   [ ] Minority owned

7. **What is the current status of the project funded by the referenced SBIR award?** *Select the one best answer.*
   [ ] Project has not yet completed Phase II
   [ ] Efforts at this company have been discontinued
   [ ] Project is continuing post Phase II technology development
   [ ] Commercialization is underway
   [ ] Products/Processes/Services are in use by target population/customer/consumers

Discontinued Projects only:

8. **Did the reasons for discontinuing this project include any of the following?** *Select all that apply.*
   [ ] Technical failure or difficulties
   [ ] Market demand too small
   [ ] Level of technical risk too high
   [ ] Not enough funding
   [ ] Company shifted priorities
   [ ] Principal investigator left
   [ ] Project goal was achieved (e.g. prototype delivered for federal agency use)
   [ ] Licensed to another company
   [ ] Product, process, or service not competitive
   [ ] Inadequate sales capability
   [ ] Other (please specify)

Discontinued Projects only

9. **Which of these was the primary reason for discontinuing the project?**
   [ ] Technical failure or difficulties
   [ ] Market demand too small
   [ ] Level of technical risk too high
   [ ] Not enough funding
   [ ] Company shifted priorities
   [ ] Principal investigator left
   [ ] Project goal was achieved (e.g. prototype delivered for federal agency use)
   [ ] Licensed to another company
   [ ] Product, process, or service not competitive
   [ ] Inadequate sales capability
   [ ] Other (please specify)

Active Projects only

10. Has your company and/or licensee had any actual sales of products, processes, services or other sales incorporating the technology developed during this project?
Select one.
[ ] No sales to date nor are sales expected
[ ] No sales to date, but sales are expected
[ ] Sales of product(s), process(es), services(s) or other sales (e.g. rights to technology, licensing, etc.)

Projects with anticipated sales only

11. You indicated that you have had no sales to date resulting from the technology developed during this project. In what year do you expect the first sales for your company? ________________________

Active Projects with Sales

12. In what year did the first sale occur? ________________________

13. What is the approximate amount of total sales resulting from the technology developed during this project? *If multiple SBIR awards contributed to the ultimate commercial outcome, report only the share of total sales appropriate to this SBIR project. Enter approximate dollars. If none, enter 0 (zero).*

13.1. Total Sales Dollars of Product(s) and/or Process(es) or Service(s) to date?: ________________________

13.2. Other Revenues (e.g., Rights to technology, Sale of spin off company, etc.) to date: ________________________

Active Projects with Sales

14. To date, approximately what percent of total sales from the technology developed during this project have gone to the following customers? *Round percentages to nearest whole number. Answers must add to 100%.*

_______Domestic private sector
_______Department of Defense (DoD)
_______Prime contractors for DoD or NASA
_______NASA
_______NSF
_______Other federal agencies
_______State or local governments
_______Export Markets
_______Other type of customers

15. As a result of the technology developed during this project, which of the following describes your company's activities with other companies and investors? *Select all that apply.*

	Finalized Agreements with US Partners	Ongoing Negotiations with US Partners	Finalized Agreements with Foreign Partners	Ongoing Negotiations with Foreign Partners
Licensing Agreement(s)	[]	[]	[]	[]
Sale of Company	[]	[]	[]	[]
Partial sale of Company	[]	[]	[]	[]
Sale of technology rights	[]	[]	[]	[]
Company merger	[]	[]	[]	[]
Joint Venture agreement	[]	[]	[]	[]
Marketing/distribution agreement(s)	[]	[]	[]	[]
Manufacturing agreement(s)	[]	[]	[]	[]
R&D agreement(s)	[]	[]	[]	[]
Customer alliance(s)	[]	[]	[]	[]

16. Employee information. Enter number of employees. *You may enter fractions of full time effort (e.g. 1.2 employees). Please include both part time and full time employees, and consultants, in your calculation.*

16.1. Number of employees (if known) when Phase II proposal was submitted: ____________________

16.2. Current number of employees: ____________________

16.3. Number of current employees who were hired as a result of the technology developed during this Phase II project: ____________________

16.4. Number of current employees who were retained as a result of the technology developed during this Phase II project:

____

17. The Principal Investigator for this Phase II Award was a... *(check all that apply or none)*

[ ] Woman

[ ] Minority

18. Please give the number of patents applied for and received based on the technology developed as a result of this project. *Enter numbers. If none, enter 0 (zero).*

18.1. Patents applied for: ___________________________

18.2. Patents received: ___________________________

19. Please give the number of copyrights applied for and received based on the technology developed as a result of this project. *Enter numbers. If none, enter 0 (zero).*

19.1. Copyrights applied for: ___________________________

19.2. Copyrights received: ___________________________

20. Please give the number of trademarks applied for and received based on the technology developed as a result of this project. *Enter numbers. If none, enter 0 (zero).*

20.1. Trademarks applied for: ___________________________

20.2. Trademarks received: ___________________________

21. Please give the number of peer reviewed scientific publications published for the technology developed as a result of this project. *Enter numbers. If none, enter 0 (zero).* ____

22. Have you received or invested any additional developmental funding in this project *(beyond funding used to as a match for Phase IIB, which is considered later)?*

[ ] Yes

[ ] No

23. To date, what has been the total additional developmental funding (excluding Phase IIB) for the technology developed during this project? *Enter dollars provided by each of the listed sources.*

Non-SBIR federal funds: ___________________________

U.S. venture capital: ___________________________

Foreign sector investment: ___________________________

Other private equity: ___________________________

Other domestic private company: ___________________________

State or local governments: ___________________________

College or universities: ___________________________

Your own company (including money you have borrowed): _______________

Personal funds: ___________________________

24. Did you experience a funding gap between Phase I and Phase II for this award?

[ ] Yes

[ ] No

25. During your funding gap between Phase I and Phase II for this award, how did your firm respond? *Please select all answers that apply.*

[ ] Stopped work on this project during funding gap.

[ ] Continued work at reduced pace during funding gap.

[ ] Continued work at pace equal to or greater than Phase I pace during funding gap.

[ ] Received bridge funding between Phase I and II.

[ ] Company ceased all operations during funding gap.

26. In executing this award, was there any involvement by university faculty, graduate students, and/or university developed technologies?

[ ] Yes

[ ] No

27. This question addresses any relationships between your firm's efforts on this Phase II project and any University(ies) or College(s). *Select all that apply.*

[ ] The Principal Investigator (PI) for this Phase II project was at the time of the project a faculty member.
[ ] The Principal Investigator (PI) for this Phase II project was at the time of the project an adjunct faculty member.
[ ] Faculty member(s) or adjunct faculty member (s) worked on this Phase II project in a role other than PI, e.g., consultant.
[ ] Graduate students worked on this Phase II project.
[ ] University/College facilities and/or equipment were used on this Phase II project.
[ ] The technology for this project was licensed from a University or College.
[ ] The technology for this project was originally developed at a University or College by one of the recipients in this Phase II project.
[ ] A University or College was a subcontractor on this Phase II project.

28. Did your company apply for NSF SBIR Phase IIB matching funds for this project? *Phase IIB is a separate application for funding, for which the applicant must show matching funds as defined by NSF. Only Phase II awardees can apply.*

[ ] Yes, we applied for and received Phase IIB funding.
[ ] Yes, we applied for, but did not receive Phase IIB funding.
[ ] No, we did not apply.

29. You have indicated that your company did not apply for NSF SBIR Phase IIB funding. Why? *Check all that apply.*

[ ] Not aware of the program.
[ ] Unable to raise matching funds.
[ ] Could not raise matching funds that qualified.
[ ] Not enough funding to be worthwhile.
[ ] Process too onerous.
[ ] Other (please specify)

30. You have indicated that you applied for but did not receive Phase IIB funding from NSF. Why do you believe your application was rejected?

____

31. What was the total amount of NSF matching funds provided for your (NSF) SBIR Phase IIB matching funds project? And what was the total amount of investment and funding you found as your part of the match (in dollars)?

31.1. NSF funding: ______________________

31.2. Your funding: ______________________

32. What sources of matching funds were used for this award? Please specify the approximate percentage of funds that came from each source *(must add to 100%).*

_______US venture capital
_______US angel investment
_______Another US company
_______Sales
_______Other internal company resources
_______Federal agency funding
_______State agency funding
_______Foreign funder(s)

33. What best describes your relationship with your funding source(s) for the matching funds for Phase IIB? *Select any that apply.*

[ ] A long time sponsor or vendor relationship
[ ] A long time technical partner
[ ] A long time purely financial partner or investor
[ ] A new financial partner or investor
[ ] Other (please specify)

34. Which of the following did you have to give up to acquire the necessary matching funds? *Select any that apply.*

[ ] Nothing
[ ] Some equity resulting in dilution
[ ] Significant equity and influence on company policy by a third party
[ ] A loan repayment agreement
[ ] Access to your intellectual property
[ ] Effective control of the company
[ ] First refusal options
[ ] Marketing and/or licensing options
[ ] Other (please specify)

35. Did Phase IIB funding accelerate or otherwise make a significant difference in bringing the product or service funded to market?

[ ] Yes
[ ] No
[ ] Too early to tell
[ ] Other (please specify)

36. In what ways did Phase IIB funding make a difference in your ability to bring the funded product or service to market?

____
____
____
____

37. To what extent did the Phase IIB funding significantly affect long term outcomes for the company?

[ ] Had a negative long term effect
[ ] Had no long term effect
[ ] Had a small positive effect
[ ] Had a substantial positive long term effect
[ ] Had a transformative effect

38. Did the Phase IIB bring investment into the company that would not otherwise have been brought in?

[ ] Yes
[ ] No
[ ] Too early to tell
[ ] Other (please specify)

39. How much investment (in dollars) was obtained by your company strictly as a direct result of your participation in the Phase IIB process and program (i.e. money that would not have otherwise been brought in)?

____

40. How difficult was it for the company to acquire the funding needed to meet the NSF Phase IIB matching funds requirement?

[ ] No additional effort needed except paperwork
[ ] Some additional effort, but less than 2 weeks Full Time Equivalent (FTE) for senior company staff
[ ] Somewhat difficult: 2-8 weeks effort FTE for senior company staff
[ ] Quite difficult: 2-6 months of effort FTE for senior company staff
[ ] Very difficult indeed: more than 6 months of effort FTE for senior company staff

41. Based on your understanding of the process, which of the following do you believe qualify as matching funds for Phase IIB purposes? *Select all that apply.*

[ ] Equity investments
[ ] Cash loans
[ ] Sales
[ ] Additional investments from founders
[ ] In-kind contributions from technical partners
[ ] In-kind contributions from marketing partners
[ ] Other (please specify)

42. In comparison to other Federal awards or Federal funding, how would you rate the process of applying for Phase IIB funding? Applying for Phase IIB funding was ...

[ ] Much easier than applying for other Federal awards
[ ] Easier
[ ] About the same
[ ] More difficult
[ ] Much more difficult
[ ] Not sure, or not familiar with other Federal awards or funding

43. To what extent did the potential availability of Phase IIB funding provide additional incentives to seek investment funds?

[ ] Not at all
[ ] Somewhat
[ ] A great deal
[ ] It drove the process

44. How closely did you work with your NSF program officer as you pursued Phase IIB funding?

[ ] Not at all
[ ] Not much
[ ] We discussed the application in detail.
[ ] The officer provided a lot of guidance during the application process.

45. Additional comments on working with your NSF counterparts?

____

46. Looking back, was the SBIR Phase IIB funding your company received worth the effort involved to get the award?

[ ] Yes
[ ] No
[ ] Not sure

47. Overall, how would you rate the impact that the NSF Phase IIB matching funds program has had on your company?
[ ] Very positive
[ ] Somewhat positive
[ ] Neutral
[ ] Somewhat negative
[ ] Very negative

48. What specific impacts on companies should NSF consider when reviewing the program?

____

49. Overall, would you recommend that the Phase IIB program be...
[ ] Expanded (with equivalent funding taken from other phases of the SBIR program at NSF)
[ ] Kept at about the current level
[ ] Reduced (with equivalent funding applied to other phases of the SBIR program at NSF)
[ ] Eliminated (with equivalent funding applied to other phases of the SBIR program at NSF)

50. How adequate was the amount of money you received through Phase IIB funding for the purposes you applied for? Was it...
[ ] More than enough
[ ] About the right amount
[ ] Not enough

51. Should NSF increase the size of Phase IIB awards even if that means a lower number of Phase IIB awards are made?
[ ] Yes
[ ] No
[ ] Not sure

52. Should NSF permit the use of "in-kind" contributions as acceptable funding for the required match, provided they can be clearly documented?
[ ] Yes
[ ] No
[ ] Not sure

53. Do you have any additional comments about or recommendations for the NSF Phase IIB program?

____

Appendix E:

Case Studies

To complement its review of program data, the committee commissioned in-depth case studies of selected companies in the period 2009-13, with the earlier studies updated in 2014. Case studies are an important part of data collection for this study, in conjunction with other sources such as agency data, the survey, discussions with agency staff and other experts, and workshops on selected topics. The impact of SBIR funding is complex and often multifaceted, and although these other data sources provide important insights, case studies allow for an understanding of the narrative and history of recipient firms—in essence, providing context for the data collected elsewhere.

The case studies are of 12 SBIR companies that all received Phase I and II awards from the National Science Foundation (NSF), with most receiving multiple Phase I and II awards, including, in a number of cases, awards from other agencies as well NSF. A wide range of companies were studied: They varied in size; two were owned by women, several others had woman managers and founders. They operated in a wide range of technical disciplines and industrial sectors. Overall, this portfolio sought to capture many of the types of companies that participate in the SBIR program.

There are multiple variables at play; the case studies provide qualitative evidence about the individual companies selected, which are, within the limited resources available, representative of the different components of the awardee population. The 12 case studies presented in this appendix have been verified by the companies that they feature, and they have permitted their use and identification in this report (See Table E-1).

TABLE E-1 Directory of Case Studies

Company Name	State	Demographic
ALD NanoSolutions	CO	
Divergence Inc.	MO	
Imaging Systems Technology	OH	W
Immersion Technologies	CA	
Intelligent Fiber Optic Systems	CA	
Learning in Motion		
Membrane Technology and Research, Inc.		
Mendel Biotechnology	CA	
Techno-Sciences	MD	
Touch Graphics	NY	
TRX Systems	MD	
Workplace Technology Research	CA	W

NOTE: The "Demographic" column describes the company as majority-owned by Women (W) or Minorities (M); these data are drawn from NSF awards data, and reflect company self-certification.

ALD NanoSolutions Inc.

Based on interview with
Dr. Karen Buechler, President and CTO
April 9, 2010
By telephone

BACKGROUND

ALD NanoSolutions ("ALD NanoSolutions") is a spin-out from work on nanoparticles and nanomaterials at the University of Colorado Boulder.[1] Incorporated in 2001, ALD has broken new ground in atomic level deposition (ALD) techniques.

The company has exclusive licensing rights for the intellectual property on ALD techniques developed at UC, and has patented its Particle ALD™ and Polymer ALD™ technology. These are broad based "platform" technologies, and are now protected by patents issued in the United States, Europe, and Japan) (see intellectual property section below).

According to ALD NanoSolutions, the company is the first to carry out atomic layer deposition on particle surfaces and on polymer surfaces (also includes non-particle surfaces). These innovations are the basis for USPTO award of broad based process and composition of matter patent claims for ALD on particles, including approximately 100 related claims. ALD on particles has been successfully demonstrated by the company on numerous substrates, such as metals, ceramics, and polymers in many different materials markets including microelectronics, defense, battery systems, consumer products, construction, and biomedical. As ALD technology matures and commercializes, ALD NanoSolutions appears in consequence to be in prime position to grow rapidly.

Dr. Buechler, the President and Chief Technical Officer (and co-founder), noted that in 2001, the technology was interesting and potentially viewed as a broad "platform" technology with many applications, but that this did not fit well with standard models for technology transfer at UC, as there were no proven applications. By 2003, the technology had been successfully licensed and a commercialization team formed, led by serial entrepreneur Mike Masterson.

The business really started in 2003, as the team did not want to seek funding before the technology was fully under their control. The first SBIR Phase I award in 2003 slightly predated finalization of the licensing agreement.

The company is now seeking further investment and partnerships, and hence is spending more on sales and development. The company has sufficient funding

[1] ALD NanoSolutions' proprietary technology is based on atomic layer deposition coating chemistries and processing methods developed at the University of Colorado by Dr. Steve George and Dr. Al Weimer. ALD NanoSolutions web site. Accessed June 6, 2010.

for the next 18 months, and expects that it will be able to demonstrate coating scalability at another order of magnitude.

TECHNOLOGY: THE ALD PROCESS

Atomic layer deposition is a gas phase two reaction process. For aluminum oxide, ALD NanoSolutions' most popular and versatile chemistry, one layer is composed of molecules containing aluminum, the second layer is oxygen-containing molecules: water molecules. The aluminum molecules react with the surface to be coated, depositing one layer of aluminum atoms on the surface (since aluminum molecules do not react with other aluminum molecules). The surface is then exposed to the oxygen-containing molecules which put down a layer of oxygen. That completes one cycle, leaving the surface ready for the next layer.

Key Attributes and Advantages of ALD Technology

ALD NanoSolutions claims that ALD technology has a number of important advantages over other coating technologies, so using ALD means the company can build coatings that other technologies cannot, or coatings that have better qualities or cost less than those of competing technologies. Among the advantages claimed are as follows:

- Close control of film thickness which is controlled by film chemistry; most ALD films grow at 0.1 nm/cycle, with the thickness being determined by number of cycles
- Near zero waste of precursor chemicals for coating particles
- The process can be reliably reproduced, because the process itself is based almost entirely on the chemistry and is not dependent on other process parameters.
- The process is easy to automate/control, and permit close monitoring of gas phase by-products.
- Improved surface wetting and interfacial adhesion of fillers and pigments means that thermal fillers, metals, sunscreens are improved as the loading of specified components is increased.
- Enhanced resistance to moisture and air provides improve stability, important for (phosphors, battery materials, medical devices).
- UV / VUV Resistance—Protect materials in space and outdoor coatings
- Surface Passivation—Improve stability, color, prevent agglomeration
- Unique Composition—Construct nanocomposites with specific properties

PATENTS AND IP

ALD NanoSolutions is, according to Dr. Buechler, the first company to carry out atomic layer deposition on particle surfaces and on polymer surfaces (also includes non-particle surfaces), and has received patents that cover broad-based process and composition of matter patent claims for ALD on particles, including approximately 100 related claims. ALD NanoSolutions, Inc. has exclusive rights to this technology.

In January 2010, ALD NanoSolutions announced that the Japanese and Canadian patent offices had issued a critical patent covering the performance of atomic layer deposition (ALD) onto polymer surfaces and particles.

ALD NanoSolutions has also received numerous recognition awards, including a 2004 R&D 100 Award for Particle ALD, and the 2006 Frost and Sullivan Award for Excellence in Technology in Advanced Coatings and Surface Technologies.

MARKET OPPORTUNITIES

Use of nanoparticles and nanomaterials is growing rapidly, as new capabilities emerge to provide improved, more customized solutions to industry. These offer improved functionalities, such as wear resistance, corrosion resistance, scratch resistance, hardness, hydrophobicity, hydrophilicity, and catalytic activity.

Commercial deployment of conformal thin coatings or films on substrates has been difficult. Conventional techniques—such as chemical vapor deposition—have been commercially employed, but suffer from drawbacks such as agglomeration of nanoparticles on the substrate (which often deactivates the coated material), nonuniform coatings, line of sight dependency, and wastage.

ALD NanoSolutions technology has applications in areas such as drug delivery, magnetic resonance imaging materials, and powdered magnetic cores. The technique can also be used to develop thermal fillers with improved properties, improved battery systems, polymer/ceramic nanocomposites, improved lighting materials, low-energy high-sensitivity sensors, thermites, dental fillers, catalytic materials, and quantum tunneling surge protection devices.

ALD NanoSolutions has also developed a low-temperature process for depositing inorganic nanocoatings on either polymer particles or substrates, independent of the chemistry, or shape of the polymer. Hermetically sealing OLED devices used in making flexible displays are just one possible application.

In short, the range of commercial applications is very wide indeed, and ALD NanoSolutions has a very well protected position in the relevant intellectual property.

FUNDING AND COMMERCIALIZATION

ALD NanoSolutions has utilized a range of commercialization strategies and approaches, aside from SBIR funding. These include the following:

- Co-developing products or application
- Licensing the technology for a specific application
- Manufacturing products for testing
- Toll-coating full scale products
- Developing full scale processes

In February 2010, ALD NanoSolutions announced that its Particle ALD™ coating platform will be used to develop advanced electronic materials and applications in partnership with Tyco electronics. The companies will work together to utilize ALD NanoSolutions' surface engineering technology to develop and fabricate thin films for certain electronic applications. Mark Ellsworth of Tyco noted that "We have identified several product development opportunities where we can potentially apply the ALD NanoSolutions technology. This collaboration provides us access to the capabilities and expertise we need to achieve the required technical solutions more effectively."[2]

ALD NanoSolutions is now actively seeking more partnerships like that recently concluded with Tyco, where the partner is prepared to fund development of potential applications, in exchange for exclusivity in elected sectors and an agreed-on level of royalties. The Tyco multi-year project addresses an entire series of products and processes, over the short, medium, and long term. The agreement acknowledges that Tyco will need to license core ALD NanoSolutions technologies before commercialization is possible. This positions Tyco as a long-term customer rather than a licensee that simply wants access to the IP.

In contrast, ALD NanoSolutions is finding that many more companies are interested in outsourced research rather than hiring in house. Consequently, the contract R&D side of the business has grown rapidly, and now accounts for more than 80 percent of annual revenues.

ALD NanoSolutions continues to receive other funding as well, including more than $7 million in Phase I and Phase II SBIR awards from National Science Foundation (NSF), the U.S. Department of Energy (DoE) and the Air Force to develop specific applications. Four Phase I awards have been converted to Phase II.

SBIR

Over time, ALD NanoSolutions has gradually grown more experienced at identifying good and bad applications. However, Dr. Buechler observed that academics are often not good at spotting commercial results. ALD NanoSolutions is now writing fewer grants, and is finding that proposals are being penalized—at least at DoD and increasingly at NSF—for a poor commercialization record (in part because ALD NanoSolutions still has no commercial product for sale).

---

[2]Dr. Mark Ellsworth, Senior Director of Technology, Tyco Electronics. Quoted in ALD NanoSolutions Press Release, February 2010. <*http://www.ALD NanoSolutionsanosolutions.com/company/news/ald-nanosolutions-inc-and-tyco-electronics-announce-collaboration-agreement/*>.

This issue with the commercialization record comes up in reviews, and Dr. Buechler noted that program managers have made it clear that the company must commercialize if it is to win more SBIR awards. They are now clearly reluctant to put more investment into ALD NanoSolutions before a product reaches the market. She noted ALD NanoSolutions as having received 2 Air Force awards, plus another recent Phase I from DARPA.

Dr. Buechler noted that the sales cycle in the materials industry is long—often 5-7 years. She believes that most of the company's early work will in the end find its way into a commercial product in some way. And she also added that ALD NanoSolutions' most promising commercial partnerships—for example with Tyco—have been based on data from NSF SBIR awards. In fact, NSF helped to find ALD NanoSolutions a commercial partner, which will become involved if the current Phase II award is successful. The commercial partner will fund Phase III, and has also provided some funding beyond P1 to broaden the research beyond the scope of the Phase I itself. The partnership is based on joint ownership of joint IP, and plans to negotiate a royalty for ALD NanoSolutions' background IP.

Dr. Buechler is a strong supporter of the SBIR program. She believes it is doing a good job in building companies in many regions around the country. It is an excellent tool for moving university technology out into the marketplace, and it is an important mechanism for bridging the early stage funding gap. She believes that the program is also operated fairly efficiently.

Dr. Buechler had some comments about the program:

- One set of rules. It would be very helpful to small businesses if a single set of rules and applications governed all agency programs. ALD NanoSolutions had found working with NSF to be very easy, with close connections to program managers and more limited auditing requirements (which did however increase the possibility of abuse).
- Funding flow. For Phase I in particular, it would be simpler, more effective and much better for companies if funding was paid two-thirds on signature and one-third on delivery of the Phase I final report.
- DoD appears to be holding small companies to higher accounting standards than large ones. DoD could consider whether lighter requirements are more appropriate for smaller firms. In 2005, ALD NanoSolutions spent $20,000 on accountants for audits.
- Some recent changes have been positive. Using grants.gov has started the shift toward a more centralized submission process. This contrasts with EPA's traditional model of single-sided hard copies stapled in the upper left hand corner. Some standardization would be useful.

Dr. Buechler further offered some opinions and suggestions:

- **The one size fits all funding model is not efficient.** For Phase I, $100-150k is more than enough to get to proof of concept. In some cases, this could be an order of magnitude too small—but in others it is twice the size needed (e.g., for some software projects). More flexibility rather than simply more money is needed. At ALD NanoSolutions, some projects would have required $80-100k in external testing—so they could not be funded through SBIR Phase I and were dropped.
- **Timelines.** People take the time allowed, so shorter timelines are preferable because they force companies to think about what is a reasonable Phase 1.
- **Selection processes** are not necessarily open and fair at all agencies. For example, ALD NanoSolutions had a lithium ion battery project rejected for a Phase 2 award at DoE, despite very good reviews.
- **External reviewers.** NSF use of external reviewers is a positive feature, in contrast, for example, to DoE where program managers appear to have substantial influence on selection.
- **Business training.** Dr. Buechler appreciated the training from NSF. She had not been invited to any DoE training sessions or others at DoD.
- **Partnership and business development funding.** Dr. Buechler felt that this was a somewhat neglected part of SBIR and did not provide sufficient funding in particular for partnership development. ALD NanoSolutions has for example partnered with A123 on batteries, but the latter is currently focused more strongly on immediate needs rather than longer term development. The ALD NanoSolutions platform is generic for all battery materials, and Dr. Buechler is actively seeking partners—for example at the University of Colorado and from offshoots. According to Dr. Buechler, even a small amount of funding might make a difference, and would have been especially important during the early years of the company when dollars were scarce.

UPDATE

As of December 2014, ALDN reported that two additional commercial partners are pursuing joint product development. The company currently only has a single Phase I STTR program that represents less than 5 percent of expected 2014 revenues. This follows more than 12 months of zero government dollars to ALDN. The Polymer ALD patent applications have been granted by the European Patent Office, which the company believes is a substantial step forward for their IP portfolio.

Divergence Inc. Case Study

Based on interview with
Derek Repp, CEO, and Dr. Jim McCarter, founder
September 21, 2010
By telephone

DIVERGENCE HISTORY

Divergence is an R&D company dedicated to the discovery of effective and ecologically sound strategies for the control of parasites and other pests. Its initial focus is on parasitic nematodes, one of the world's major pest groups. Nematodes are roundworms that cause billions of dollars in damage annually to numerous crops, including soybeans, corn, cotton, strawberries, and bananas. Nematodes also cause widespread human diseases including hookworm, whipworm, roundworm (Ascaris), and the filarial worms responsible for lymphatic filariasis and onchocerciasis.

The company was founded by Dr. James McCarter at Washington University's Genome Sequencing Center in St. Louis, Missouri, to use genomics to help control parasitic nematode infections in plants, veterinary animals, and people. Divergence argues that safer, more efficient agriculture is critical to our future, and control of pests including nematodes is an important part of the equation.

The company now employs 23 full-time staff, which includes scientists trained at Washington University School of Medicine in computational biology, molecular biology, genomics, and biochemistry. Divergence also attracted former Monsanto Company employees, who brought complementary skills in business and product development. Other scientists have been recruited from leading academic institutions.

In 2009, Divergence moved into a new building in the Bio-Research & Development Growth (BRDG) Park, a life sciences lab and office park located on the Danforth Center campus in suburban St. Louis County. Divergence also leases greenhouse facilities and uses core labs in advanced proteomics/mass spec and microscopy at the Donald Danforth Plant Science Center.

According to Dr. McCarter, very little was known about the genomes or even the molecular targets of parasites. On the basis of his long career focused on nematodes and roundworms, he saw the need for the ability to detect and treat parasitic illnesses, in plants, animals and humans.

The primary opportunity at the time was based on genomics. Washington University in St. Louis had been heavily involved in the sequencing of the *C. elegans* genome, and did considerable work on sequencing the human genome after that. NIH funding had been substantial for this work. Starting in 1999, university staff began sequencing the genomes of parasites.

BOX E-1
Divergence Inc. Milestones

Divergence has

- developed cutting-edge discovery platforms in RNAi-based functional genomics, agrochemical discovery (Harvest™), and transgenic plants (STEM™);
- demonstrated the efficacy of its novel nematicidal chemistry against plant parasites in multiple field trials;
- validated approaches to plants with built-in nematode resistance which are being developed for soybeans and other crops;
- raised more than $20 million from investors and received more than $10 million in grant funding for research and development;
- created a significant intellectual property portfolio, including multiple pending and 35 issued patents as well as trade-secrets, around our discovery platform, molecular targets, chemistries, and transgenic methods for parasite control;
- established multiple important research and commercial relationships, including a collaboration with Monsanto Company to develop nematode-resistant soybeans and a relationship with a leader in animal health.

SOURCE: Divergence Inc.

Dr. McCarter expected many investigators in industry and academia to take advantage of this work, and believed that the best opportunity lay in the formation of a small company. He believed that while PIs often did excellent work operating academic labs, he did not see sufficient real world applications and impacts. In contrast, a small firm could focus, take risks, and assemble a multidisciplinary team. That would be difficult to do as new employees even in a large firm.

By February 2001, Divergence had 7 employees, and had completed its first financing round, comprising $1.4 million from 20 individual angels, in addition to $770,000 in family investments. This allowed Divergence to hire professional management, and Derek Rapp joined as CEO.

Divergence's work began with the results from published university research. This meant that no technology licensing was involved—Divergence was unencumbered, in Mr. Rapp's words. Exploiting genomics information, Divergence aimed to address the high toxicity inherent in the use of organophosphates as pesticides. Genomics made it possible to focus on molecular targets that were essential for nematodes, but divergent from humans.

BUSINESS STRATEGY

The company's philosophy is focused on the following:

- Building expertise around a particular area—parasitic nematodes (including microbiology, human health, and plant biology). Divergence has acquired world class expertise in these areas.
- Identifying practical applications for knowledge. Not only does this generate revenues and act as a market for the research side of the company, it provides validation of the company's technical approach

In essence, Divergence is aiming to build a company that has both high levels of research knowledge and effective development know-how. One core question is the balance between licensing and product-based market strategies. The core project of Divergence has been focused on discovery, and, according to Dr. McCarter, that will likely remain the case. But the company has been deliberately slow to license out product candidates, in order to retain significant value and to ensure that the maximum value is generated for the company. And where it does license, Divergence often imposes geographical or sectoral limitations.

Overall, Divergence does not expect to directly commercialize products except in specific circumstances. The marketing costs and regulatory challenges are seen as too prohibitive for a company the size of Divergence. Instead, Divergence is working with a number of potential collaborators, including, for example, Monsanto.

Divergence can afford to pick and choose among collaborators because it raised a substantial amount of investment funding. It has generated a total of about $40 million in income and investment, of which about 50% came from angel and venture investors (the Divergence C round closed in 2009), about 25 percent came from grants (including SBIR awards), and about 25 percent came from corporate relationships.

Strategically, Divergence has worked through the initial identification of targets, and continues to move downstream all the way to small molecule targeting and drug development. It is now shifting its focus from bioinformatics to a cheminformatics approach.

Finally, it should be noted that Divergence utilizes contract research organizations (CROs) and universities for aspects of its research beyond its core capabilities. It currently has over 20 contracts and collaborations in areas including synthetic chemistry, toxicology, and animal health.

PRODUCTS AND MARKETS

Nematodes are one of the world's major agricultural pests, causing an estimated $80 billion in worldwide crop damage annually. Traditional nematicides are environmentally dangerous, expensive, and difficult to apply. Nematode-infested crops with major economic losses include soybeans, potatoes, bananas, cotton, corn, citrus, strawberries, tomatoes, coffee, carrots, peppers, turf, and greenhouse ornamentals.

Divergence's innovations include new nematicides and nematode-resistant crops, offering improvements in both parasite control and environmental safety. These constitute the primary market targets for Divergence.

Nematicides

Nematode control has traditionally depended on highly toxic pesticides now restricted or eliminated in the United States. Similar restrictions are being implemented in other countries. The current global market for nematicides is estimated at $0.7-1 billion annually, but Divergence believes that improved control methods could expand that market several fold. Overall damage caused by nematodes and insects are similar in value, and worldwide insecticide sales are approximately $8 billion annually.

Divergence has targeted markets for better and safer nematicides, and its nematicides are currently in 150 field trials. The EPA Reduced Risk Initiative may permit accelerated regulatory review timelines for these products.

Nematode Resistant Crops

Aside from nematicides, nematodes can also be controlled by developing plants with internal resistance to nematodes. Internal resistance could provide highly-specific season-long protection from nematode damage—without the need for nematicide treatment. This approach is especially attractive in some widely planted row crops—such as soybeans, corn, and cotton—where the costs of nematicide treatment are especially high.

Other High-Potential Markets

Divergence has identified several other markets where its core technologies could be applied commercially.

Veterinary Medicine: Animal Parasites

Livestock and companion animal parasites include internal worms such as nematodes (endoparasites) and external fleas, ticks, and flies (ectoparasites). These cause a number of diseases in domestic and commercially-raised animals. While global sales of antiparasitic compounds account for approximately $3.5 billion annually, resistance to all major drug classes is now widespread in sheep and goats, and is emerging in the North American cattle market.

Human Health: Human Parasitic Nematodes

Nematodes infect nearly three billion people worldwide, mainly in developing countries. According to Divergence, diseases caused by nematodes include

- Hookworm infection, a major cause of anemia and stunted growth in children in tropical countries;
- Ascariasis, a gut roundworm infection, which affects more than one billion people and results in decreased quality of life;
- Filariasis or elephantiasis, an infection of the lymphatic system resulting in grossly swollen and scarred extremities.[3]

No vaccines are available for these diseases, so there is an urgent need for a compound that will be effective.

DIVERGENCE TECHNOLOGY

Divergence utilizes its expertise in the application of comparative and functional genomics to the control of parasitic nematodes. The last decade has seen revolutionary progress in both the generation of sequence information and methods for rapid gene knock-down. Divergence was an early adopter in applying advances such as RNAi to gene target validation for nematicides and to the generation of plants that were resistant to parasitic nematodes. This focuses research on targets that are biochemically distinct and vital for the life cycle of the infecting organism.

Publicly available DNA sequences increased from fewer than 50 million nucleotides in 1990 to more than 200 billion in 2008, and hundreds of genomes have been or are being completed. Washington University's Genome Center in St. Louis has played a leading role, and has now published more than 500,000 expressed sequence tags (ESTs) from 32 nematode species. Divergence has applied bioinformatics mining approaches to select promising targets from this basic genomic information, and with its collaborators the company has also directly generated genome sequences from key parasites of interest such as soybean cyst nematode. Divergence in-house expertise also includes a cross-species gene discovery approach that can rapidly clone gene orthologs from parasites of interest.

A Divergence scientific advisory board member was part of the team that discovered how to silence genes by degrading the corresponding messenger RNA, a process called RNA interference (RNAi). Divergence began RNAi-based work in 1999, and now leads its application to the control of parasitic nematodes.

Divergence has developed other proprietary technology platforms with potentially wide application, including Harvest™, which allows more rapid discovery and improvement of novel chemicals, and hence shortens the timeline from project conception to lead selection and reduces research and development

[3]Divergence web site. Accessed October 17, 2010. <*http://www.divergence.com*>.

costs. STEM™ is another proprietary protein-engineering platform technology. Divergence is currently applying STEM™ to nematode control.

PARTNERSHIPS

Divergence has developed a considerable range of partnerships relationships with both academic and commercial partners. The company has had a licensing agreement with Monsanto since 2004, focused on imparting nematode resistance into soybeans. The company also has a research partnership with the National Corn Growers Association aimed at developing nematode-resistant corn since 2003, and a close relationship with the Donald Danforth Plant Science Center since 2001. Divergence's laboratories are located next to the Danforth Center, and Divergence collaborates with multiple Danforth investigators, utilizes core laboratories in analytical chemistry and microscopy, and has received joint research grants with the Center.

DIVERGENCE AND SBIR

Divergence has received more than 33 awarded grants totaling more than $8 million, most by competitive peer review through the SBIR program.[4]

SBIR has, according to Mr. Rapp and Dr. McCarter, had a huge impact on Divergence. It was particularly helpful as the company prepared to offer a B round to venture investors in 2002—SBIR awards were seen as important factors in validating the company's research capability. Mr. Rapp said that he believed very strongly in the SBIR program. Divergence is a very strong supporter; without SBIR funding he did not know where the company would be: it might not exist, and it certainly would not be the company that it is.

In addition, these awards provided a significant influx of non-dilutive funds, which added to the company's attractiveness to professional investors. Mr. Rapp observed that it made life much easier when talking to investors if he could show that more than 50 percent of income and investment flows came from non-dilutive sources.

Divergence has received more than 29 SBIR awards (out of a total of 33 grants), which provided $8.8 million in funding. This funding has been absolutely critical in moving projects forward.

Since the addition of VC funding, the role of SBIR funding has shifted somewhat. It is now used more for projects which are more speculative and have less data to support them, where VC funding would not likely be forthcoming. This allows projects to mature and prove the design to both company management and reviewers to the point that they can reach product development. Even here, though, the company is disciplined in ensuring that SBIR funds are only used for

[4]SBA Tech-Net database. Accessed June 10, 2011.

projects that fit the company's broad strategy. Divergence is careful not to apply for awards on projects that do not fit the core strategy.

Divergence has also used SBIR to apply core knowledge to new areas. For example, initial work partly funded by NSF SBIR awards focused on soybeans, in partnership with Monsanto. Knowledge drawn from the project has since been applied more generally to root crop nematodes, where Divergence is currently seeking SBIR funding from NSF. Similarly, NSF has funded work applying Divergence technologies to corn and sugar cane.

Finally, Divergence also notes that SBIR awards are a powerful help in the recruitment of high-level scientists. They provide funding for projects but also generate excitement within the company. Eight different PIs have been in charge of projects at Divergence.

For Divergence, one of the biggest challenges in working with the SBIR program is timing. Grant applications require that the company look ahead to where the project might be 12 months in the future. Often, a number of the specific milestones to be addressed under a proposed award have been completed by the time funding arrives. It is therefore critical that program directors and TPOCs have flexibility to work with PI's in adjusting objectives, such as by adding more advanced milestones.

Some program directors have been very flexible—indeed, NSFs have been especially so over the years and have been very strong on personal contact between NSF program directors and company PIs. According to Divergence, personal contacts at NIH are harder to manage, although they are now improving. USDA is also improving rapidly, and is now getting closer to the NSF model. Mr. Rapp and Dr. McCarter both stressed how important it was for the program director to maintain close personal relations with the company PI and company management, so that the PD could fully understand the project and could therefore provide active support as needed.

Divergence sees the NSF SBIR program as somewhat different from those at other agencies.

- Submission requires use of FastLane, which is completely different than other application processes, and requires that the company provide more detail.
- Funding—for Phase I, NSF often provides two-thirds of total funding on signature, with one-third on completion. Most others provide a steady stream of ongoing funding.
- NSF requires more detail in both submission/tracking science report, including time report by individual person, and a financial report that identifies funds spent on each category.

NSF is also different in the degree of pressure it exerts toward commercialization. This has become even stronger in the past two years, possibly coinciding with the move to LARTA's commercialization support program (see below for more on commercialization).

This has had some unanticipated consequences. At the start of this new approach, in Divergence's opinion, some grant applications had been dismissed apparently simply because they were in the biotech sector, where regulatory timelines imposed significant delays, which apparently pushed the projects out of the timeframe for commercialization that appeared acceptable to NSF. The rejection of these applications has led to indefinite delays on projects that the company sees providing a potentially powerful range of applications.

Recommended Improvements

- NSF should—like most other agencies—provide more than one annual opportunity or deadline for each topic.
- NSF should also consider adopting at least in part the NIH model of open solicitations, where topics indicate areas of agency interest but proposals outside those areas are not automatically excluded. (According to NSF, it already has adopted this approach.)
- NSF should find ways to permit companies to rebut reviewers. In one case, according to the company, a reviewer who completely misunderstood a Divergence proposal "torpedoed it." According to Divergence, "reviewer comments included five misstatements and one or two complete misunderstandings." In contrast, USDA reportedly already uses a system whereby the program director emails Divergence a list of up to 10 questions arising from review. This gives the company an opportunity to make its case in more detail and to clear away misunderstandings.
- Resubmission. As with rebuttal, Divergence sees value in allowing applicants to improve their applications in response to review. The NIH resubmission approach responds to this need.
- Splitting commercial and scientific review. Divergence saw the need for this in a recent NIH application for funding to work on diagnostics for human parasites. Review comments noted that "poor people have no money to buy anything" and hence there could be no market for these diagnostics. USDA is reportedly now splitting commercial and scientific review.
- Commercialization support programs. Mr. Rapp participated in commercialization support programs conducted both by Dawnbreaker and by LARTA. He sees both as helping inexperienced scientists and engineers understand and prepare for the business world. However, being so strongly encouraged to participate is in his view a tax on the company's executive resources. For example, Divergence is at a stage where it is looking at world markets for corn seed treatment. Generic business plans are of no use here—instead, the company hired an industry insider with

more than 30 years of experience as a consultant. Mr. Rapp noted that his team has more than 120 years of experience running Divergence, and that he has more than 20 years as a senior marketing and management executive.

COMPANY UPDATE (NOVEMBER 2014)

Divergence achieved a successful exit in 2011 when it was acquired by Monsanto Company for $76 million. Three products from the company are being commercialized. The Divergence nematicide tioxazafen is now in phase III in Monsanto's technology pipeline as a seed treatment for corn, soy, cotton and eventually other crops with an anticipated 2017 launch. Divergence's collaboration with IDEXX Laboratories has resulted in commercialization of a revolutionary new test for whipworms in dogs with additional tests in development. An antiparasitic compound is in development for hornfly control in cattle with a pharmaceutical company licensee. All twenty-five Divergence employees joined Monsanto, and nearly all research scientists and technicians remain with the company. Divergence CEO Derek Rapp led M&A for Monsanto until 2014 when he joined the Juvenile Diabetes Research Foundation (JDRF) as President and CEO. Divergence founder Jim McCarter was an Entrepreneur in Residence (EIR) with Monsanto Growth Ventures until 2014. He is now a Senior EIR with BioGenerator, a St. Louis seed-stage venture group, working on his next start-up.

Imaging Systems Technology

Based on interview with
Carol Wedding, CEO and founder
October 12, 2010
By telephone

HISTORY

Imaging Systems Technologies is a privately held woman-owned firm located in Toledo Ohio, founded in 1997. The firm was a spin-off from a previous family business focused on display technologies, and is still family owned and operated. The previous firm—a technology innovator in displays—focused on military displays in the 1970s and 1980s, but could not compete effectively against globalized sourcing in the late 1990s.

IST has developed customized touch screens which can be mounted over standard computer displays. These screens fit around any flat panel display, and do not reduce viewing area, distort, obscure or dim the image in any way. The system allows for full mouse emulation via the touch screen input device. IST also sells a specialized video controller, and suite of test products for the auto industry. IST now focuses on research and development of large area distributed electronic networks including flexible displays and flexible sensors.

IST management has been working on flexible plasma displays since the late 1990s. The company's strategy was based on pairing high tech glass display capabilities, which were located in the Toledo region, with low tech "bubbles" to create large, flexible displays at various sizes without the need to invest in a multi-billion dollar manufacturing plant.

Currently, the smallest bubbles in use are 250 microns. This means that viewers have to stand 5-6 feet away to get an acceptable view. It is a large display technology. IST believes that cost and deployment advantages will allow bubble-based technologies to capture a substantial share of the large markets for outdoor advertising billboards and large wall displays. The strategy is to develop technologies that are comparable in quality but much lower cost and much easier to transport and deploy than the current-standard LED based products. LED displays are \$1-20k/sq. ft.; bubbles in contrast cost pennies per bubble to build even at relatively low volume.

IST has maintained an extensive consulting practice, on displays, imaging, and optical technology. Display technology includes AC plasma, DC plasma, LCD, and EL and their related drive electronics. The consulting practice has provided ongoing staff funding support while flexible plasma display technologies were developed. IST maintains a worldwide consulting business with large scale display manufacturers. However, this is insufficient to fund product development or the marketing push needed to find the first major customers.

IST TECHNOLOGIES

IST is now focused on the application of bubble technologies to flexible displays. Hollow spheres formed of glass, ceramic, and metal are used in many industrial, scientific, and medical applications, including structural applications, insulation, imaging, solar collectors, and transducers.

IST has developed the capacity to fabricate high quality custom hollow spheres with uniform wall thickness and uniform diameter in quantities consistent with research and development projects. These range in size from 500 microns to 3mm, with wall thicknesses from 40 microns to 300 microns. IST uses a variety of shell materials, including glass, metals, and ceramics. Custom techniques include layering various shell materials, and developing customized shapes.

This capacity is the basis for an existing consulting practice, and could be the basis for a wide range of applications. However, the company is primarily focused on the production of flexible monochrome and color plasma displays using Plasma-spheres. Plasma-spheres are hollow microspheres encapsulating an ionizable gas mounted on rigid or flexible substrates.

A conventional PDP is composed of two glass substrates that enclose an ionizable gas. Electrodes are deposited on the two substrates and covered with dielectric. Barrier ribs are formed on one substrate. Phosphor is deposited on walls of the barrier ribs.

In contrast, a Plasma-sphere is a hollow sphere composed of a glass shell encapsulating an ionizable gas. The Plasma-sphere display is formed on a single substrate. The substrate is made from a variety of materials.

Plasma-sphere technology can be deployed through a continuous flow process, instead of a costly batch process. Some of the more costly steps involved in manufacturing standard plasma displays can be eliminated. These include blasting, vacuum deposition, gas processing, and numerous screen-printing cycles. The elimination of these various process steps and the cost advantages of flow production yield substantial cost advantages, according to IST. Plasma-spheres are also produced much more rapidly, as the 12-16 hour gas-processing step is bypassed.

According to IST, plasma spheres have substantial advantages over standard plasma displays. While Plasma-sphere displays are comparable in terms of color pallet, viewing angle, video speed, and ability to scale to large sizes, Plasma-sphere displays are much more durable, temperature tolerant, and above all flexible compared to standard plasma displays.[5]

[5]C. A. Wedding, W. W. Olson, D. K. Wedding, O. Strbik, J. Guy, R. Wenzlaff, "Flexible AC Plasma Displays Using Plasma-spheres," *SID Symposium Digest* 35:815, 2004.

IST has now developed displays that are highly flexible and can be manufactured in rolls rather than batches. Ms. Wedding noted that its flexible displays have substantial advantages in terms of shipping. They represent an 80 percent reduction in weight, are not fragile and hence do not require crates and pallets, and can be rolled for convenient shipping. Finally, they do not crack at high altitude.

MARKETS

Plasma-sphere display technology is initially being applied in the digital billboards, which is a multi-billion dollar market, and the company is developing relationships with a number of large scale display manufacturers—Christy Display, Barco, and Diamond, all of which make digital billboards. The market for digital billboards is, according to Ms. Wedding, growing at 40 percent annually.

The critical strategic problem for the company is the lack of capital needed to fund market entry. IST would be prepared to license its technology or (perhaps preferably) to provide proprietary components for display manufacturers. IST believes its optimal strategy is to market large strip displays (2ft × 6ft), which can be seamlessly integrated into displays that contain multiple strips. Currently, IST has prepared a demonstration swatch of flexible display, 1ft × 1ft. but expects to be able to build 2ft × 6ft swatches. This approach would not require radical change to existing IST production facility.

Companies in this market segment have, according to IST, adopted fairly conservative approaches to new technology. In particular, they have focused attention primarily on the home theater market for Plasma TVs, rather than on larger scale displays.

Ms. Wedding believes that IST is in a very good strategic position. No significant competitors appear to be working on bubble-based technologies, although one Japanese firm is utilizing tube structures in a somewhat similar way. That firm is somewhat better funded than IST and is further toward entering the market. She noted that it has a 2m × 3m display that is up and working, although this technology is considerably heavier than the IST equivalent would be and is also not bendable.

Market entry will not be inexpensive. Ms. Wedding estimated that costs would be on the order of $10 million, and that a number of technical problems would also have to be addressed, including the provision of a market-ready power supply.

Bubble-based display technologies have substantial opportunities in a wide range of areas aside from large-scale digital billboards. Ms. Wedding noted potential markets in several programs in DoD. IST recently received an inquiry and subsequent contract from the US Air Force for adapting bubble technologies for use as programmable antennae. Other potential DoD projects have focused on adapting the technology for development of a large-scale radiation detector and for shielding on a stealth project. In each case, the same core technology is applied—putting gas into a bubble and the placing the bubble onto a substrate. Bubbles do not have to be made out of glass—IST has made bubbles out of ceramics and metals. For example, IST recently developed metal bubbles for a lightweight buoyancy application.

The range of possible applications is so wide that IST spun out another company to focus on non-glass structural applications, where large quantities of material are involved, and no electronics. IST might be interested in spinning off the display company.

IST AND SBIR

IST has a long history with SBIR, including a Tibbett's award in 2001. The company is a strong supporter of SBIR. Without it, Ms. Wedding notes, "IST could not do the necessary research to develop its innovative products."

IST has positive views on the NSF SBIR program in particular, which IST believes does a particularly good job of selecting high quality projects for funding.

According to Ms. Wedding, IST's experience with other agencies has not always been so positive. The company found it particularly frustrating that DoD applications did not automatically receive a detailed debrief, which was provided only verbally rather than in writing.

Ms. Wedding had participated in more than one commercialization training program. She believed that overall commercialization training did help, and that the Dawnbreaker program was the most useful. She observed that Dawnbreaker provided more customized support.

IST has not made serious efforts to attract venture funding, according to Ms. Wedding. The company has presented at two venture gatherings, but has attracted little interest there.

However, other marketing efforts have been more fruitful. IST purchased a booth at the Society of Information annual conference in 2010, and this generated a considerable amount of interest. The 1ft × 1ft demonstration display was enough—for this expert audience—to attract attention, specifically from digital billboard manufacturers, though not plasma display firms.

Improvements to SBIR

Ms. Wedding and IST offered a number of suggestions for possible improvements to the SBIR program at NIST.

More NSF SBIRs

Ms. Wedding did not believe it to be good policy to focus funding on a smaller number of awards. She belied that $150,000 was reasonable for Phase I and $500,000 was appropriate for Phase II. She observed that it was probably better to give smaller awards to more projects. Larger awards would lead to too great a focus on hot topics, and potentially good projects would be ignored.

More Flexibility on Matching Funds for Phase IIB

In particular, IST saw the need for more flexibility on the timeframe for acquiring matching funds. Currently, funding has to be obtained during the exact 18-month period from the start of the Phase II award to the last point of reasonable application for Phase IIB. She found it hard to believe that, starting on a new project, it would be possible to get to an investable point so quickly. At a minimum, she believed that NSF should accept matching funds acquired any time after the start of Phase I. Aside from timing, she thought the limits on the acceptable sources of matching funds were appropriate.

Topics

NSF topics are now very broadly defined—particularly in the materials sector—so it is possible for companies to find a topic in almost every solicitation. This used not to be the case at NSF and is an improvement in the program.

Application Process

IST was particularly complimentary about the feedback received from NSF SBIR applications. There were a substantial number of reviewers even for Phase I applications, and NSF provided both a summary review and individual reviewer comments.

Rebuttal

IST strongly supported the idea of providing companies with the opportunity to respond to reviewer comments within the framework of the selection process. Ms. Wedding observed that the ATP application process had provided exactly that opportunity by providing companies with a preliminary review and follow up questions to applicants.

Resubmission

NSF has in the past been flexible when an element of the application was not completed properly, but IST remains concerned about the possibility of having an application removed for administrative reasons, particularly as NSF does not permit resubmission of rejected applications.

Electronic Submission through FastLane

The FastLane process is not well designed for SBIR. Formatting can be very time-consuming, and as FastLane is used for all NSF applications, there are additional sections that are not relevant to SBIR applications. At a minimum, FastLane should identify application elements that are mandatory or not relevant for SBIR.

Recognition Received by the Company

- R&D 100, 2005
- Edison Emerging Technologies Award, 2005
- Roland Tibbett's Award, 2001

COMPANY 2014 UPDATE

IST continues to develop the Plasma-shell technology that was previously funded by NSF under several SBIR awards. IST has teamed with a manufacturing partner in Japan to assist in integrating the Plasma-shell technology into display and lighting products. In conjunction with its manufacturing partner, IST is developing several prototype products, including a UV lighting product for a U.S distributor. This development is being funded internally.

IST considers itself a successful small business. IST sells three commercial products, and engages in various engineering consulting projects. Technology funding by NSF has resulted in 60 patents; licensing revenue; and the creation of two new spinout companies.

Since 2010, the company believes that the NSF SBIR program has evolved from providing assistance in commercialization (via DawnBreaker) to demanding commercialization within a time frame of about two years. Small businesses that do not fall into this accelerated growth trajectory are considered unsuccessful. In this respect, much more is being expected of small businesses than is of either large businesses or universities with federally funded research projects.

In Ms. Wedding's view, the SBIR program was established to allow small businesses to access federally funded research opportunities. However, NSF seems to be focused on funding "start-ups" over established small businesses. In view of this, IST offered the following suggestions:

- *Reduce focus on outside investment and venture capital.*
- *Broaden the definition of a success.*
- *Be mindful of the fact that established small businesses tends to have infrastructure and resources (facilities, equipment, personnel, IT, and proper accounting and project management) to apply to a research project that a "start-up" might lack.*

Immersion Corporation[6]

Based on interviews with
Christophe Ramstein, CTO
August 17, 2009
In person

Chris Ullrich, Senior Research Director[7]
October 2, 2009
By telephone

BACKGROUND

Immersion Corporation ("Immersion") is a publicly owned company headquartered in San Jose, California. Founded in 1993, Immersion went public in 1999 and is now publicly traded on the NSADAQ, with a market value of about \$115 million.[8] Immersion has focused on the provision of "haptic" technologies. These technologies allow users to engage their sense of touch when operating digital devices including touch screens, gaming devices, and other tools where touch adds a further dimension of connection for the user, as well in many cases as enhanced functionality.

Immersion currently focuses marketing and business development activities on five major sectors:

a. automotive,
b. gaming,
c. 3D CAD systems for industrial devices and controls,
d. medical simulation, and
e. mobile communications.

Depending on the market, Immersion sometimes licenses technology for inclusion in products branded by other manufacturers (e.g., video console gaming, consumer electronics, mobile phones, and automotive controls). In other markets (notably medical simulation), Immersion sells products directly under its brand name (see "Lines of Business" section below for details).

---

[6]Note that Immersion Corporation was also included in a series of NSF case studies in an earlier study of NSF SBIR grant companies, published in 2008. Then the focus was on the endoscopy applications of the technology.

[7]Unless otherwise specified, information in this report is drawn from the interview with C. Ramstein and the Immersion Corp. web site (<*http://www.immersion.com*>).

[8]NASDAQ stock quotes, <*http://quotes.nasdaq.com/asp/SummaryQuote.asp?symbol=IMMR&selected=IMMR*>, accessed September 30, 2009.

Immersion is currently in the middle of a period of steady revenue growth, driven by the widespread adoption of its haptics technologies in cell phones. Overall revenues reached $36.5 million in 2008, up from $23.8 million in 2004.[9] Along with some long-time agreements in this area (e.g., with Samsung), Immersion recently concluded a licensing agreement with Nokia. The successful conclusion of a patent infringement case against Sony appears likely to lead to further licensing agreements with Sony.[10]

INTELLECTUAL PROPERTY (IP)

Immersion is somewhat unusual in the extent to which its resources are expended on IP protection. The company claims to hold more than 700 patents,[11] a large number for a relatively small company not yet breaking even. However, the need for these patents is clear in the context of the license-based business strategy for non-medical devices, and the Sony patent infringement settlement shows that this approach has in some ways already been successful.

HAPTICS TECHNOLOGY

Immersion haptics systems typically include five kinds of elements:

- One or more sensors
- Actuator (motor) control circuitry
- One or more actuators that either vibrate or exert force
- Real-time algorithms (actuator control software, which we call a "player") and a haptic effect library
- Connecting software that includes the application programming interface (API) and often a haptic effect authoring tool

Immersion's technical advantage is focused on actuators, actuator control software, and the API and authoring tools. Mechanisms used to convey forces to the user's hands or body include vibrotactile actuators; direct-, belt-, gear-, or cable-driven mechanisms; and other proprietary haptic devices that supply textures and vibration, assistance, resistance, and damping forces to the user. The Immersion API is used to program calls to the actuator, specifying which effect in the haptic effect library to play.

When the user interacts with the product's buttons, touchscreen, lever, joystick/wheel, or other control, this control-position information is sent to the OS, which then sends the play command through the control circuitry to the actuator, which in turn physically translates the command into touch-based effects.

---

[9] Immersion Corporation, Annual Report 2008, p.32

[10] Interview with Christophe Ramstein, Immersion CTO, August 17, 2009.

[11] Immersion Corporation, Annual Report 2008, p. 7.

LINES OF BUSINESS[12]

Immersion operates five lines of business. The gaming/mobile communications/automotive/CAD systems lines are based on licensing haptic technology to brand-name or OEM manufacturers. The medical devices business in contrast uses haptics technology to provide competitive advantage in the market for simulated surgical training.

The Licensing Business

Key components of the licensing business include the following:

Gaming Devices

Immersion began to license products in 1996, starting with the gaming devices sector. Clients include Microsoft (for use in its gaming products), Apple Computer (operating system), and Sony Computer Entertainment (PlayStation products), as well as more than a dozen gaming peripheral manufacturers and distributors such as Logitech and Mad Catz.

Mobile Communications and Portable Devices

Immersion's TouchSense technology covers haptic touchscreens and programmable haptic rotary controls. In early 2009, Samsung announced its new P3 personal media player, which uses Immersion haptic feedback technology for touchscreen interactions. In 2008 Cue Acoustics announced and began shipping a premium AM/FM radio and iPod docking station that includes a TouchSense rotary control module as its primary control mechanism.

Immersion currently licenses TouchSense technologies to the top three makers of mobile phones by volume in the world: Nokia, Samsung, and LG Electronics, plus others such as Pantech Co., Ltd. and KTF Technologies Inc. In 2008, approximately 40 million handsets with TouchSense technology were shipped by Immersion licensees.

Automotive

Immersion began licensing TouchSense for use in vehicle controls in 2002. Licensees include Siemens VDO Automotive (now Continental) (for use in the high-end Volkswagen Phaeton sedan and Bentley cars); ALPS Electric (Mercedes-Benz S—Class sedans and Lexus RX 350/450h). Other automotive industry licensees include Methode Electronics, Inc., Visteon Corporation, Volkswagen,

[12]Ibid. p. 9.

and SMK Corporation of Tokyo. Since 2001, over 2.4 million vehicles have included TouchSense technology.[13]

The Medical Simulation Business

Immersion has developed numerous simulation technologies used for medical training and testing. By more fully engaging the sense of touch, Immersion's technologies support more realistic simulations. In turn, this improves the training of medical students, doctors, and other health professionals. Simulators allow these professionals to practice in a risk-free environment where mistakes have no dire consequences and animal or cadaver use is unnecessary.

Specifically, Immersion has developed four lines of medical simulation products covering

a. needle-based procedures such as intravenous catheterization and phlebotomy;
b. endoscopic procedures, including bronchoscopy and lower and upper GI procedures;
c. endovascular interventions including cardiac pacing, angiography, angioplasty, and carotid and coronary stent placement; and
d. minimally invasive procedures involving abdominal and pelvic organs.

Each product line is designed to maximize the number of procedures that can be simulated with minimal additional customer hardware investment. These systems may then generate additional sales of relatively inexpensive software modules. Immersion currently offers more than 25 software modules that replicate such medical procedures as intravenous catheterization, laparoscopy, bronchoscopy, colonoscopy, cardiac pacing, and carotid and coronary angioplasty.

In addition, Immersion has developed simulation technology for other medical device companies, such as Medtronic.

IMMERSION'S FINANCIAL SITUATION

Immersion is not yet profitable; in fact operating losses increased during 2008 to about $25 million excluding special items (up from approximately $12 million in 2007). However, the company has substantial liquid assets (more than $85M) apparently drawn to a significant degree from its successful lawsuit against Sony, which resulted in a $134.5 million settlement in 2007.[14]

[13]Ibid. p. 8.
[14]Ibid. p. 14.

R&D

After an early period focused on defining and testing its core technologies in the 1990s, which resulted in the patents that form the core of the company's IP, Immersion switched its attention toward the exploitation of its existing technologies, and the rate of technical innovation within the company slowed somewhat.

This strategic choice has to some extent been recently reversed. Additional technical research staff have been hired, and the company's R&D budget has doubled to $12 million since 2005. The CTO indicated that this expansion was expected to continue.[15]

SBIR AWARDS

Immersion won a series of SBIR awards starting in the late 1990s (See Table E-2).

ROLE AND PURPOSE OF SBIR AWARDS AT IMMERSION

The SBIR awards outlined in Table E-3 can be divided into four categories:

- Three early awards, including two DoD Phase 2 awards, supported development of the company's core haptics technology.
- Starting in 1997, a number of DoD and NIH (HHS) awards supported development of medical applications, which now account for more than 40 percent of company revenues (see Table E-3).
- Subsequent NSF awards supported development and adaptation of the core technologies to the medical simulations business.
- The most recent Phase II award was in 2003; the most recent Phase I was in 2005.

Dr. Chris Ullrich, Senior Director of Research, provided further insights into the role and value of the NSF awards.[16] He was originally part of the team developing virtual reality CAD systems for a very small company called Virtual Technologies, which was acquired by Immersion.

The NSF awards were originally made to support development of the VR CAD technologies, but Immersion's research interests did not in the end support the long term research required for this technology. Instead, Immersion discovered that the VR CAD technology could be adapted for use within their medical simulator business.

The NSF Phase II and Phase IIB awards were, according to Dr. Ullrich, instrumental in funding this important development. The Phase IIB match was

[15]Interview with Christophe Ramstein, August 17, 2009.

[16]Telephone interview with Chris Ullrich, October 2, 2009.

TABLE E-2 Immersion SBIR Awards

	Number of Phase I Awards	Phase I Funding (Dollars)	Number of Phase II Awards	Phase II Funding (Dollars)
NSF	6	439,455	3	1,395,600
HHS	3	299,888	2	1,236,452
DoD	9	878,334	5	3,459,764
Total	18	1,617,677	10	6,091,816

SOURCE: SBA Tech-Net SBIR awards database. Accessed September 15, 2009.

TABLE E-3 Distribution of Immersion Sales by Sector

	Percent of Sales		
	2006	2007	2008
Gaming devices	18	21	23
Mobile communications	1	7	13
Automotive	7	10	9
3D CAD systems	17	14	13
Medical business	51	44	41
Misc.	6	4	1
Total	100	100	100

SOURCE: Immersion Corp., *Annual Report*, 2008.

provided through an existing long-term relationship with Medtronic. Medtronic funded research aimed at creating virtual demonstrators of their surgical tools to be deployed to surgical centers as a sales tool.

The NSF funding thus allows Immersion to develop the core technology that continues to underpin the medical simulation business. Dr. Ullrich indicates that the NSF award came at a critical time for his group and provided the funding that allowed them to transition from VR CAD to medical situations.

SBIR AT NSF

Dr. Ullrich noted that Immersion was ideally suited for a Phase IIB, given the existing relationship with Medtronic. As with other NSF Phase IIB awards, this suggests that in some ways the Phase IIB program may simply reward existing relationships.

Dr. Ullrich said that he was positively impressed by the flexibility provided by the NSF project liaison, who understood the technical and market shifts underway at Immersion, and approved adjustments to the Phase II award that were critical to its eventual success.

Immersion participated in the Awardee conferences organized by NSF. Dr. Ullrich commented that although they provided useful networking opportunities and helped to build valuable connections between companies and NSF program managers, commercial training was not especially important to a publicly traded firm like Immersion. However, he thought the commercialization training was probably very useful for smaller companies with weaker commercial experiences.

Intelligent Fiber Optic Systems (IFOS®) Corporation

Santa Clara, CA
Based on interview with
Dr. Behzad Moslehi, CEO/CTO and founder
November 14, 2010

HISTORY AND BACKGROUND

IFOS designs, manufactures and markets advanced photonic sensing systems to monitor and control high-value assets in harsh and demanding environments. It is a privately held company, based in Santa Clara, California. The company was incorporated in the State of California in 2001 by its current CEO/CTO, Dr. Behzad Moslehi.

In its early years, IFOS was a classic Silicon Valley startup—working out of a garage. It did not really gain traction until 2000-2001, when initial Phase II SBIR awards with NASA and NSF allowed the founder to commit himself fulltime to the company and lease its own 4000-square-foot facility initially in Sunnyvale, CA. SBIR funding was key to equipping the facility and hiring a dedicated founding staff. This development coincided with the final stages of the telecom boom.

Subsequent SBIR/STTR awards allowed IFOS to strengthen its technical team and better-equip its facilities, delivering across a wider range of disciplines and subsequently to move (in 2006) into a larger facility in Santa Clara, CA, with 10,000 square feet of research and development space.

Since the company's inception, IFOS has worked toward implementing its vision of end-to-end fully flexible and scalable networked optical sensing systems that can detect and monitor a wide variety of physical and chemical data over a dynamic fiber network. IFOS is taking the leading role in optical sensing and will continue to supply innovative, advanced technologies for this market. The goal of the company is to meet a wide range of sensing needs cost effectively and competitively.

In IFOS products, optical fibers are not merely a transmission medium, but rather an active intelligent medium with simultaneous sensing, processing, and transmission capabilities. The company began to sell commercial products in 2006.

In recent years, IFOS has expanded its marketing efforts internationally and is working harder to connect to potential customers. It has been attending key industry trade shows and taking part in exhibitions and industry days. The company's products are often of interest to clients who require substantial customization for applications in harsh and demanding environments, so it is not feasible to develop simple distributor-based marketing strategies at this time.

IFOS has developed a sensor platform that could be applied to a number of industries. This array of options presents marketing challenges and requires judgment and prioritization. IFOS addresses the needs of these emerging markets opportunistically. Following an extensive research effort, IFOS is presently focusing on three core markets:

- aviation and safety
- energy
- life sciences

The life sciences sector illustrates in part how IFOS works. Initially, the company worked with practitioners and surgeons at Stanford University Medical School, in cardiology, radiology, and oncology. Through its Stanford connections, the company then moved on to partner with several local medical devices companies who will manage FDA market testing and industry-specific marketing.

Dr. Moslehi completed his PhD in Electrical Engineering at Stanford University, where the company retains important connections. Prior to founding IFOS, Dr. Moslehi helped to commercialize fiber optic gyroscopes, known as FOGs, for avionics and towed hydrophone sensor arrays in submarine applications as a member of the fiber optic development team at Litton Industries (part of Northrop-Grumman). Dr. Moslehi holds numerous patents and has many peer-reviewed publications in this area. He played a part in the design of the first commercial WDM multiplexers based on diffraction gratings at Physical Optics Corporation (POC). His work also contributed to the founding of ONI Systems, a spin off from Optivision, which went through a successful IPO in 2000 and was later acquired by Ciena.

IFOS revenues are derived from numerous grants and contracts from a range of federal agencies and from commercial clients. The IFOS I*Sense® Product Family has been sold to a number of clients worldwide, through in-house sales and marketing involving representatives and selected strategic partners.

IFOS retains close ties with several prime contractors, notably Lockheed Martin, Boeing, Raytheon, and Pratt and Whitney. Further connections have been made with customers in France, Germany, Italy, Japan, and Korea. Inquiries have also been recently received from several medical centers of research excellence in the United States, including UC San Francisco and Harvard medical schools.

In the energy sector, IFOS has worked with medium-sized oil/gas services companies (in Bakersfield, CA, Wallingford, CT, and Houston, TX) on down borehole well temperature profiling (based on Raman Distributed Temperature Sensing or DTS) and on tools for angled directional drilling and measurement-while-drilling (MWD), which would be highly applicable to deep sea drilling. As with other market activity sectors, however, there are challenges facing innovative technologies. Oil companies tend to operate in boom-bust cycles and are inconsistent in funding innovative research or products.

Partnering is again a cornerstone of IFOS' strategy in this market and a partner typically contributes field-testing facilities while adding a new technology-enabled service to its own portfolio. One of the company's partners has recently completed initial borehole testing of the IFOS DTS oil well temperature profiling system to prove the economics of using the IFOS solution, showing that the technology can work reliably in the field both on hot days and cold nights. IFOS is also planning a trial for the MWD tool incorporating its FOG technology next year. The project is now seeking partners to fund more advanced trials.

Related to these activities, IFOS is now in discussions to develop an advanced pipe monitoring system for a leading US infrastructure pipe maintenance company. Aging infrastructure in the US, including bridges, tunnels, railways, and pipes necessitate new intelligent means of planned maintenance and life expectancy modeling. Intelligent sensors and sensor interrogators are essential to support decision making in these situations and optical fiber sensors are among the most cost-effective new solutions emerging in this nascent market. Dr. Moslehi believes optical sensors will become ubiquitous due to their competitive properties and attractive economies.

PRODUCTS AND SERVICES

Dr. Moslehi believes that a key barrier confronting photonic-based applications is the level of discomfort users may have with these new and emerging technologies. As a result, both IFOS and the customer must expend considerable effort, cost, and time in establishing field reliability, and in educating and convincing customers to embrace the transition from Industrial Age processes. Unlike standard electronic-based wired and wireless transmission systems, the integration and insertion of photonic-based components into a system presents technical difficulties for many users who could otherwise be likely customers for IFOS. A symbiotic relationship that combines IFOS expertise in the technology and the client's expertise in the field application engineering is often a key success factor, but such relationships take time and investment to establish.

In addition, Dr. Moslehi observed that photonics face the classic difficulties encountered by many highly innovative technologies in that they face underdeveloped markets if they are early to the game: while customers may *want* a new product, he notes that they may not feel they *need* it immediately. This sometimes causes substantial problems for innovative companies working in emerging markets. They must either invest to create the market—like Apple—or wait for the market to emerge while other companies catch up. While there have been substantial US government investments in photonics and optical networks—from DARPA and other agencies—the US has not really replaced the infrastructure of dedicated world class research behemoths like Bell Labs, AT&T, XEROX PARC, etc., conducting costly fundamental long-term research. So sustainable innovation is facing challenges in the United States.

Accordingly, IFOS has worked to develop end-to-end solutions for particular applications. This has required development of additional company capacities, branching out from the original core competencies surrounding sensors to include electronics, algorithms, software, and proprietary electronic/mechanical/optical user interfaces. Unlike telecom/datacom, standards are not yet developed in the optical sensing field. This is a barrier to faster market growth at this stage and may, in the future, cause new players with larger financial backing to hijack the direction of the technological standards to benefit their own versions of products.

Aside from lower cost, IFOS' innovative products provide a number of important benefits to end-users:

- **High Speed.** IFOS has demonstrated speeds of 1MHz on multi-channel-count IFOS sensing platforms and expects that the technology will yield higher speeds still in the future.
- **Easy to use.** Advanced decision-aid software for data management provides support for automated or programmed tasks.
- **Economical and reliable.** Highly innovative photonic designs reduce costs and enhance system scalability and reliability.

The I*Sense® family of commercial products includes several core elements:

Sensors

Optical sensors are the nerve endings of any monitored system and IFOS provides many different types of sensors as described below:

- Off-the-shelf sensors packaged for industrial usage
- Bare fiber Bragg gratings (FBG) that can be adhered to various structures for strain, temperature and other types of measurements. These FBGs can be embedded, enclosed or placed in highly protective casings.
- Customized packaging for user-specific applications. IFOS works closely with the user to define the system requirements and sensing needs.

IFOS has also developed several customized physical and chemical sensors using its proprietary and patented fiber half-coupler (FHC) technology (referred to as FyberSpace) to build some of its custom sensors. The technology also has potential for biochemical sensing.

Optical Interrogators

At the heart of the I*Sense® Interrogation System is a proprietary state-of-the-art, solid state, high-speed photonic spectral processor (PSP) with sub-picometer resolution (0.01 picometers demonstrated). This system supports integration of multiple optical sensing systems for physical, chemical, and biological applications, ranging from a single sensor to a multi-sensor system with auto-calibration and highly customized end-user display. The base version of I*Sense® can presently support up to 16 high-speed sensors per fiber on four fibers (total up to 64), the multisource version supports up to 64 per fiber over four fibers (total up to 256). IFOS expects to expand this to approach to several thousand sensors. Each wavelength (or color) is fully flexible and sensor independent. Different sensing elements can be deployed on a single optical fiber using wavelength (color) multiplexing, enabling users to mix sensors, such as strain, vibration, acceleration, displacement, tilt, pressure, temperature, moisture, gas concentration, and other variables. These sensors can be deployed at one or multiple locations several kilometers apart. A unique feature of the IFOS interrogator design is the possibility to configure them for *simultaneous* examination of up to 16 sensors at rates up to 1 Mega-sample-per-second measurement of high frequency phenomena such as acoustic emissions as indicators of structural damage. Support of higher sampling rates and larger sensor numbers are in development.

User Interface

IFOS provides a proprietary Windows-based user interface to monitor sensor outputs. This interface is highly configurable, offers multiple display options for concurrent monitoring of multiple sensors. Because the interface is proprietary, it is tuned to meet the specific needs of IFOS customers, which include features related to decision-aid intelligent algorithms, data export and storage, varied monitoring modes, and extensive online support. While this component was not originally part of the IFOS core technical competency, it is now an important element in the end-to-end options that IFOS provides.

FyberSpace™

FyberSpace™ optical components (such as couplers, polarizers, modulators) are based on the well-established geometry of standard fibers, but add precision side polishing. This eliminates many device interfacing challenges and permits the adoption of further technical advances in integrated optics. One initial FyberSpace™ product is a simple precision family of fiber half-couplers used as a sensing platform. By depositing overlay specialized materials (including biomaterials and nanomaterials), one can build novel optical (bio) chemical and physical sensors. IFOS is presently negotiating a licensing agreement with a well-established company.

APPLICATIONS AND MARKETS

Traditional electronic sensors can meet many current marketplace needs but often fail to operate in harsh and demanding environments, have insufficient sampling speeds, limited resolution, or inflexible user interfaces, and have high costs. IFOS' innovative product families address these limitations by providing tools that can monitor a flexible number of sensors in harsh and demanding environments cost-effectively. A modular design allows precision signal measurement by a scalable series of multiplexed optical sensors sharing interrogation hardware.

Primary application areas include

- Avionics & Safety (including Civilian Infrastructure and Transportation);
- Energy (including Oil & Gas, Geothermal and Wind);
- Life Sciences (including Smart Surgical Tools and Robotic Surgery).

IFOS anticipates that market dynamics are pointing toward more ubiquitous use of advanced sensors. In preparation for this, IFOS has recently hired a fulltime President with substantial experience and expertise in commercializing high-technology products.

Current applications for IFOS technologies include infrastructure health monitoring and condition monitoring for energy production.

Weigh-in-Motion (WIM) measures the weight and speed of trucks as they move along roads, without requiring them to stop. The system can collect statistical traffic data, support commercial vehicle regulation enforcement, and help analyze funding allocations for the repair and upkeep of roadways, tunnels, and bridges. Fiber-optic sensors provide significant advantages in this sector; they are not affected by electromagnetic interference (including lighting strikes), can withstand harsh environments, and have low power requirements.

IFOS has also demonstrated the use of FBG-based sensors to monitor the structural health of a composite marine pile under static and dynamic loading, large span wind turbines, exhaust temperatures of jet engines, and other high-value assets and structures.

In oil and mining applications, the operating and loading environments are harsh. As recent events in the Gulf of Mexico have shown, it is critically important to monitor the performance and structural health of drilling equipment, down borehole conditions, transport pipelines and production equipment. Fiber optic sensors are used because they are small, highly sensitive, immune to electromagnetic interference, and reduce the risk of explosion.

To take one example, optical sensors can be used in sub-sea environments to monitor dynamic pipeline strain levels, provide real-time detection of impacts by hydrate movements, and to monitor pressure, temperature and strain in deep-water pipelines at water depths greater than 2,000 feet and at temperatures around 400 degrees F.

IFOS is also applying its sensing technologies for use on the emerging smart electric grids, for measuring temperature, strain, vibration, and other critical parameters.

IFOS AND SBIR/STTR

IFOS is a very strong supporter of the SBIR/STTR program and believes the program should be expanded, as it "really fosters innovation," according to Dr. Moslehi.

IFOS has developed good working relations with several NASA, DoD, and DoE centers. Dr. Moslehi noted that many of NASA/DoD/DoE/NIH Technical Points of Contact (TPOCs) are highly technical and understand the difficulties encountered in developing cutting edge technologies, do not require instant commercialization results, and appreciate companies that work on challenges worth addressing.

The first commercial product from IFOS was based on work completed for NASA Langley Research Center (augmented by NSF funding), where the head of the relevant research group was a well-respected scientist in the photonic field and had a deep understanding of the technical/commercialization issues and challenges. Subsequent work completed in collaboration with the Jet Propulsion Laboratory (JPL) and Stanford University led to other commercial developments, and IFOS currently has two awards with NASA Dryden to develop exciting new technologies on structural health monitoring and damage detection of complex avionics structures.

By contrast, some NSF program directors do not entirely appreciate the challenges involved in developing highly innovative technologies for emerging markets. Unlike other agencies, which have an end use in mind for their own needs and are therefore happy with a working system for their use, the NSF struggles to demonstrate a wider impact and consequently expects a broader commercial success. Dr. Moslehi noted that this is often unrealistic, considering that venture backed companies are often funded with tens of millions of dollars to succeed commercially and a $1 million program cannot expect the same results.

As a result, NSF sometimes has unrealistic expectations for its awards and tends to treat all technologies as though they could be commercialized on the same time line and with the same resources. This is clearly nonviable. Currently, IFOS is being discouraged from applying for NSF SBIR/STTR funding because, according to Dr. Moslehi, its commercial sales are not in the many millions of dollars. He observed that as a result, NSF is no longer a good source of funding for certain medium size businesses working on challenging emerging technologies.

Overall, Dr. Moslehi is concerned that NSF's approach needs to be reviewed so that it can have a realistic impact on the competitiveness of small innovative businesses over the long term. He believes that NSF should ideally focus on industry-university partnerships organized around long-term high-risk high-reward projects with the *potential* for substantial commercialization.

IFOS supports the Phase IIB program, in which it has participated in the past. Matching funds have been primarily generated through purchase orders from customers as part of IFOS efforts to garner interest in its products and partner with larger primes to act as clients or OEM manufacturers. IFOS has also recently worked with the DoD through the MDA, where it has received some Phase 2 Transition (P2T) funding for its G*Sense® (fiber optic gyroscope) platform with matching funds through a purchase order from Lockheed Martin. IFOS has also attended the Navy Forum and has worked with the Dawnbreaker programs.

Overall Dr. Moslehi is concerned that agency SBIR programs do not distinguish clearly enough between emerging and non-emerging technologies or between areas where incremental improvements can be marketed and others that are potentially transformative but markets are limited in the short term.

SBIR/STTR RECOMMENDATIONS

TPOCs need to be better trained and empowered to champion the technology they manage. In addition, once TPOCs exhaust their allocation of time for an SBIR project, they have no way to charge time on their timesheets. This effectively means their effort ends, often when it is most needed (this problem is most closely related to DoD/NASA TPOCs). The diffusion of technological innovation is not a straightforward linear process nor is it a short-term sprint. It is more akin to a long running marathon with several sprinting episodes. Managing the early phases only is not sufficient.

TPOCs at most agencies are not trained by the contracting/acquisition officers to be fully aware of Phase II transition programs and support mechanisms beyond Phase II, so there are often disconnects along the road to commercialization. TPOCs are often technically skilled but can have a relatively limited understanding of commercial realities and barriers.

Dr. Moslehi believes that small businesses which have successfully obtained traditional VC funding should not be allowed to participate in the SBIR/STTR programs, as these funds should not be used to protect the interests of financial players. Dr. Moslehi also believes that commercialization prospects of a well-funded company should not be compared to one that is bootstrapped (growing organically) and needs SBIR funding to reduce technology risks.

Likewise, larger companies, including major primes that benefit from large government orders, should be obligated proactively to partner with the SBIR/STTR companies, according to Dr. Moslehi. By providing resources (cash and in-kind) in Phase II for technology transition/transfer without taxing the SBIR system with their large overheads and other internal accounting surcharges when they partner with small companies, larger companies often make desirable partners.

Learning in Motion, Inc.

Based on interview with
Marge Cappo, CEO and co-founder
Santa Cruz, CA
August 19, 2009

Learning in Motion (LIM) is a privately held company located in Santa Cruz, California. LIM now focuses mainly on providing consulting services to the education publishing industry, covering both software and print materials.[17]

LIM was founded in 1993 by Marge Cappo and Mike Fish, who were employees of Sunburst Communications which was relocating to New York, NY. As Vice President of Sunburst Communications (1982-1992), Ms. Cappo's division had developed and marketed over 350 educational software programs.

Initially, LIM had about 10 employees, and was focused on a business model very similar to Sunburst—becoming a profitable software publisher. However, initial revenues—like those of many SBIR firms—were drawn primarily from consulting contracts with other publishers.

LIM was, however, still focused on pursuing its original business plan. Ms. Cappo notes that "at the time, I did not believe we could be successful just as a consulting company. I thought we needed products as well."[18] Development of products would, however, require resources, and at such an early stage in the company's development, additional resources for product development were not available.

As a result, LIM decided to apply for NSF funding. LIM received its first Phase I award from NSF in 1994, and this was followed by a Phase II in 1995 (LIM eventually converted all three of its NSF Phase I awards into Phase II awards).

These first awards were used to develop the company's first major educational software product, AssessMath! This product aims to provide teachers with tools for accessing a database that included more than 1,000 problems and exercises, which could be selected using a range of criteria (see Box E-2 for detailed description).

Subsequent Phase I awards in 1997 and 2000 were used to build a tool for developing mathematical stories and a subsequent project to adapt the story tool for use with deaf children that includes technology for generating sign language animations. The latter allowed LIM to address a market which was too small to attract substantial attention from large publishers—perhaps analogous to the market for orphan drugs in the health sector.

Each of these three Phase I awards were converted to Phase II, and each resulted in the development of software titles that reached the market.

---

[17]Unless explicitly noted otherwise, all information in this case study is derived from an in person interview with Ms. Cappo in Santa Cruz (August 19, 2009) and from the company's web site.

[18]Interview, Santa Cruz, August 19, 2009.

BOX E-2
Assess Math

Designed by Learning in Motion and the Freudenthal Institute, a renowned international center for mathematics and science education based in the Netherlands, AssessMath! allows teachers to customize tests to meet the needs of particular students and curriculum goals. Over 1,000 mathematics tasks in grades K-8 are available.

Teachers can set criteria used to select problems aligned with their teaching and learning goals. Six different criteria can be set: grade level, mathematical content, level of difficulty, item format, use of context, and time.

Problems cover number, algebra, measurement, geometry, statistics and probability, discrete mathematics, and integrated mathematics. Selected problems have also been included from trigonometry and calculus. These problems can be selected at different levels of difficulty, ranging from basic skills and computations to challenging open-ended questions.

More than half of the problems have an essential context—real-world situations where a student must understand and interpret the context to solve the problem. Since a key goal is to develop students who can use their mathematical knowledge in various contexts, context-embedded problems are essential. Once a test is created, AssessMath! can check the balance of content, skills, and levels of thinking to ensure that appropriate variety is included.

AssessMath! can also serve as a shared resource within a school or district. It is a resource that is intended to grow and to make an ongoing contribution to a balanced program of instruction and assessment in mathematics. By aggregating results a more systematic picture of instructional outcomes can be presented.

SOURCE: Adapted from LIM, *<http://www.learn.motion.com>*. Accessed September 10, 2009.

THE INTERNET REVOLUTION OF THE LATE 1990s

Reaching the market is insufficient for commercial success. The widespread advent of the Internet in the late 1990s changed educational software dramatically; the provision of online tools and web-based environments generally meant that publishers had to be prepared to offer larger system able to handle much larger numbers of simultaneous users. Web-based publishing required a new set of authoring tools; more important, instead of simply delivering a CD which simply plays an application within the PC or Mac operating system, web-based applications run remotely on company-owned and -maintained servers. And at the same time, textbook publishers began to branch out to provide web-based tools and resources of their own.

LIM found that selling individual titles to school systems presented multiple challenges. Understandably, buyers preferred to acquire complete solutions that covered multiple years of math via multiple successive courses. And school systems were somewhat reluctant to buy titles that did not explicitly fit with their curricula but simply provided support for it (like AssessMath!). These changes would necessitate a change in LIM's business model and require investment to compete with major text book publishers.

NEXT STAGE

Because there was little evidence that LIM could compete effectively as a software publisher in this new environment, it returned to focusing on its consulting business.

Today, LIM sometimes acts as a publisher for independent projects that cannot attract support from the larger software publishers. For example, LIM published the Voyages Through Time series, which includes information about evolution and had been rejected by other potential publishers.

LIM acquires varying shares of ownership for these projects. In some cases, LIM has acted in ways similar to OEM manufacturers. LIM has helped to develop products based on the latest research, which they monitor both in the US and internationally. This in itself provides a valuable service for clients.

LIM's titles are, according to Cappo, well regarded and continue to sell. They will remain available as long as the technology still supports them. However, they have accounted for a steadily declining share of company revenue and now in aggregate generate less than 2 percent of total revenue.

Today, LIM is an educational product consultant that provides a wide range of services needed to develop educational products. Perhaps surprisingly, for a period of time from 2000 to 2010, a large percentage of LIM revenues has been drawn from print-based products. In 2011-2013, Learning in Motion did a major project with Pearson to develop a K-12 math curriculum delivered on tablets.

IMPACT OF SBIR

SBIR funding from NSF was critical for the company during its early years. Cappo stated that the company likely would not be in existence without this funding, even though SBIR awards did not directly support the company's foundation. NSF funding provided a critical revenue stream that supported staff while the consulting business grew.

Over time, LIM stopped applying for NSF SBIR awards, for two primary reasons. First, the company shifted toward consulting and away from self-generated stand-alone projects. Second, according to Ms. Cappo, NSF SBIR topics themselves increasingly focused on testing and electronic student records, which were not of interest to LIM.

COMMENTS ON SBIR

Ms. Cappo identified a number of important issues related to SBIR awards at NSF and elsewhere.

- ***Funding Gaps.*** LIM experienced a significant gap—up to a year—between Phase I and Phase II. This presented substantial problems, even though the company was able to fund project staff through other work. These gaps and lags need to be addressed, especially given the importance of stable funding for small firms.
- ***Overambitious requirements at the Department of Education.*** The Department of Education effectively requires that Phase I applications include information covering Phase II commercialization. This is often not feasible.
- ***Erratic selection.*** LIM applied again in 2009 to the Department of Education for a highly promising reading project in Eugene, Oregon, based on an award-winning initiative from an elementary school teacher to provide 30-minute readings delivered via student IPods. This application was rejected for reasons that seemed obscure to LIM.
- ***Topics.*** The NSF focus on testing has narrowed the range of potential projects. Broader topics would be very welcome and would support a wider range of innovation.
- ***Annual deadlines.*** Given the speed with which market conditions change, a single annual deadline seems unnecessarily inflexible.

Ms. Cappo views NSF grant administration as highly professional. The project manager at NSF provided pre-application advice and early feedback on applications She views the NSF team as highly supportive and focused on making the project successful.

Membrane Technology and Research Inc.

Based on interview with
Dr. Hans Wijmans, Director of Research[19]
August 19, 2009

BACKGROUND

MTR is a privately held company headquartered in Menlo Park California. The company has sales offices in Houston and Brussels. It has focused primarily on providing membrane technology and products in the oil and gas industry, and has applied its technologies in other industries as well.

The company was founded in 1982 by Richard Baker, previously a co-founder of Bend Research. Mr. Baker believed that a company based only on contract research would not be viable in the long term. He also believed that the standard alternative for research companies, a model based on licensing proprietary technology, was equally fraught with difficulties. Thus from its foundation, MTR has been concerned with the direct commercialization of its own technologies.

Originally all revenues were from contract research, and MTR grew very gradually with revenues on the order of $5 million by 1997. Of this, about $1 million per year came from SBIR. According to Dr. Wijmans, during this time MTR was a technology-driven company that focused on addressing interesting technical problems, and pioneering membrane applications for the petrochemical and natural gas industries.

In the early 1990s, a turning point occurred for MTR when the company was approached by a PVC company to solve their problems with carbon tetrachloride, which was both very dangerous and on the verge of being outlawed by the EPA. The company was, according to Dr. Wijmans, so desperate for a solution that MTR was able to sell a system with limited testing and no warranty for $200,000. This constituted a large step up in pricing for MTR but more importantly provided funding for a prototype system which could be deployed into the market at no risk to the company. Today, according to Dr. Wijmans, about two-thirds of all PVC plants worldwide use MTR membrane technology to address carbon tetrachloride issues.

After the success of the PVC product, MTR moved on to address industrial markets related to polypropylene and polyethylene manufacture, a market two orders of magnitude larger than the PVC market. MTR delivered its first commercial system in this area in 1996, and MTR now sells several systems a year in the

[19] All data for this case study, unless otherwise attributed, was provided through the interview with Dr. Wijmans or from the MTR web site at <*http://www.mtrinc.com*>. Accessed on September 17, 2009.

$6 million range. This finally transformed the company from a contract research organization into a company dominated by the manufacture and sale of commercial products. The impact of this transformation is captured in MTR revenues.

With a strong and growing position in these markets, MTR was able to explore other commercial opportunities. VOCs—volatile organic compounds—represented a new opportunity. Membranes allow the collection of VOCs during various kinds of industrial processing, and can therefore be used to remove potentially harmful or banned contaminants before they reach open airways. At the same time, the membrane approach can also collect valuable compounds that can generate an additional revenue stream for the client.

One example of this approach relates to chlorofluorocarbons (CFCs). As CFCs were phased out in the 1990s, users of heavy duty industrial refrigerators faced difficult issues: these refrigerators represented a major capital investment with substantial remaining product life but required increasingly expensive inputs as CFC prices rose. CFCs generated emissions that were subject to increasingly higher legal penalties. MTR developed new applications that focused on VOC recovery. These applications were sold at about $50,000 each. More importantly, MTR designed and built the whole system and gained critical experience in product design manufacturing while building ties to customers.

Today, MTR generally sells complete systems, built using outside fabricators, but completely engineered by MTR. Two-thirds of sales are outside the US, although MTR sees a growing domestic market, especially for gas separation and a new biofuel business.

Aside from its substantial and growing presence in the petrochemical sector, MTR is increasingly focused on green energy including a carbon sequestration project for which MTR has won a $4 million Department of Energy contract. MTR announced that it would conduct a six-month field test of its membrane process to capture CO_2 from flue gas at the Arizona Public Service's (APS) Cholla coal-fired power plant near Phoenix, Arizona. The system was scheduled for startup in the first quarter of 2010 and was designed to process 250,000 scfd of flue gas, separating about 1 ton CO_2/day. The field test validated the potential for MTR's membrane process to efficiently capture up a substantial portion of the CO_2 from coal-fired power plant flue gas.[20]

SBIR AWARDS

Clearly, SBIR played a critical and ongoing role in funding the development of MTR's core membrane technology, while providing the equivalent of stable contract funding to underwrite a portion of MTR's overall budget during the 15 years before commercial products finally began to dominate.

[20]Tim Merkel, Xiaotong Wei, Jenny He, Bilgen Firat, Karl Amo, Saurabh Pande, Steve White, and Richard Baker, "Membranes for Power Plant CO Capture: Slipstream Test Results and Future Plans.

While MTR has won SBIR awards from seven agencies, and NSF and EPA have provided significant SBIR funding, the bulk of MTR's awards have come from DoE.

According to Dr. Wijmans, SBIR was an integral part in the evolution of the company, and supported the development of most of the company's early core technologies:

- early DoE awards supported development of the CFC technology
- EPA SBIR awards addressed PVC manufacturing
- DoE SBIR awards funded work on polyolefin applications and demonstrations of the application of existing technologies to wider applications

Most awards supported the work well beyond simple feasibility studies.

NSF awards came later, and supported work on hydrogen (for green energy) and perfluoropolymers. In general, NSF awards focused on development of novel or improved materials, which appeared to be critical for success in the NSF SBIR competition according to MTR. This contrasts with DoE, where reviewers were more open to funding process innovations.

TECHNOLOGY

Since its beginnings in 1982, MTR has grown continuously as industry embraced membranes as an effective gas separation technology. After MTR sold its first commercial system to the petrochemical industry in 1992, the portfolio of applications expanded quickly. MTR now provides a full range of gas separation solutions for petrochemical plants, refineries, and gas processing facilities. Systems for these demanding applications need to be effective, economical, reliable, safe, and conforming to industry standards. MTR's systems are based not only on state-of-the-art membrane know-how—they are also custom engineered to fit the application and the industry.

MTR received the Kirkpatrick Chemical Engineering Achievement Award for the successful commercialization of the original VaporSep® membrane technology. The award, which is sponsored by *Chemical Engineering* magazine, honors the best chemical engineering technology commercialized during the preceding two years.

Vapor Separation Technology

Membrane-based vapor separation systems are used in the petrochemical, refining, and natural gas processing industries. Current applications include the following:

- Recovery of olefins from resin degassing vent streams in polyolefin plants
- Recovery of liquefied petroleum gas (LPG) from refinery vent streams
- Fuel gas conditioning (removal of heavy hydrocarbons from fuel gas)
- Recovery of natural gas liquids (NGLs) from natural gas streams

The VaporSep process works by separating hydrocarbons from mixed liquids or gases. A mixture of hydrocarbons in nitrogen, hydrogen, or methane is compressed and cooled, condensing some of the hydrocarbons which are recovered as a liquid. The remaining gas is fed to the VaporSep membrane. The membrane separates the gas into a hydrocarbon-rich permeate stream and a hydrocarbon-depleted residue stream (the purified gas). The permeate is recycled to the compressor; the residue stream is vented or reused.

MTR's competitive advantage is based on the membrane unit, comprising one or more VaporSep modules. Each module contains proprietary membranes which are manufactured as flat sheets and then rolled into spiral-wound modules. The feed gas enters the module and flows between the membrane sheets. Hydrocarbon vapor passes through the membrane and flows inward to a central collection pipe. Lighter gas (e.g. nitrogen or hydrogen) is excluded by the membrane, and exits as the residue.

This membrane-based approach has been applied in other industries, including

- Recovery of flavor compounds from food industry process streams.
- Recovery of ethanol from fermentation and food industry process streams.
- Removal of organic contaminants from wastewater streams.

POLYPROPYLENE PRODUCTION

During the production of polypropylene (PP), a portion of the propylene feedstock is lost. The value of the lost feedstock is substantial, ranging from \$1 million to \$3 million per year for a typical polypropylene plant. Propylene losses occur primarily in resin degassing vents.

For resin degassing applications, the vent stream is compressed and then cooled to condense the propylene. The gas leaving the condenser still contains a significant amount of propylene. This gas is fed to the membrane unit, which separates the stream into a propylene-enriched permeate stream and a purified nitrogen residue stream. The permeate is recycled to the inlet of the compressor and then to the condenser, where the propylene is recovered. The purified nitrogen stream is recycled to the degassing bin.

For C_3 splitter overhead applications, the VaporSep unit is very simple, consisting of membrane modules only, with no moving parts. The stream leaving the column overhead is primarily propylene, mixed with light gases such as nitrogen or hydrogen. The VaporSep unit splits this stream into a propylene-enriched stream and a light-gas-enriched stream. The propylene-enriched stream is returned to the distillation column, where the propylene is recovered, and the light-gas-enriched stream is vented or flared.

VaporSep units are currently used by major polypropylene producers including Formosa Plastics, Ineos, SABIC, Sasol, and Sinopec.

POLYVINYL CHLORIDE (PVC) PRODUCTION

Polyvinyl chloride (PVC) is produced by polymerization of vinyl chloride monomer (VCM). Unreacted VCM is pumped out of the reactor and condensed, and non-condensable gases are vented from the condenser. Depending on the temperature and pressure of the condenser, the vent stream also contains from 50 to 2,000 lb./h of VCM. As VCM emissions are tightly regulated, the vent stream must typically be incinerated and scrubbed before release.

The vent stream from the existing VCM condenser is sent to the VaporSep system. VCM passes through the membrane at a greater rate than inert gases, producing a VCM-enriched permeate and a VCM-depleted residue. The permeate is recycled to the inlet of the existing compressor and the residue is incinerated. The VCM recovered by the VaporSep system is condensed in the existing condenser.

VaporSep systems allow PVC producers to recover 90 percent to 99+ percent of the VCM currently lost in vent streams, providing a significant economic benefit. VaporSep systems are currently used by major PVC producers including Oxyvinyls, Westlake, Solvay, and Aiscondel.

IMPROVING SBIR

Dr. Wijmans believes the NSF SBIR program is too focused on the VC funding model. In his opinion the DoE model with a DoE-funded demonstration phase after Phase II is more realistic.

NSF is, according to Dr. Wijmans, not interested in counting partnering agreements as the matching funds needed for Phase IIB, even though such an agreement is important to commercialization. NSF does not count in-kind contributions, only cash from third parties or sales revenues. In contrast, DoE counts cost share letters from partners. Wijmans says cash in is almost impossible—"investors want to play further downstream."

The NSF approach can force companies into agreements with VC funders, which may not make strategic sense. In particular, these agreements narrow the strategic options open to the company, focusing it on the specific product identified for VC funding. Dr. Wijmans contrasted the rigidity of VC funding with the need to make strategic changes as opportunities grow and shrink. Many companies need to switch commercialization direction during research, but VC agreements can make this difficult. VC funding is best suited to projects that are designed to grow very rapidly and very big.

SBIR awards are very good for small companies and support the foundation of new companies. Barriers to entry are low and reporting is limited and manageable. MTR still uses SBIR to look at new things. The company sees Phase II awards as an important stepping stone toward bigger things, such as new applications of existing technologies or new membranes. Now that MTR has manufacturing and a steady stream of sales in place, it is much easier to add products incrementally.

COMPANY UPDATE 2014

MTR's commercial sales were flat in 2010 through 2012, but have been picking up since 2013. Sales and profits will be at record highs in 2014; MTR now has more than 90 employees. The U.S. shale gas boom has significantly increased MTR sales in the natural gas industry.

MTR's participation in the SBIR program is significantly reduced compared to 5 years ago, particularly in the NSF program. MTR's impression is that the company is "too big" for the SBIR program, even though it employs fewer than the 500 employee limit used to define a small business. According to MTR, proposals submitted jointly with universities are received more favorably, and generate a higher success rate.

Mendel Biotechnology

Based on interview with
Neal Gutterson, CEO
Hayward, CA
August 21, 2009

Mendel is a privately owned biotechnology company located in Hayward CA. The company focuses on biotech for agriculture. From its foundation in 1997 until 2003, the company was funded largely through a research partnership with Monsanto.

Until recently, the company focused on developing its core technology, largely via a contract based research relationship with Monsanto, aiming to develop technologies that can be licensed either within or outside the partnership with Monsanto. In line with this, the company developed a number of partnerships focused on different applications of its core technology.

Since 2005, the company has increasingly focused on a new set of markets and a new business model. By applying its technologies to the biomass needed for the biofuels industry—which is projected to grow dramatically in the US and worldwide in the coming years—Mendel believes it can add substantial value to the sector. It has therefore decided that it will seek to be a direct operator in this sector, developing genetically enhanced feedstocks and pre-processing them to the point of purchase by Mendel customers—the utilities. This ambitious new approach is being developed in partnership with BP (see Box E-3).

HISTORY

With the human population set to reach 9 billion by 2050 and the ever increasing environmental pressure on agriculture, there is an urgent need to develop crops with enhanced productivity and yield stability. In short, global agriculture will need to produce nearly twice the current amounts of food, feed, fiber and fuel with less energy and with an improved carbon footprint.

Mendel was founded in early 1997 based on the idea that controlling gene expression would create new opportunities to improve plant productivity and quality. From 1997-2003, Mendel focused on a class of genes encoding products known as "transcription factors" (TFs) given these proteins act as master regulators of gene networks. During this period, Mendel identified essentially all of the TF genes from a model plant species (Arabidopsis thaliana), and systematically analyzed the function of each of the encoded proteins by producing experimental plants that had increased or decreased amounts of the target protein.

Mendel discovered individual TFs that control complex valuable traits such as freezing tolerance, drought tolerance, intrinsic growth rate, photosynthetic output, plant form, disease resistance, water use efficiency, and nitrogen use effi-

BOX E-3
Vision

During the next 20 years, with an expanding global population, improved diets and rapidly growing energy demands, society will need to develop plants enhanced for food, feed, fiber, and fuel production in order to limit the demand for increased production acreage. With the growth of energy generation from agricultural feedstock, agricultural and energy supply chains serving the needs of a growing carbohydrate economy are expected to become integrated. Agricultural systems have had a major impact on the global environment; Mendel's technologies will contribute substantially to minimizing environmental consequences of agriculture for future generations.

SOURCE: Mendel Biotech *Annual Report,* 2008.

ciency. Many of these discoveries are now patented by Mendel. These discoveries were in testing internally in 2000-2001, and then to Monsanto.

From 2003 through 2009, in collaboration with partners, Mendel showed that TF technologies can generate commercially meaningful improvements in crops, producing corn and soybean varieties with improved yield, drought-tolerant corn varieties, freezing-tolerant eucalyptus trees, and drought-tolerant ornamentals.

In 2004, Mendel started work on the regulation of valuable traits in plants by direct chemical application to the plant. Starting again with Arabidopsis, Mendel identified molecules that improve tolerance to freezing, drought, and cold. That led to a collaboration with Bayer CropScience to identify commercially valuable chemistries that enhance stress tolerance.

BOX E-4
Example of genetic modification: Enhancement of Photosynthetic Output

Mendel has discovered novel gene networks and multiple transcriptional regulators of those networks that control core photosynthetic output in a number of different ways. One acts by directly regulating the chlorophyll content and density of chloroplasts, the organelles within plant cells that capture solar energy. Mandel has filed patent applications on the TFs that control such networks.

SOURCE: Mendel Biotechnology.

In 2005, Mendel made a strategic decision to pursue biofuels. This industry will require dedicated sustainable energy crops with high biomass yields and low inputs. In collaboration with BP, Mendel is beginning to create elite, proprietary varieties of energy grasses. This will be the basis for a new BioEnergy Seeds and Feedstock business that will provide seeds and services directly to farmers, and refineries.

Business Strategy

For the years up through 2003, Mendel appears to have been largely dependent for funding on its relationship with Monsanto. In exchange for funding, MB delivered packages of information to Monsanto, which then decided whether the technology would be commercialized. This close relationship made it relatively simple for Mendel to meet the requirements for NASF SBIR Phase IIB without any change to its standard business operations (see SBIR below). It is worth noting that Monsanto is by far the largest company worldwide involved in the genetic modification of plants for agriculture. According to Dr. Gutterson, it accounts for more than 80 percent of the global market for genetically modified seeds. So Mendel's partnership with Monsanto is potentially of tremendous long-term commercial significance.

That strategy was complemented starting in 2003 with a growing emphasis on the acquisition of additional marketing and development partners beyond Monsanto able to address different markets (see partnerships). This expansion has led to several commercially significant developments in a range of areas, including ornamental plants and the use of eucalyptus as a feedstock for wood products.

More recently, this expanded strategy has again been complemented by an ambitious effort to build a completely different business focused on biofuels. Rather than relying on licensing and royalty payments, Mendel intends to become a physical supplier of biomass to end users.

This new approach represents a major shift in strategy for Mendel. Instead of relying entirely on R&D, developing IP, and licensing for long term revenue, the company is prepared to make the investments necessary to become a feedstock provider itself, with facilities in several locations. Obviously, the success of this strategy will likely impact the overall success of the company itself.

Mendel's successful development and expansion of its partnership strategy is reflected in its employment growth, more than 60 percent in 2007-2008, with over 100 employees in 2010.

MENDEL AND SBIR

Starting in 1999, Mendel has received a series of SBIR awards primarily from NSF and USDA. Over the course of seven Phase I awards and four Phase II awards (the most recent of which being in 2005), Mendel received a total of approximately $2.3 million in SBIR Phase I/Phase II funding. These awards were used primarily to further validate Mendel's TF technology first in soybeans and then in ornamental plants.[21]

The freeze-tolerant TF technology developed with the help of NSF awards has been at the heart of Mendel's commercial strategy to date, according to Dr. Gutterson. The ArborGen eucalyptus project is specifically based on enhanced freeze tolerance.

SBIR also helped Mendel return to genetic approaches to disease resistance, for example addressing the rapidly growing threat of Asian rust in soybeans. SBIR funding and the award itself helped Mendel attract Monsanto's attention to a possible solution to this problem.[22] This reflects Gutterson's view that "SBIR is all about leveraging to build off the core Mendel platform into new areas."

Mendel also received some Phase IIB funding from NSF. In 2006, it received an award of $500,000 based on a match against $2 million in funding from Monsanto for a project entitled "Engineering Broad-Spectrum Disease Resistance in Crop Plants." In 2007, it received another $500,000 for a project on "Developing Crop Plants with Wide-Spectrum Disease Resistance," again based on a match against (unspecified) funding from Monsanto.[23]

It is not clear whether Mendel had to adjust its existing business plan or operations in order to qualify for the Phase IIB funding. The partnership with Monsanto fits the Phase IIB model so closely that it is entirely possible that Mendel could receive Phase IIB funding without any adjustment at all.

This is not necessarily a bad thing: it is not clear whether NSF sees Phase IIB as a tool for enhancing and rewarding commercialization activities, or as a tool for encouraging firms to undertake or expand those activities and attract outside funding, or both. Mendel's entire business model is predicated on attracting funding from companies like Monsanto.

Moreover, Gutterson notes that the Phase IIB did have some significant effects on Mendel in terms of shifting the internal balance of power between the business and technical sides of the company, and as a result, enforced the need for scientists to learn more about the business side.

Looking forward, Mendel will again be seeking SBIR funding, although not from USDA where Mendel believes the process is too burdensome to justify the effort.

PARTNERSHIPS

Mendel has maintained a growing number of partnerships, starting with its founding relationship with Monsanto.

---

[21] Gutterson interview.

[22] Gutterson interview.

[23] Phase IIB Information provided privately by NSF.

Monsanto

Mendel's long-term technology collaboration with Monsanto was initiated in 1997, and provides Monsanto with exclusive licenses to Mendel technology for application in some large-acreage crops and vegetables. Essentially, Monsanto's biotechnology/genomics group funds Mendel to develop and license to them technologies for incorporation into their R&D pipeline. Mendel receives royalty and milestone payments on the developed products—a fairly typical biotechnology industry relationship.

Monsanto's most advanced soybean yield trait product, based on a Mendel-developed technology, has now entered into Phase III advanced commercial development based on excellent field results.[24] According to Dr. Gutterson, some other technologies are also showing promising results in the Monsanto commercial development pipeline.

According to the Mendel Annual Report, "the companies have (also) initiated a systems biology program to develop an integrated framework for predictive control of plant gene expression that is anticipated to streamline future discovery and product development activities."[25]

ArborGen

According to Mendel's Annual Report, ArborGen has deployed Mendel's freezing tolerance technology (which was supported by the first of the NSF Phase II awards) to create eucalyptus varieties that are tolerant to the periodic freezes that occur in the Southeastern United States. Eucalyptus is a fast growing and valuable tree used for a variety of wood products.

ArborGen submitted a regulatory dossier in December of 2008 for approval of a freezing-tolerant Eucalyptus variety. The submission of a regulatory dossier represents a major step toward final commercialization.[26]

Bayer CropScience

In early 2008, Mendel announced a new research partnership with Bayer CropScience, which continues previous joint activities focused on stress responses generated by Bayer agrochemicals like Imidacloprid and Trifloxystrobin. The program aims to discover and develop further chemical products that regulate plant stress tolerance, leveraging Mendel's knowledge of plant transcription factor pathways with the expertise of Bayer CropScience as a leader in agricultural

[24] See Mendel Biotechnology Press Release, "Mendel Biotechnology Yield Trait Reaches Phase III for Monsanto Soybean Products," January 9, 2009.

[25] Annual Report 2008, p. 7.

[26] Annual Report 2008, p. 8.

chemistry. This collaboration follows the discovery of chemicals that can be applied to crops to enhance their tolerance to a range of stresses.

BP

In May 2007, Mendel and BP entered into a strategic long-term collaboration for the development of a BioEnergy Seeds and Feedstock business. This reflects Mendel's new focus on biofuels as a pillar of the company's future.

According to Mendel, BP is developing advanced biofuels and will need the biomass feedstock that Mendel's crops will produce. So BP is funding development of Mendel's new business which will provide energy crops to refinery customers including BP. Mendel's new BioEnergy Seeds business will have BP as a preferred customer, but expects to attract other customers throughout the biofuels and power generation industries. Because the grasses used as feedstock are low density, Mendel envisages a regionally organized delivery structure, with up to 40,000 acres serving a 50 million gallon/year bio-refinery.

Selecta Klimm

In 2006, Mendel formed a joint venture with Select Klimm called Ornamental Biosciences, Inc. for the commercialization of ornamental crops with improved growth and survival characteristics. Research facilities are located in Stuttgart, Germany.[27]

MMR Genetics/Richardson Seeds

In 2008, Mendel partnered with MMR Genetics/Richardson Seeds to develop sorghum varieties for the bioenergy industry, with maximized cellulosic biomass rather than starch or protein. MMR Genetics is a leading sorghum breeding company, associated with Richardson Seeds, one of the largest sorghum seed producers in the United States.

IMPROVING THE SBIR

Gaps

Mendel is concerned about the existence and potential size of financing gaps between Phase I and Phase II for a company of this size. While other funding can usually be found, the gaps do present problems, such as challenging a company's ability to retain key staff.

[27] See Ornamental Biosciences web site. Accessed September 22, 2009.

Training

Gutterson said that the SBIR symposia he attended were of little use, although they were likely to be helpful to smaller and younger companies. Mendel did use Foresight's support and found it useful.

Application Deadlines

Gutterson strongly endorsed the need for multiple deadlines; a single annual deadline is no longer sufficient given the rapidly accelerating speed of technical change.

Topics

Mendel would like to see more broad topics, where firms can decide which technologies fit the agency's requirements.

COMPANY UPDATE

In December 2014 the research business of Mendel Biosciences was purchased for an undisclosed amount by Koch Agronomics Services LLC.

Techno-Sciences Inc.

Based on interview with
Professor Gilmer Blankenship,
CEO June 1995-June 2009, Chairman June 1995-May 2014

BACKGROUND

Techno-Sciences, Inc. (TSi) is a high technology company headquartered in Beltsville, MD. Lee Davidson, a Professor of Electrical Engineering at the University of Maryland who specialized in information theory, founded the Company in California in 1975. The Company was created to provide systems engineering services to the U.S. government and prime contractors in communications, signal processing, and search and rescue. In 1988 Techno-Sciences merged with Systems Engineering, Inc., a company founded by Gil Blankenship and Harry Kwatny.

Until the late-1980s, Techno-Sciences was largely a contract research house that used government contracts, including SBIR awards, as way of funding investigator-initiated research, and as a basis for research and development in the U.S. SARSAT program. In the mid-1990s, the company underwent a major shift of emphasis. Professor Davidson retired, and Professor Blankenship[28] became CEO and Chairman.

In 1988, the company had developed its first significant product—search and rescue command centers satellite ground stations for international search and rescue programs. The new product line formed the basis for a new kind of company. Since then, TSi changed to a company with a global market, selling ground stations and mission control centers in more than twenty countries, most of which have retained TSi for ongoing management and maintenance, often for decades.

In the early 2000s TSi rolled out a second major product line, the Trident Integrated Maritime Surveillance System (IMSS). This was sufficiently successful to create a new operating division for the company, called Trident Maritime. The Trident IMSS is now deployed on more than 3500km of coastline in Southeast Asia and North Africa—one of the largest such deployments in the world.

As a result of these successful products, TSI transitioned from a contract research house to a company primarily concerned with the development, deployment, and support of new products.

In May 2014 Techno-Sciences was acquired by the Orolia Group.

[28]Dr. Blankenship is also professor and associate chairman of the Department of Electrical and Computer Engineering at the *University of Maryland*, College Park.

COMPANY STRUCTURE

Prior to its acquisition, TSi had three divisions:

- SARSAT, which provides ground stations for search and rescue at sea and over land. TSi's SARSAT products are now mature systems, backed by an experienced staff that has a well-developed process for scoping projects, deploying systems, and following up with effective maintenance and support. In short, the Division has a smoothly operating ISO 9000 certified model of what it takes to deploy and support these systems on an international basis. Working with the US NASA and NOAA, the SARSAT Division developed the next generation SARSAT ground systems based on MEOSAR satellite technology. TSi has sold these important new systems in the US, Australia and New Zealand, and in Algeria. Many additional sales are expected, as the COSPAS-SARSAT community changes to this next generation technology.
- Trident, which sells coastal and ship-based surveillance and security systems, is active in Southeast Asia and North Africa. It has installed about 35 coastal stations, several command centers, and multiple shipboard systems. The coastal station network in Indonesia and Malaysia covers more than 3000km of coastline along the Strait of Malacca and around the Celebes Sea. Trident has also installed surveillance and security systems on oil platforms in the Middle East. The Trident coastal stations include dual band radars, AIS, long range day and night vision cameras, and command and control systems and communications systems. Trident Maritime Operations Centers feature remote access and control functions and extensive cyber security systems. Since most of the stations are installed in extremely remote regions, the Trident Division also manufactures and installs grid free power systems using solar, wind, and generator units.
- The Advanced Technology Division, which undertakes both contract research and supports TSi's products and services. The Division has worked in software, sensors, control systems, and active materials, including magneto-rheological fluids for semi-active dampers. Supported in large part by the SBIR program, the Division has investigated a wide range of areas, some leading to new products for TSi (elements of the coastal stations), and two spinoff companies. The Division has strong ties to universities and has funded several million dollars of university based research and development. Innovital Systems, Inc. acquired the Advanced Technology Division in 2013.

SPIN-OFFS

TSI has spun off three companies: TRX systems, which focuses on a specific application of TSI tracking technologies: the ability to track personnel in GPS denied areas; Innovital Systems, which designs novel medical devices, including an implantable respirator for persons with impaired diaphragm function; and E14 Technologies, Pvt. Ltd. a Mumbai based company that produces custom electronics for a wide range of applications.

TRX's personal location and tracking products are based on years of research following the disaster of 9/11 in which hundreds of firefighters were among those lost in the collapse of the World Trade Center buildings. From the outset TRX's research focused on meeting stringent operational requirements for first responders. The system had to be low cost; highly portable (i.e. laptop based "command center"); built largely from off the shelf components; and able to work in 3-D without building maps.

TRX Systems met these requirements. Its products are deployed in several countries with firefighters and the military. TRX is now working on location and mapping services for consumer applications using handheld technology.

Innovital Systems has leveraged TSi's defense based technologies to design novel medical devices, including an implantable respirator for people with diminished diaphragm function. The Innovital DADS system employs pneumatic muscle technologies to move the diaphragm to support breathing. As a small business, Innovital has made use of the SBIR program to fund its basic research.

TSi Products and Markets

Satellite-based Search and Rescue (SARSAT)

A wide array of information is available to search and rescue (SAR) personnel. Integrating and managing the data from Mission Control Centers (MCCs), for SAR crews on land and in the air, and other sources is crucial to saving lives. The faster SAR resources are mobilized, and the more efficient the response, the greater the potential for saving lives. TSI's SARSAT system automates the coordination of SAR information and resources.

The COSPAS-SARSAT system generates distress alert and location data for SAR operations. Emergency transmitters (distress beacons) are detected by polar orbiting, geosynchronous, and medium earth orbiting satellites, and these signals are relayed to ground facilities, where they are processed for location and identification and ultimately distributed to Rescue Coordination Centers (RCCs), which perform the actual search and rescue missions.

SAR personnel require accurate, concise, information that can be accessed quickly and easily. SAR missions may involve high-risk rescuers and costly resources. So accurate, reliable, and timely data is critical. The TSi SARSAT system links information from the international search and rescue system (COSPAS-SARSAT) via MCCs that have database, communications, and 3-dimensional graphical information systems (GIS). Data drawn from comprehensive digital maps of the world help rescuers understand the search requirements in specific locality (roads, rivers, lakes, population centers, airports, geographic elevations, ocean currents, etc.).

TSI's RCC maintains an extensive, automated database that manages all received alert information. New alert information generates alarms, and the map display highlights recently updated locations. Users can easily access data by time (most recent) or for a specific incident. Messages are tracked and archived automatically.

The TSI MCC is a command and communications system based on a client server structure, which gathers data from satellite ground stations (Local User Terminals), aggregates and manages the data through its server and proprietary software, and delivers the data for display and analysis in a graphical interface and 3D GIS. By using a standard client-server architecture based on standard Microsoft/Intel technologies, costs are reduced and reliability enhanced. Proprietary software provides the competitive edge needed by TSI.

International sales have always been important to TSi, since search and rescue (S&R) systems are sold on a national (or sometimes regional) level. The company's record as a highly trusted supplier of SARSAT systems has allowed it to penetrate other markets including those for maritime safety and security and the personal location technology developed by TRX Systems.

TSI has worked to limit the cost of initial installation with the objective of developing long-term maintenance and upgrade contracts and customer retention. This approach has been successful, with almost all SARSAT and Trident customers purchasing long-term contracts from TSI. Some have been customers for more than twenty years.

Trident

The Trident Division provides TSI's Integrated Maritime Surveillance System. It is designed for governments and other authorities that need to manage the complex flow of traffic and information around crowded, vital coastal regions. The system "deploys a tightly integrated network of ship and shore based sensors, communications devices, and computing resources that collect, transmit, analyze, and display a broad array of disparate data including automatic information system (AIS), radar, surveillance cameras, global positioning system (GPS), equipment health monitors, and radio transmissions of maritime traffic in a wide operating area. Redundant sensors and multiple communications paths make the system robust and functional even in the case of a major component failure."[29]

The system can be sold as an integrated package or in component elements. In 2004, the Indonesian Navy bought the first TSI coastal radar system. This was the result of $7.5 million in R&D investments, primarily from the US government

[29]TSI: the Trident Maritime Integrated Marine Surveillance System, <*http://www.technosci.com/trident/imss.php*>. Accessed October 30, 2009.

and to a considerable degree from the NSF and DoD SBIR programs. Specifically, the core technologies for the Trident system were derived from a single NSF SBIR (Phase 1 and 2) award.

The NSF awards allowed TSI to develop the technology that would go into a ship-based system. A subsequent SBIR award from US Special Operations Command (SOCOM) supported the adaptation of the system for use by Navy Seal operations, to track the precise location and status of Seal boats.

The sole source advantage conferred by these SBIR awards had a significant effect on the subsequent decision by US Space and Naval Warfare Systems Command (SPAWAR) to deploy the technology in the United States. Overall TSi received more than $70 million in contracts to install coastal systems as SBIR Phase 3 awards, and has received over $100 million in contracts in this business area.

Other Advanced Technologies

In the Advanced Technology Division, TSi worked on a wide array of technology areas including software engineering, operations scheduling (for maintenance operations), sensors and actuators, wireless networks, and many others. One particularly interesting application area involved the use of magneto-rheological (MR) fluids for (semi-) active dampers for vehicles and occupant safety. Using MR dampers for soldier seating, TSi and its partners at the University of Maryland were able to demonstrate dramatic improvements in occupant safety when the vehicle was subjected to a dramatic shock such as an IED explosion. Both SBIR and BAA funding supported this research.

In parallel, TSi used SBIR funding to develop solutions using flexible hoses and air to provide air driven mechanical operation of flaps on aircraft wings. The air driven hoses ("pneumatic muscles") can deliver 300lbs or more of force, while avoiding the weight penalties of hydraulic systems. SBIR projects, joint with the University of Maryland, were used to support research on pneumatic muscle applications. One project funded by the US Army, as part of the development of a robot for battlefield rescue of wounded soldiers, led to the development of a powerful robotic arm. The pneumatic muscle powered arm could easily pick up a 300lb person (including their equipment) using 90psi of air pressure.

In other applications Bell helicopter has tested pneumatic muscle controlled wing flaps in the University of Maryland wind tunnel. If adopted, this technology would revolutionize helicopter design. However, it is has other potentially important applications as well. Wind turbine efficiency could be substantially improved through the adoption of automated flaps; the weigh and cost of hydraulic systems have made this impractical thus far.

SBIR and TSI

Prof. Blankenship stated that SBIR awards had played a pivotal role in several different ways at different times in the Company's life cycle. Initially, SBIR awards had provided funding for investigator-initiated research and an important funding stream that allowed for the growth of the company and its personnel during its early years.

As the Company transitioned toward a product-driven model, SBIR funded the research that led to both of the Company's core product lines—SARSAT search and rescue, and Trident ship based monitoring. It also supported the creation of two of TSi's three spinoff companies: TRX Systems and Innovital Systems.

The Advanced Technologies Group is now part of Innovital Systems, where several SBIR proposals are submitted each year. SBIR projects are now helping to fund Innovital's push into new technologies and new markets for next generation medical devices.

SBIR and Advanced Staff Training

According to Prof. Blankenship, SBIR awards played a critical role on the human resources side of TSi. He observed that SBIR projects provided an ideal training ground for certain classes of project managers. TSi's research groups typically hired PhD researchers soon after graduation—at which point they are technically trained but have little understanding of how to manage projects, support clients, or work to fixed schedules.

SBIR projects at TSi were treated as stand-alone projects, and were often handed off to staff not yet ready for major commercial projects. In the course of managing one or two SBIR awards, Dr. Blankenship strongly believes these staff acquire critical management skills, which can then be applied to commercial projects and eventually to the management of entire product lines.

For example, TRX Systems is a spin-off from TSi. Its CEO—Dr. Carol Teolis—was hired by TSI as a new PhD from the University of Maryland. She was assigned to several SBIR projects before entering senior TSI management as Vice-President of Engineering. Her experience at TSi—which included complete management responsibility for a research project for the US Mint, and other key customers—allowed her to develop skills in customer development and support. Her skills have translated into several million dollars of research contracts that have supported the development of TRX Systems. Two other TSi employees have followed a similar path and now lead their own companies (Innovital Systems, and E14 Technologies).

SBIR and Skills Acquisition

Prof. Blankenship also sees the SBIR program as means of acquiring technical skills and know-how that, while not necessarily directly commercialized, may have significant uses downstream on other projects.

For example, TSi won an SBIR award to build high performance gun turrets. As part of the project, TSi built a prototype that required a high-performance gimbal. Commercially available gimbals were not suitable, so TSi learned to build its own high performance gimbal. While the DoD ultimately did not eventually pick up the gun turret technology for acquisition, the gimbal design knowledge was later applied to coastal surveillance systems, supporting the Trident long-range cameras. Similarly, TSI now builds high performance cameras, which are also sold as part of its integrated systems, and grid free power systems for installations in remote areas lacking in reliable power.

Phase IIB

TSI's spinoff company TRX Systems won one of the first Phase IIB awards from NSF. This $500,000 award matched $1 million in investments by strategic partners and sales of the company's products to key customers. This project helped to create what is now TRX Systems main line of business.

SBIR Improvements

Prof. Blankenship indicated that the current size of awards is acceptable, although he is confident that TSi would not have suffered if the size of SBIR awards were to be increased and the number of awards reduced. He noted that the gap between Phase I and Phase II awards had been a problem for many smaller companies; however the introduction of optional tasks to bridge the gap has remedied this.

Prof. Blankenship was somewhat concerned about what he called Phase I SBIR mills, which win numerous Phase I awards but in general fail to convert them into Phase II awards or to commercialize the research. TSi focused heavily on converting Phase I awards, and according to Prof. Blankenship, the Company typically matched a Phase I award with an additional 50 percent internal company money to ensure that the result is good and that TSI has a strong case for a Phase II award. TSi's commercialization record for SBIR projects achieved and sustained the maximum rating.

Prof. Blankenship also observed that larger small businesses—those with more than 100 employees for example—had a smaller need for SBIR awards, which should be focused primarily on very small firms (those with less than 10 employees), and then on smaller and mid-size small firms. The Government is often the only investor willing to take a chance on a company just starting out. Indeed, as TSi grew, SBIR contracts supplied about 5 percent of revenue.

TECHNO-SCIENCES TRANSITIONS

In May 2014 the Orolia Group, a rapidly growing French group, acquired the SARSAT and Trident Divisions of Techno-Sciences. This acquisition followed a period of sustained rapid growth for the Company. Over the period beginning in 2005, the Company grew rapidly both in revenue and number of employees. In June 2009 Jean-Luc Abaziou joined the Company as CEO, with the mission of managing growth and increasing the value of the company. (Prof. Blankenship continued as Chairman of the Board and Principal Scientist.) Mr. Abaziou had led Torrent Networks prior to its acquisition by Sony-Erickson. He later worked at Highland Venture Capital. Under his leadership, TSi was among the Deloitte Fast 500 Technology companies for 3 years in a row. In 2010 the Company was named as the High Tech Company of the Year in Maryland. Several companies expressed interest in acquiring the TSi. The Company entered in to negotiations with the Orolia Group in 2013, and the deal closed in May 2014. Since the acquisition, the SARSAT Division was merged with the McMurdo subsidiary of Orolia. McMurdo is one of the world's leading manufacturers of emergency beacons for the COSPAS-SARSAT program. The merged company is "vertically integrated," offering beacons, ground stations, and rescue planning systems to a global market.

Prof. Blankenship retired from Techno-Sciences in June 2014. He has since started two new companies, one working in sleep health, and the other in medical devices. Both have received SBIR funding.

TOUCH GRAPHICS

Based on interview with
Steven Landau, founder and CEO
October 11, 2010

BACKGROUND

Touch Graphics ("TG") is a privately held company located in New York, New York. It focuses on providing tools for building tactile images so that educational materials can be used by visually impaired students. TG is a world leader in this field, with commissions from a wide range of educational and government organizations.

Mr. Landau started in a completely different profession. He was an architect, and was recruited by the university academic (at City University of New York) initially as a consultant to help develop tools. As a consultant, Mr. Landau created a CAD-CAM system for making tactile maps for the NY subway system. Once that project was completed, the idea was to move the technology into a more universal platform, based on audio-tactile graphics where raised line pictures responded to touch with an audio response.

HISTORY

TG's technology derives from a patent held by Mr. Landau in partnership with his co-inventor, by an academic researcher who is an advisor and collaborator with TG, but not a stockholder. She has no interest in commercialization. In fact, the university paid patent legal fees in exchange for 5 percent royalty payments until the $30,000 debt is eventually retired, after which the university will receive a 1 percent royalty thereafter.[30]

TG was founded in 1997, and started selling its flagship product, the Tactile Touch Tablet (TTT) in 2002. the TTT required development of both hardware and software elements. For example, while touch screens are now a stock component for many products, a standards screen is transparent. To persuade a manufacturer to customize the product, TG had to make a large purchase (much of which is still on hand).

Over time, Mr. Landau indicated that the market for TTTs in U.S. schools—and especially in Schools for the Blind—has largely been saturated in the United

[30]Today, TG's technology is protected by two patents, issued in 2006 to Mr. Landau and his academic partner, Professor K. Gourney. See S. Landau, "System for Guiding Visually Impaired Pedestrian Using Audio Cues," U.S. Patent No. 7,039,522, Issued May 2, 2006; and K. Gourgey & S. Landau, "Tactile Graphic-Based Interactive Overlay Assembly And Computer System for The Visually Impaired," U.S. Patent No. 7,039,522, Issued September 12, 2006.

States. TG has in some cases sold 20-30 units to a single school (such as the Perkins School).

TG is now developing a completely different version of the tablet which uses the touchscreen in a new housing. However, the TTT currently retails for $699, which is far too much for a developing country. The new version, designed to reduce cost, can be sold for about $300, and is much lighter weight as well, which will reduce shipping costs.

TTT has helped found a sister company in Spain (TG Europe). It has no ownership stake in TG Europe, seeing it as a distribution vehicle. However, the difficult economy in Europe has limited sales to date. More generally, TG finds that partnering is not so easy because it operates in a very small niche market. Standard distribution agreements are not appropriate, as distributors simply want to move product, but the TTT requires much more contact and ongoing relationships with customers.

The TTT and related products have found a place in classrooms across the US and Canada. These tools remain an effective way to address legal requirements to serve disadvantaged students, without requiring a large local staff.

MARKETS/PRODUCTS

Products

In 2003, Touch Graphics introduced the Talking Tactile Tablet (TTT), now the company's flagship product, which adds audio annotation that is accessed by the user pressing his or her finger on any part of a picture. Combining tactile images with user-triggered speech improves comprehension and independence.[31]

The TTT is currently in wide use around the world, and TG now provides additional capabilities related to the TTT, such as the TTT Authoring Tool, which allows teachers and other users to develop their own talking tactile materials, and are now also being used to publish illustrated digital talking books.

In addition, the TTT has been used to deploy standardized tests to visually impaired students, for example the MCAS 8th grade math assessment.

For more sophisticated users who need to probe materials more deeply—such as high school and college students—TG has developed the Talking Tactile Pen. The Pen has allowed development of a scientific calculator as part of the STEM binder developed by TG. According to TG, "*A camera near the pen's tip "sees" clusters of dots and the on-board computer performs rapid calculations that are transparent to the user. Once the computer knows its location, it speaks the name of the element that was touched.*" TG is now working on a library of fundamental STEM illustrations for use with the TTP.

[31]See S. Landau, Tactile, "Audio-tactile and Other Multi-Sensory Curricular and Assessment Materials," Pearson Accessibility and Innovation Conference, Upper Saddle River, NJ, <*https://docs.google.com/present/view?id=djngnsb_870dcqmswfh*>. Accessed October 25, 2010.

TG's technology can also be applied outside the classroom, in the form of museum guides and tools for other environments.

TG has developed an audio-tactile version of the "Getting in Touch with Ancient Egypt" self-guided tour at the Metropolitan Museum of Art in New York city.[32] The technology has been applied to development of talking maps (e.g., for the Carroll Center for the Blind in Newton MA),[33] as well as portable tactile maps (e.g., Lincoln Center for the Performing Arts, NY),[34] and Talking Kiosks (for New York City).[35]

Following the 2008 settlement of a Justice Department suit under the Americans with Disabilities Act (ADA) against the International Spy Museum in Washington DC, which required the deployment of tactile map options,[36] advocates for more tactile map deployment hope that these tools will be deployed more widely in museum settings.

Strategically, TG seeks to develop a pipeline of projects at different stages of development. Some are on the market, others are just entering the market, and others still are in the very earliest stages of concept development. So TG has 5-6 separate product lines at different stages of development.

Some of these products can be understood as extensions of the TTT—for example, TG is well along in deploying Braille courseware which runs on the TTT. This product is the result of partnering with other SBIR companies—in this case, Exceptional Teaching Inc., which developed the course independently, and then teamed with TG to bring the course out as an accessory for the TTT. This successful partnership is based on a profit share approach.

The Braille courseware project and others reflect the company's view that it would be unwise to place too large a bet on any single product or product line, and also that with the accelerating pace of technological change, it is not feasible to bank on any single product indefinitely.

The Talking Tactile Pen (TTP) is now on the verge of market deployment as the next generation of audio tactile products. The TTP is more suitable for high

---

[32]This project is described and explained in detail in S. Landau, "Exhibit design relating to low vision and blindness: tactile mapping for cultural and entertainment venues," 2010 LEAD Conference, San Diego, CA. <*http://www.touchgraphics.com/publications/Tactilemapping-Landau.pdf?id=djngnsb_827cmfkq3kd*>. Accessed October 25, 2010.

[33]S. Landau, *An Interactive Talking Campus Model at Carroll Center for the Blind: Final Report*, 2009, <*http://www.touchgraphics.com/downloads/carroll center talking campus model final report low.pdf*>. Accessed October 25, 2010.

[34]S. Landau, "Multi-sensory way-finding and orientation tools for cultural and entertainment venues," 2010 LEAD Conference Workshop on Exhibit Design for Visitors with Print Disabilities, San Diego, CA USA, <*https://docs.google.com/present/view?id=djngnsb_827cmfkq3kd*>. Accessed October 25, 2010.

[35]S. Landau, "*New York City's Growing Network of Talking Kiosks*," Access and the City Conference, Dublin, Ireland 2008, <*http://docs.google.com/Present?docid=djngnsb_539gxcj74cp*>. Accessed October 25, 2010.

[36]Settlement agreement between the United States of America and the International Spy Museum under Title III of the Americans with Disabilities Act, DJ 202-16-130.

school and college students, and hence allows TG to expand its markets into other groups that do not currently use the TTT.

TG has also sought other vectors for deploying its technologies. Wayfinding information kiosks utilize information maps and 3-D models and other displays in public places. Users can interact with them in a multi-sensory way. Chicago has deployed two new Wayfinding kiosks at the Lighthouse for the Blind, which serves hundreds of visually impaired users every day. Building directory. TG has deployed kiosks for many institutional clients, for travel systems and rehabilitation facilities. The company has developed a methodology for doing it cheaply and well, so this is a growth area for the company. It is in discussion with the Veterans' Administration to deploy at facilities for wounded returning veterans. Museums are another substantial market.

More recently, TG deployed another new product—a cane based around a Wii controller, which can be used to train blind people how to walk safely using a cane. This product could be useful for preparing people who are newly blind to use their canes in a safe manner, doing the training in a safe indoors/ lab setting.

STRATEGY

TG is in the midst of a long term shift out of funded research, and is focused on taking concepts into more mainstream setting. R. Landau believes that there may be wider opportunities for presenting multi-sensory information—for example, auto controls use a lot of tactile marking.

In part, this shift is driven by the view that there is an increasingly poor fit between the timeline for a completed SBIR project and the faster moving markets in which TG operates. Mr. Landau points out that from Phase I application to Phase II completion is on the order of three to four years, which is much too long for commercial applications in his market sector.

Collaboration remains apart of TG's strategy, as long as there is a strategic fit. The partnership with Exceptional Teaching (ET) in California was in fact engineered by the NSF program director, Ms. Sara Nerlove, whose hands-on approach generated some additional work, but also led her to make a connection that neither company was initially very interested in making.

In this case, ET had received a grant from NSF to develop speech assisted learning, a way to present Braille in a more multisensory way. ET was also working with Freedom Scientific, a larger company that made much more expensive assistive devices. The result was a hardware device to use with Braille courseware which cost $5,000 each. The product did not sell at this price point, so was pulled from the market, leaving ET no deployment vector through which to sell their software.

TG in contrast already had the hardware device (the TTT) which cost about 10 percent of the Freedom Scientific device, but needed software courseware to run. This partnership is referenced further on the NSF web site.[37]

ROLE OF SBIR AT TOUCH GRAPHICS

SBIR provided initial funding for company formation. The niche market for these assistive devices is not suitable for venture financing.

There was a learning curve for TG with regard to SBIR. Their first application failed—as did their first phase II application. But after the first Phase II, TG had a remarkable run of successful applications—according to Mr. Landau, their next ten applications were successful. The five or six Phase II awards made it possible to develop most of the TG products now for sale. These came directly from SBIR, but not always by the most direct route.

According to Mr. Landau, SBIR was very important for TG: The company would never have developed any commercial products without the SBIR funding. There is no money in the assistive technologies field for new technologies.

However, the transition out of SBIR funding is now under way. In part this is driven by the mismatch with SBIR timelines (see above). The fixed funding amounts are also a problem for TG.

Mr. Landau observed that TG is competing fiercely with other companies, and this requires that TG be very nimble in bringing products to market quickly when needed. While there was initially only one competitor, now there are several. This makes it all the more important that TG retain its technological edge over the competition. SBIR would slow development to the point that the company would likely be doomed.

More importantly , the nature of technology development for TG has shifted. Previously, the company was developing core technology that required a considerable amount of technological innovation. Now it is focused on developing applications for its flagship technology which is already in use worldwide, and which requires less fundamental research.

RECOMMENDATIONS FOR IMPROVING THE SBIR PROGRAM

1. **Higher standards of due diligence for proposals.** Different standards of evidence are needed in the commercial world. Reviewers coming from academia may be unaware that in the business world it is normal to emulate your competitors or to develop a similar and competing model. This would be unethical in academia. Mr. Landau notes that another company used SBIR funds from a different agency to compete with the TTT—in his

[37] See NSF Discoveries, "Electronic Braille Tutor Teaches independence," *<http://www.nsf.gov/discoveries/disc_summ.jsp?cntn_id=105832&org=IIP>*. Accessed November 10, 2010.

view, a waste of government money and counterproductive. The competing company eventually left the market. It is important that applicants be required to be honest about prior art.

2. **More flexible funding amounts and timelines.** Increased flexibility in these areas would permit TG to return to NSF for further SBIR funding.
3. **Better focus on proposed budgets.** These are often an afterthought, and reviewers in his experience rarely pay much attention to budgets (at NSF and other agencies where Mr. Landau has reviewed proposals).
4. **More open solicitations.** NSF is clear about their funding priorities. Some other agencies are less targeted (for example DARPA and NIST). This would open the door for a wider range of projects.

PRIZES AND RECOGNITION

- Louis Braille Prize, National Braille Press, 2007.
- IDEA Competition, Computer Equipment category. Gold Medal, 2006.
- Tech Awards Laureate (Microsoft Education Award), Tech Museum of Innovation, 2004.
- US National Inventors' Hall of Fame, Invent Now America Finalist, 2004.

COMPANY UPDATE (NOVEMBER 2014)

Since this case study was completed, the company has entered into an important new relationship with the premier publisher of educational material for the blind, American Printing House for the Blind (APH). Touch Graphics developed the US Map for Talking Tactile Pen for APH, the first of a series of new products based on the TTP platform, a technology pioneered through multiple SBIR Phase 1 and 2 grants.

TRX Systems

Based on interview with
Carol Politi, CEO
February 2, 2013
Greenbelt, MD

TRX Systems is a privately held company headquartered in Greenbelt, Maryland. Founded by Carole Teolis, Gilmer Blankenship, and Ben Fun, TRX established a focus on indoor location in part because of its success in winning an SBIR award from NSF in 2007, which provided critical seed capital.

The company focuses on developing new tools for geo-locating personnel in locations where existing technologies (such as GPS) do not work or work poorly—for example, indoors or other areas where there is considerable signal interference.

BUSINESS MODEL

TRX is the developer of NEON®, an indoor location system that delivers precise, infrastructure-free tracking of personnel inside buildings where GPS is not available and in outdoor urban centers where GPS is unreliable. NEON greatly improves situational awareness and command effectiveness through the use of advanced sensor fusion, time of flight ranging, and mapping algorithms that deliver precise, real-time location of personnel in GPS-denied locations.

The TRX business model focuses on selling NEON, including a TRX developed accountability system, to federal and public safety markets as well as in licensing the NEON technology on an OEM basis for integration into third-party products.

CORE TECHNOLOGIES

GPS, Wi-Fi, and ultra-wideband technologies are all used for geo-locating, but they have significant shortcomings. They work less effectively in certain environments, particularly indoors where GPS and compasses are unreliable, and in circumstances where building maps are not available.

TRX NEON is a software suite that integrates data from numerous and disparate sources to create accurate 2D and 3D maps and to track personnel across them. These patent-pending sensor fusion and mapping algorithms integrate data from a broad range of sensors including compasses, GPS, ranging, inertial, light, and pressure sensors to deliver accurate tracking of personnel paths.

NEON determines when a degraded sensor (e.g., compass or GPS) is providing accurate estimates and when it is not. Poorly functioning or degraded sensors are isolated and de-emphasized or eliminated in the navigation solution. As a result, NEON works well even when a sensor's data is more than 50 percent erroneous.

NEON uses received sensor information to build site maps, building features, and other landmarks dynamically as people move about an area or building. Information from multiple people is merged to deliver "team-wide" location estimates. Essentially, by managing the data flow from multiple sensors NEON can determine the likelihood that any one data stream is erroneous, and act accordingly.

Ranging information, if available, can also be used to constrain location results. Examples include people operating within 50 meters of each other or working within range of a fixed RF node.

Accuracy is further increased where there are known features or floor plan information, because NEON can also match location estimates and inferred maps to known features and floor plans. User corrections can also be incorporated into NEON's data stream.

PRODUCTS

NEON Location Services

The NEON Location Services are the core product for TRX. NEON uses an open architecture that is easily implemented with sensors from partner-provided hardware systems. The NEON Engine software includes application programming interfaces APIs for integrating input or constraints from partner systems and for providing indoor location data to third-party applications.

System Components

In addition to delivering location information in an API, TRX delivers an application into the public safety market (called NEON Tracker Command Software) that supports rapid and easy 3D building mapping, clear 2D and 3D views of personnel operating in and around buildings, and a record/history of personnel activities for after action review. System configuration can be performed in advance or immediately prior to an event or training session. Personnel equipped with NEON tracking units, or devices running NEON software, are automatically detected and monitored. NEON's Tracker Software allows commanders to visualize the location of personnel outfitted with NEON Tracking Units in both 2D and 3D as they operate indoors.

NEON Tracking Units are waist-worn devices (about the size of a deck of cards). They include a number of sensors: temperature compensated triaxial accelerometers and gyroscopes, triaxial magnetic sensors, barometric pressure, light sensor, Time of Flight (TOF) RF ranging, and GPS. These Tracking Units interface to radios or smartphones to transmit location information back to the NEON Tracker Command software. TRX is also now implementing its location algorithms on Android smartphones which now have many of the same sensors that exist in the NEON Tracking Units.

NEON Multi-Sensor Anchor Nodes can be used in fixed site applications to enhance precision. Anchor Nodes do not require networking and can be added during operations. It's also possible to use both Bluetooth and Near-Field Communications to support location initialization using cellular devices.

Markets

There are numerous potential markets that require an efficient means to geolocate people in circumstances where standard GPS does not work effectively. TRX has focused on three core areas: first responders, defense, and security.

First Responders

First responders often work inside buildings, in dangerous circumstances, where GPS is unavailable and where the location of personnel is a critical need for commanding officers.

NEON's key feature is that it can work in areas that are currently unmapped and that are not equipped with networked beacons: it does not require a building plan or pre-installed infrastructure to constraining the routes through which people move. This differentiates NEON sharply from many competing approaches, which rely in part on existing building maps and often require installation of beacons to deliver location indoors.

Defense

Dismounted war fighters increasingly rely on location for navigation and to deliver the situational awareness required for optimal command effectiveness. In some cases, GPS may be either unavailable or insufficiently precise. NEON is currently being adapted for use specifically for training, where trainers benefit from immediate review of exact personnel movements in near real time, as well as information on personnel orientation and proximity to other personnel or entities to enhance training realism.

Security

Security applications require easily deployed systems to support monitoring and tracking of essential security personnel. Event security requires highly portable systems that can be rapidly deployed with a minimum of facility integration, reliance, or impact. Many fixed facilities need to incorporate monitoring and tracking of security personnel in harsh environments, where infrastructure cannot be relied upon, or where networking of infrastructure is difficult or expensive.

NEON greatly improves situational awareness and command effectiveness for these applications, delivering real-time 2D and 3D locations for personnel operating in and around buildings. Key features include

- Real-time 2D and 3D location of personnel,
- Clear situational awareness, indoors and out,
- Effective after action review,
- Portable, lightweight, and rapidly deployed.

Partners and Business Model

TRX has focused primarily on infrastructure-free applications for which it has a substantial competitive advantage and on government applications in particular. It is now expanding to include mobile applications. Primarily, sales are made through partner organizations, which include Motorola, Globe Manufacturing, and ST Electronics.

In general, partners integrate TRX's NEON geo-tracking system into their own solutions, thus becoming in effect a distribution channel for TRX, which can then focus on R&D and partner management. Partners often bring an extensive suite of tools in the form of a fully integrated solution, such as Motorola's radio systems or Globe Manufacturing's fire suit, and may also have expandable existing contracts and a sales and support organization already in place.

In addition, TRX has deployed some systems directly both in the United States and internationally. Such sales typically involve sale of a system for evaluation, followed by customization to integrate the system with existing radio networks or other situational awareness tools. This has allowed TRX to deploy very rapidly, providing a further competitive advantage.

Finally, sometimes TRX directly licenses its algorithms for use on other hardware, which provides greater form factor flexibility for the partner.

Over the long term, the management team at TRX foresees a much wider range of potential uses. According to Carole Teolis, CTO, "While TRX Systems started with a focus on firefighters, it has become clear that there are many situations that would benefit from precise indoor locations without relying on pre-installed infrastructure for support. In places like malls and office buildings, this technology would allow a person to navigate to the exact restaurant where a friend is waiting, to a store with a favorite item is on sale, or to an office cubicle to meet a colleague."[38]

[38] A. Rote, TRX Systems, Inc., "Taking a new Perspective, researchers develop new locating tech," *NSF Livescience* online magazine, December 12, 2012.

R&D

TRX R&D programs include the deployment of NEON location services in a "software as a service" or SAAS environment, and on mobile devices. Current research projects include the following:

National Science Foundation: Collaborative Indoor Mapping Technologies

TRX is developing a smart phone application that creates indoor maps through sensor fusion and crowd-sourcing. The resulting indoor map database changes dynamically as individuals move about indoor spaces, using data gathered from sensors in Android smartphones. Building features and navigable passageways are detected and displayed, while accuracy increases as the number of users increases.

Federal Highway Administration: Navigation Aid for the Blind and Visually Impaired

TRX is developing a navigation aid for the blind and visually impaired, to track the location of a blind person anywhere, including areas where GPS is not available or reliable (for example, indoors or in urban areas with tall buildings). The application also aims to plan and adaptively update a route based on recognized obstacles to be avoided (for example, people or construction within the path). A third objective is to take gestural input and provide natural route guidance based on tactile stimulus instead of relying solely on auditory or visual instructions.

Army: Distributed Navigation

The goal of high accuracy and robust navigation for mobile soldiers requires a flexible system design that uses all available information. A network of soldiers must be able to move seamlessly from operating individually to navigating as a team. TRX is building a soldier-worn device that shares location information and leverages available communications (to other squad members and optionally to ground sensors and vehicle-based navigation systems and command), generating dynamic and timely information for improving navigation.

Department of Homeland Security

The Department of Homeland Security Science and Technology Directorate has sponsored Honeywell, with team members Argon ST and TRX Systems, to develop the Geo-spatial Location, Accountability and Navigation System for Emergency Responders (GLANSER). GLANSER provides accurate and reliable location of emergency responders (ERs) in all types of environments. It aims to

BOX E-5
Army Contract Enhancement to Develop Urban Training

TRX was awarded an enhancement contract with the U.S. Army Research Laboratory's (ARL) Human Research and Engineering Directorate (HRED), Simulation and Training Technology Center (STTC). The contract is to enhance and deliver a flexible, low-cost training system to support real-time location and tracking of war fighters during urban training operations.

TRX will deliver a small, Android-based application that war fighters can use to the view location of themselves and their teammates, as well as maps of the surrounding environment. A software application provides visualization of the urban terrain and allows for rapid creation of 3D urban environments. By implementing the system STTC aims to deliver flexible training any time and in any geographic location; effective after-action review will support real-time assessment of individual and team learning.

provide indoor/outdoor precision navigation, robust communications, and real-time position updates for commanders.

PRIVATE INVESTMENT

In 2012, TRX received a $2 million A round of venture funding from Motorola Solutions Inc. (NYSE: MSI), New Dominion Angels, the Maryland Department of Economic Development, and inside investors. It is using the investment to fund integration of NEON with Motorola Solutions' radios and to expand sales and marketing operations more generally.[39] (Since that time, NEON was approved for use with the Motorola APX radio and is now available to Motorola customers through its catalog).

"The ability to locate personnel operating indoors and often in hazardous situations improves command effectiveness, increases personnel safety and ultimately saves lives," said Mel Gaceta, investment manager, Motorola Solutions Venture Capital. "The TRX NEON Indoor Location System clearly complements Motorola Solutions' capabilities to improve safety for mission-critical users."[40]

TRX AND SBIR

TRX can already be viewed as an SBIR success story. Only 5 years after its founding, it received a Tibbetts award in 2012. TRX founder Carole Teolis was

[39] J. Clabaugh, "TRX Systems gets $2 million in VC funding," *Washington Business Journal*, November 13, 2012.

[40] Motorola Solutions invests in TRX Systems Inc., *PRWeb*, November 12, 2013.

well aware of the SBIR program; Teolis had previously been a PI at TSI, another successful SBIR recipient in the Maryland suburbs of Washington, DC.

The company's focus on indoor location was critically enabled by its first SBIR award from NSF. It has since received several awards from DoD, including a recently expanded award from DoD's Army training command. Three of TRX's four Phase I awards have been selected for Phase II, providing total committed SBIR funding as of year end 2012 of about $3 million.

Important early support was also provided by a TEDCO grant from the state of Maryland, which together with the SBIR program provided critical early funding to deliver proof of concept. Carol Politi notes that this early support was very important to the company's success and allowed it to file its first patents in 2007 and 2008. (Since that time, TRX has had 7 patents awarded, four of these in the US).

TRX worked with Army's Simulation and Training Technology Center (STTC) on group training and simulation technologies, focused on developing an application to help train soldiers in urban areas. Army is required to develop effective urban training and particularly needed a tool for after-exercise review in near real time. Existing solutions required expensive networking technologies such as ultra wide band or the introduction of numerous cameras for video review. A better approach would be lightweight and rapidly deployable, and it would require no pre-existing infrastructure or network, while still providing a means to track the location and path of all soldiers during an exercise. The Army also sought integration of interior maps where available.

TRX received substantial support from a program manager, Tim Roberts, at SSTC, who linked the company to staff conducting live training exercises. This provided important feedback for improving NEON, as well as access to testing venues.

Eventual take-up in DoD, according to Politi, must be based on end-user support and establishment of the right partners. TRX has recently partnered with General Dynamics to integrate NEON within the Army CTIA training architecture, and to extend the NEON capabilities to further enhance training realism.

For TRX, key competitive advantages include the following:

- low costs (no infrastructure required, which means that training organizations can simply buy the technology without any significant prior planning or authorization, or need for integration with current systems);
- easy interface with programs of record, but no requirement for integration;
- multiple sales options (more than 100 military training organizations are potential buyers);
- light weight both physically and technically (which means high degree of portability, so systems can be deployed for training within theater).

TRX primarily markets its product by looking for partners. Ms. Politi notes that in many cases, "TRX is an important piece of a much larger program." As a result, partnership is inherent to TRX's business strategy.

Similar solutions and approaches are used to address the training needs of other organizations, notably law enforcement, first responders, and others in the wider field of security. Here partnerships such as that with Motorola and Globe, and the development and potential licensing relationship with Honeywell, are the primary conduits for sales.

TRX has a flexible business model. Although Ms. Politi expects to make most sales through partners, TRX is set up to make direct sales where necessary or to offer OEM services where it provides the product but not fulfillment.

Ms. Politi observed some angel and VC funders are concerned that companies will become dependent upon SBIR funding, and apply for programs that become distractions from developing a product business. TRX frequently rejects opportunities to pursue SBIR funding in order to stay focused on its core business of location and mapping.

SBIR Matching Funds and Enhancements

TRX has found enhancement programs within SBIR to be of considerable value and would recommend expanding them, particularly at DoD where they can be used to help fund company efforts to traverse the difficult and demanding DoD validation process. Developing hardened products is expensive, and enhancement programs can provide key funding in that area.

DoD funding in this case required matching funds, which TRX was able to raise from a strategic partner (Motorola) as well as other investors.

TRX was also the recipient of an NSF Phase IIB award, which provided another important contribution. NSF support was central in helping the company raise its first angel funding: the ability to point to a federal contribution that effectively doubled the money of investors was "a huge benefit in raising outside money."

More generally, Ms. Politi observed that "matching programs give you a reason to reach out to people, and the double-your-money offer is very well received."

Recommendations

TRX is not woman-owned but it is woman controlled. Both the CEO and CTO are women. Because TRX was successful in raising outside funds, its time as a woman-owned business, according to SBA's definition, was limited. So, although the company is still well below 50 percent venture owned, it is more than 50 percent owned by outside funders—and therefore is no longer woman-owned. This change suggests a significant weakness in efforts to track the engagement of women (and minorities) within the SBIR program: successful companies quickly fail to meet the standard SBA definition of woman-owned.

Ms. Politi observed that through NSF, TRX had received commercialization support from LARTA, whose process was especially helpful in relation to a new collaborative mapping initiative. LARTA's method focuses on business planning and partnerships from the start of Phase I, which could also help to support a new initiative within an existing company. TRX has also used the method to train new PIs.

In addition, through NSF, TRX has received invaluable marketing support. This support included the development of publicly available spotlights of TRX founders and technology, as well as videos showcasing TRX developments.

Workplace Technologies Research Inc. (WTRI)

Based on interview with
Dr. Lia DiBello, Research Director and Founder
September 21, 2010, and October 30, 2014
By telephone

COMPANY HISTORY

WTRI (Workplace Technologies Research Inc.) is a woman-owned firm located in San Diego, CA and Brooklyn, NY. Originally based in academic research—first at City University of New York and then at UC San Diego, Dr. DiBello responded to customer demand by moving her research and work on workplace cognition into the private sector (her first major client, Amtrak, preferred a private sector base).

Dr. DiBello has retained strong connections to the academic community, and remained primarily academy-based until WTRI received its first SBIR award from NSF in 2000. She sees WTRI as continuing to serve two missions: growing a commercial enterprise, and continuing its commitment to academic research. According to Dr. DiBello, this dual focus draws criticism from some parts of the academic community; however, other senior academics have been strong supporters of WTRI's work.

WTRI's business to date can perhaps best be understood as providing sophisticated and highly customized war gaming capacity to business organizations, focused on improving the efficiency of their internal processes within the business environment that they face. WTRI builds a highly sophisticated model of business processes, populated in part by data drawn from public business databases, and then provides either a physical or Multi-Player 3-D Virtual World environment in which company executives—usually at the C-level—can run the model to identify bottlenecks and inefficiencies, and review the impact of alternative strategies for addressing them.

This approach has attracted an impressive roster of clients, including

- Amtrak
- Brigham and Women's Hospital
- ComEd
- EdNet
- IBM
- Invitrogen
- Kellogg

- Monroe Plan for Medical Care
- NASA
- New York City Transit
- NSF
- Siemens
- Most of the major mining companies, such as Kenross Gold Corp., Rio Tinto
- Merck
- Number of major schools such as UCSD, Yale University

The roster of clients reflects Dr. DiBello's view that the role of C-level managers and especially the CEO is changing, and that a more dynamic perspective is required. She believes that the age of careful management has in some senses been replaced by CEOs who can respond rapidly to accelerating change. This can be technical, via the introduction of disruptive technologies, but it can also be political or commercial, as the rapid collapse of major corporations such as Pan-Am or TWA suggest.[41] In this more dynamic environment, the ability to war game new strategies is increasingly useful. Companies are prepared to pay premium prices for the capabilities embedded in WRTI's approach.

WTRI struggled for some time with the highly labor intensive character of the initial approach (the "OpSim" model). Customers however insisted that the company continue to provide these services, and eventually, as noted below, much of the process was automated (in part using technologies developed with funding from SBIR), so OpSim is now a profitable enterprise for the company.

WTRI is comfortably profitable, and aside from seeking to develop mass market applications of its work for training and assessment purposes, and now has a global network of alliances, with offices in Sydney and London and partners in Africa, Europe and elsewhere. It is considering alliances with larger consulting organizations.

WTRI'S TECHNOLOGY BASE

While WTRI is based on academic research that describes how individuals and organizations change and can be changed, the technologies it uses play a crucial role in its simulations. Dr. DiBello observes that by breaking business processes into discrete elements, a substantial amount of the simulation design and build process—perhaps 90 percent—can be automated and scaled.

[41] See Clayton Christenson, the Innovator's Dilemma when new technologies cause great firms to fail. Harvard Business Press. 1997.

Dynamic Strategic Modeler

WTRI has developed a modeling tool that provides expert system support for applying the theory of constraints to the value proposition of the business, and determining what aspects of the business process to work on. Models are based on detailed analysis of trends evidencing themselves in business environments. As a result, the modeler has automated substantial parts of the strategy development and review process, which now takes 15 minutes instead of weeks of staff effort. The resulting charts still require expert review, but the process itself is operated by student interns.

Cognitive Agility Assessment Tool

This tool supports the elicitation of embedded knowledge. Using it, WTRI staff can expose the underlying mental model of the expert—which could be a mental model for clinical interventions, operating a lathe, or marketing a business. The point of the tool is to automate the interview and the scoring process. As participants in a simulation complete tasks, the tool automatically scores their efforts. This rapid feedback is valuable, as it permits companies to see whether their executives can handle change at the pace and depth required. For example, a chemical company planned to introduce radical change into their business model, and wanted to determine whether C level managers can could work with the new strategy. Building on expertise focused on the specific needs of C-level managers in the chemical industry, the tool is now being used by the CEO and other C-executives to make mission critical decisions. WTRI believes that the company is now seeing information in the tool that even WTRI cannot identify. In this case, WTRI did not even have to meet the client in the course of the project.

WTRI PRODUCTS

WTRI provides three main products, which can be combined in different configurations: OpSim, Modeling, and Profiling.

OpSim

WTRI's high end product is OpSim, a live immersive environment where companies and organizations can war game solutions to problems and issues facing the organization. According to WTRI, "OpSim™ Live" provides a "safe" environment in which to rehearse strategies to address current challenges or crises (e.g., losing revenue based on significant backorders), or to anticipate future uncertainties (e.g., a potential decline in the value of your core products)."

OpSim focuses on helping businesses achieve core outcomes, designed in conjunction with the company. WTRI then builds the artifacts and materials that support achievement of this goal in a simulated environment. So for example an

OpSim involving a foundry company included development of real molds at a miniature scale. Participants actually run the company under accelerated time pressure (20 minutes often represents one month of real time (according to WTRI), so that participants face the real pressures and competing demands experienced in actually running the business. By performing the exercise multiple times, participants can experience directly the impact of different decisions and strategies.

Critically, the OpSim approach draws on WTRI's work on what it calls "cognitive agility," defined as the extent to which an individual's thinking is flexible when data indicate the situation has changed. This approach incorporates cutting edge brain science.[42] The first component of each OpSim exercise encourages participants to manage the company in line with status quo procedures and strategies, while measuring outcomes. The experiential nature of the exercise helps to break down resistance to new approaches, as participants experience for themselves the disastrous consequences of some current strategies (which is of course why WTRI has been engaged by the company—to address critical problems). Once participants are open to new possibilities, OpSim allows for scenario-based efforts to address newly identified difficulties and bottlenecks.

WTRI has also developed versions of OpSim using virtual world operational simulation, using environments such as Second Life.[43] The virtual world OpSim provides more flexibility in design, and is used by WTRI in particular for rehearsal of business strategies that are heavily technology supported, such as logistics distribution or network management. In 2008, WTRI collaborated with IBM to developed 3-D environments for training IBM employees in the managing enterprise IT engagements.[44]

WTRI is now heavily involved in using OpSim to work with leaders of the project management and mining industries, and will likely sell its lower end capabilities in this area to a buyer in the project management sector.

Modeling

OpSim works, according to Dr. DiBello, in part because it is closely integrated with WTRI's FutureView™ modeling software. FutureView™ can perhaps best be understood as an expert system, based on modeling the insights and approaches of highly experienced business strategists.

WTRI initially populates the model with publicly available data (drawn from standard business database subscriptions). Once the initial model has been populated, users—normally C-level company executives—can use the model to

---

[42]WTRI "What is Cognitive Ability?" <*http://www.wtri.com/documents/What_is_cognitive_agility.pdf*>.

[43]An online multi-player environment with significant analogs to real work processes including virtual money, trading, and business development. See <*http://www.secondlife.com*>.

[44]Made in IBM Labs: IBM Develops a "Rehearsal Studio" to Let You Practice Your Job in a 3-D World <*http://www.wtri.com/documents/IBM-WTRI_press_release_v2.pdf*>.

simulate the effect of changes in resource distribution and other company policies that affect critical constraints (including for example capital structures) as measured against the result of an "ideal" competitor company.

FutureView™ dynamically models the impact of bottlenecks in an organization's operations on company performance, and conversely, the net present value of moving or removing the bottleneck. For example, it can be used to calculate the impact of specific changes in R&D to the organization's pipeline of new products, and then calculate the subsequent value that can be realized (gross revenue, profits, stock price).

It can also use a mathematically derived profile of financial to benchmark company performance against competitors, which can be used to help with strategic planning. For C-level executives, the modeling tool can be used to move around constraint points to visualize the impact of changing resource allocations, or the impact of adjusting major financial variables on company performance.

Scaling and Automation

Based on the expanding library of business cases completed using OpSim and the Cognitive Assessment toolset, WTRI is now moving to build a set of decision-assessment tools that would be available at a much lower price-point, targeted as assessing executive decision-making and also at training to improve skills and hence outcomes.

These new tools confront executives with cases of actual companies. Using information that would be available to management, users must make the decisions that will lead the company into the path required. Review of decisions made in this environment can help companies assess executives' capabilities (and indeed those of potential executive hires). Executives can also improve skills by completing simulated cases online through the Profiling toolset, and receiving feedback on their strengths and weaknesses. According to WTRI

> *"As the situations are real companies, the candidate's performance can be compared with the actual outcomes under varying needs, allowing the tool to profile both their accuracy and the suitability of their mental models for interpreting situations and changing needs. In addition, companies looking to assess a more specific problem in their team can request a customized profiling and assessment layer."*[45]

SBIR AT WTRI

Dr. DiBello remains overall a strong supporter of the SBIR program in general, and the program at NSF in particular. SBIR has been central to the development of WTRI technology—and has been used to fund development of

[45]WTRI <*http://www.wtri.com/profiling.html*>. Accessed September 28, 2010.

each of the WTRI tools. The initial NSF SBIR award in 2000 was, according to Dr. DiBello, successful in part because the NSF program director was familiar with the relevant academic work.

The projects that NSF funded in early 2000s were targeted as different markets entirely. However, they were used to make virtual worlds "immensely powerful," leading to a WRTI's world leading position as a provider of virtual worlds for war gaming for business, according to Dr. DiBello. Overall, she believes that NSF sees its role as funding innovative ideas, without being too concerned about how exactly those ideas will be commercialized.

This funding was important not least because some of the projects funded by SBIR in the early 2000s would not have been funded by private sector sources because they were simply too risky, with too much uncertainty about the eventual market.

WTRI believes it was the first company permitted to use revenues from sales as the matching funds for a Phase IIB award at NSF. This was at the time controversial within NSF, although this is now apparently the preferred option within the agency.

WTRI's own experience suggests that the initial NSF focus on encouraging firms to acquire VC funding might have been misplaced. Dr. DiBello notes that WTRI's experience in fixing broken companies indicates that many of them got VC funding too early, at too high a price. In addition, many VC companies failed after the financial crash in 2008 and others have become more conservative in their investments.

SBIR funding helped WTRI transform the OpSim business, from a homemade product that is almost entirely customized for each client, to a much more sophisticated product with substantial automated and reusable components. Dr. DiBello also noted that NSF's support for the company's financial modeling capability transformed the value of the overall service, making it much more valuable.

Similarly, the SBIR funding supported automation of the cognitive assessment tool. WTRI did not have enough trained staff to continue expanding if it continued to use a paper questionnaire and an in-person interviewer. In addition, the interviewer introduced variables into the process that affected observed outcomes. The SBIR funding helped WTRI develop an assessment tool that clients could complete on their own—reducing costs and improving quality. Very large scale assessments are now routine.

Overall, Dr. DiBello has a positive view of the NSF SBIR program, and of the division with which WTRI has been working. WTRI's strong commercialization track record and effective use of SBIR funds in the past has, she believed, helped to support successful application record.

NSF's program has changed over the years. When WTRI initially interacted with the program, the focus was on very innovative ideas that might have the potential for significant commercial success. This underpinned initial WTRI awards in 2000-2001.

This has gradually changed.

Today, Dr. DiBello sees a contradiction between the demand for innovation and growing requirements that projects be far along toward commercialization, even before the start of the first Phase I award. It is not much of an exaggeration, according to Dr. DiBello, to suggest that NSF is seeking projects that are more or less ready for Phase IIB at the time of the Phase I.

Yet on the other hand, Dr. DiBello has experienced negative reviews because the technology and/or the project were too mature, too close to market-ready. It is possible that a current application for a project in partnership with IBM may be declined for that reason.

In short, Dr. DiBello believes that NSF now wants a much more completed idea for Phase I than is reasonable, but at the same time screens out projects for being too mature.

There are also growing problems with the Phase I review process, which has become somewhat less helpful than her initial experience. Dr. DiBello is concerned that WTRI may be facing a competitor on the review panel, but she is unable to determine whether this is the case, as NSF uses anonymous reviewers. She believes it is too easy for a competitor to find something wrong with an application. Further, the consistency of quality and qualifications of reviewers themselves is not what it once was; some reviewers do not seem to have the background to follow the proposal details while others do an exemplary job and offer important insights, even if they are being critical. This is a problem only for Phase I review; her experience is that Phase II reviews separate commercial and technical review, and that the resulting reviews are better quality.

WTRI's experience with NSF program directors has varied. Some, like the first program director with whom WTRI worked, were very helpful, seeing their mission as aiding the company. Others have been much less helpful, and have shown little understanding of how companies like WTRI work with their customers. This has in some cases been a problem, especially in relation to the oral defense part of the application process for Phase IIB.

RECOMMENDATIONS

Dr. DiBello was primarily concerned with what she saw as two core issues for the NSF SBIR program:

1. NSF should review the balance between innovation and commercialization in the review of applications. While it was important to ensure that projects were commercially focused, it was also important to allow sufficient room for innovation.
2. NSF should ensure that Phase I remained focused at the level of feasibility studies, and did not demand too much in terms of completed research. To a lesser degree this was also true for Phase II.

Appendix F

Bibliography

Acs, Z., and D. Audretsch. 1988. "Innovation in Large and Small Firms: An Empirical Analysis." *The American Economic Review* 78(4):678-690.

Acs, Z., and D. Audretsch. 1990. *Innovation and Small Firms*. Cambridge, MA: MIT Press.

Adelstein, F. 2006. Live Forensics: Diagnosing Your System Without Killing It First, http://frank.notfrank.com/Papers/CACM06.pdf. Accessed July 17, 2014.

Advanced Technology Program. 2001. *Performance of 50 Completed ATP Projects, Status Report 2*. National Institute of Standards and Technology Special Publication 950-2. Washington, DC: Advanced Technology Program/National Institute of Standards and Technology/U.S. Department of Commerce.

Alic, J. 1987. "Evaluating Competitiveness at the Office of Technology Assessment," *Technology in Society* 9(1):1-17.

Alic, J. A., L. Branscomb, H. Brooks, A. B. Carter, and G. L. Epstein. 1992. *Beyond Spinoff: Military and Commercial Technologies in a Changing World*. Boston, MA: Harvard Business School Press.

American Association for the Advancement of Science. R&D Funding Update on NSF in the FY2007. http://www.aaas.org/spp/rd/nsf07hf1.pdf.

American Psychological Association. 2002. "Criteria for Evaluating Treatment Guidelines." *American Psychologist* 57(12):1052-1059.

Archibald, R., and D. Finifter. 2003. "Evaluating the NASA Small Business Innovation Research Program: Preliminary Evidence of a Tradeoff Between Commercialization and Basic Research." *Research Policy* 32:605-619.

Archibugi, D., A. Filippetti, and M. Frenz. 2013."Economic crisis and innovation: Is destruction prevailing over accumulation?" *Research Policy* 42(2):303-314.

Arrow, K. 1962. "Economic welfare and the allocation of resources for invention." Pp. 609-625 in *The Rate and Direction of Inventive Activity: Economic and Social Factors*. Princeton, NJ: Princeton University Press.

Arrow, K. 1973. "The Theory of Discrimination." Pp. 3-31 in *Discrimination in Labor Market*. Orley Ashenfelter and Albert Rees, eds. Princeton, NJ: Princeton University Press.

Audretsch, D. B. 1995. *Innovation and Industry Evolution*. Cambridge, MA: MIT Press.

Audretsch, D. B., and M. P. Feldman. 1996. "R&D Spillovers and the Geography of Innovation and Production." *American Economic Review* 86(3):630-640.
Audretsch, D. B., and P. E. Stephan. 1996. "Company-Scientist Locational Links: The Case of Biotechnology." *American Economic Review* 86(3):641-642.
Audretsch, D., and R. Thurik. 1999. *Innovation, Industry Evolution, and Employment.* Cambridge, MA: MIT Press.
Baker, A. No date. "Commercialization Support at NSF." Draft.
Baker, J. A. K., and K. J. Thurber. 2011. *Developing Computer Systems Requirements.* Ithaca, NY: Digital Systems Press.
Barfield, C., and W. Schambra, eds. 1986. *The Politics of Industrial Policy.* Washington, DC: American Enterprise Institute for Public Policy Research.
Baron, J. 1998. "DoD SBIR/STTR Program Manager." Comments at the Methodology Workshop on the Assessment of Current SBIR Program Initiatives, Washington, DC, October.
Barry, C. B. 1994. "New Directions in Research on Venture Capital Finance." *Financial Management* 23 (Autumn):3-15.
Bator, F. 1958. "The Anatomy of Market Failure." *Quarterly Journal of Economics* 72:351-379.
Biemer, P. P., and L. E. Lyberg. 2003. *Introduction to Survey Quality.* New York: John Wiley & Sons.
Bingham, R. 1998. *Industrial Policy American Style: From Hamilton to HDTV.* New York: M.E. Sharpe.
Birch, D. 1981. "Who Creates Jobs." *The Public Interest* 65 (Fall):3-14.
Bouchie, A. 2003. "Increasing Number of Companies Found Ineligible for SBIR Funding," *Nature Biotechnology* 21(10):1121-1122.
Branscomb, L. M., K. P. Morse, M. J. Roberts, and D. Boville. 2000. *Managing Technical Risk: Understanding Private Sector Decision-Making on Early Stage Technology Based Projects.* Washington, DC: Department of Commerce/National Institute of Standards and Technology.
Branscomb, L. M., and P. E. Auerswald. 2001. *Taking Technical Risks: How Innovators, Managers, and Investors Manage Risk in High-Tech Innovations*, Cambridge, MA: MIT Press.
Branscomb, L. M., and P. E. Auerswald. 2002. *Between Invention and Innovation: An Analysis of Funding for Early-Stage Technology Development.* Gaithersburg, MD: National Institute of Standards and Technology.
Branscomb, L. M., and P. E. Auerswald. 2003. "Valleys of Death and Darwinian Seas: Financing the Invention to Innovation Transition in the United States." *The Journal of Technology Transfer* 28(3-4).
Branscomb, L.M., and J. Keller. 1998. *Investing in Innovation: Creating a Research and Innovation Policy.* Cambridge, MA: MIT Press.
Brav, A., and P. A. Gompers. 1997. "Myth or Reality?: Long-Run Underperformance of Initial Public Offerings; Evidence from Venture Capital and Nonventure Capital-Backed IPOs." *Journal of Finance* 52:1791-1821.
Brodd, R. J. 2005. "Factors Affecting U.S. Production Decisions: Why Are There No Volume Lithium-Ion Battery Manufacturers in the United States?" ATP Working Paper No. 05-01, June.
Brown, G., and J. Turner. 1999. "Reworking the Federal Role in Small Business Research." *Issues in Science and Technology* XV(4 Summer):51-58.
Bush, V. 1946. *Science—the Endless Frontier.* Republished in 1960 by U.S. National Science Foundation, Washington, DC.
Cahill, P. 2000. "Fast Track: Is it Speeding Commercialization of Department of Defense Small Business Innovation Research Projects?" In National Research Council, *The Small Business Innovation Research Program: An Assessment of the Department of Defense Fast Track Initiative*. Washington, DC: National Academy Press.
Carden, S. D., and O. Darragh. 2004. "A Halo for Angel Investors." *The McKinsey Quarterly* 1.
Caves, R.E. 1998. "Industrial Organization and New Findings on the Turnover and Mobility of Firms." *Journal of Economic Literature* 36(4):1947-1982.

Ceulemans, S., and J. K. Kolls. 2013. "Can the SBIR and STTR Programs Advance Research Goals?" *Nature Immunology* 14(3):192-195.
Christensen, C. 1997. *The Innovator's Dilemma.* Boston, MA: Harvard Business School Press.
Christensen, C., and M. Raynor. 2003. *Innovator's Solution*, Boston, MA: Harvard Business School.
Clabaugh, J. 2012. "TRX Systems Gets $2M in VC Funding." *Washington Business Journal.*
Clinton, W. J. 1994. *Economic Report of the President.* Washington, DC: U.S. Government Printing Office.
Clinton, W. J. 1994. *The State of Small Business.* Washington, DC: U.S. Government Printing Office.
Coburn, C., and D. Bergland. 1995. *Partnerships: A Compendium of State and Federal Cooperative Technology Programs.* Columbus, OH: Battelle.
Cochrane, J. H. 2005. "The Risk and Return of Venture Capital." *Journal of Financial Economics* 75(1):3-52.
Cohen, L. R., and R. G. Noll. 1991. *The Technology Pork Barrel.* Washington, DC: The Brookings Institution.
Congressional Commission on the Advancement of Women and Minorities in Science, Engineering, and Technology Development. 2000. *Land of Plenty: Diversity as America's Competitive Edge in Science, Engineering and Technology.* Washington, DC: National Science Foundation/U.S. Government Printing Office.
Cooper, R. G. 2001. Winning at New Products: Accelerating the Process from Idea to Launch. In Dawnbreaker, Inc. 2005. "The Phase III Challenge: Commercialization Assistance Programs 1990-2005." White paper. July 15.
Council of Economic Advisers. 1995. *Supporting Research and Development to Promote Economic Growth: The Federal Government's Role.* Washington, DC: Council of Economic Advisors.
Council on Competitiveness. 2005. *Innovate America: Thriving in a World of Challenge and Change.* Washington, DC: Council on Competitiveness.
Crane, G., and J. Sohl. 2004. "Imperatives for Venture Success: Entrepreneurs Speak." *The International Journal of Entrepreneurship and Innovation* May. Pp. 99-106.
Cutler, D. 2005. *Your Money or Your Life.* New York: Oxford University Press.
Cycyota, C. S., and D. A. Harrison. 2006. "What (Not) to Expect When Surveying Executives: A Meta-Analysis of Top Manager Response." *Organizational Research Methods* 9:133-160.
Czarnitzki, D., and A. Fier. 2002. "Do Innovation Subsidies Crowd out Private Investment? Evidence from the German Service Sector." ZEW Discussion Papers, No. 02-04.
Dalton, A. B., S. Collins, E. Muñoz, J. Razal1, V. H. Ebron, J. Ferraris, J. Coleman, B. Kim, and R. Baughman. 2003. "Super-Tough Carbon-Nanotube Fibres." *Nature* 423(4):703.
David, P. A., B. H. Hall, and A. A. Tool. 1999. "Is Public R&D a Complement or Substitute for Private R&D? A Review of the Econometric Evidence." NBER Working Paper 7373. October.
Davidsson, P. 1996. "Methodological Concerns in the Estimation of Job Creation in Different Firm Size Classes." Working Paper. Jönköping International Business School.
Davis, S. J., J. Haltiwanger, and S. Schuh. 1994. "Small Business and Job Creation: Dissecting the Myth and Reassessing the Facts," *Business Economics* 29(3):113-122.
Dawnbreaker, Inc. 2005. "The Phase III Challenge: Commercialization Assistance Programs 1990-2005." White paper. July 15.
Dertouzos, M. L. 1989. *Made in America: The MIT Commission on Industrial Productivity.* Cambridge, MA: MIT Press.
Dertouzos, M. L., R. Lester, and R. Solow. 1989. *Made in America: The MIT Commission on Industrial Productivity.* Cambridge, MA: The MIT Press.
Dess, G. G., and D. W. Beard. 1984. "Dimensions of Organizational Task Environments." *Administrative Science Quarterly* 29:52-73.
Devenow, A., and I. Welch. 1996. "Rational Herding in Financial Economics." *European Economic Review* 40(April):603-615.
Dillman, D. 2000. *Mail and Internet Surveys: The Tailored Design Method.* 2nd Edition. Toronto, Ontario: John Wiley and Sons, Inc.

DoE Opportunity Forum. 2005. "Partnering and Investment Opportunities for the Future." Tysons Corner, VA. October 24-25.

Eckstein, O. 1984. *DRI Report on U.S. Manufacturing Industries.* New York: McGraw Hill.

Eisinger, P. K. 1988. *The Rise of the Entrepreneurial State: State and Local Economic Development Policy in the United States.* Madison, WI: University of Wisconsin Press.

Ernst and Young. 2007. "U.S. Venture Capital Investment Increases to 8 percent to $6.96 Billion in First Quarter of 2007." April 23.

Evenson, R., P. Waggoner, and P. Ruttan. 1979. "Economic Benefits from Research: An Example from Agriculture," *Science* 205(14 September):1101-1107.

Feenstra, D. 2014. "Public Support of Innovation in Entrepreneurial Firms." *Journal of Applied Management and Entrepreneurship* 19(2):135.

Feldman, M. P. 1994. *The Geography of Knowledge.* Boston, MA: Kluwer Academic.

Feldman, M. P. 1994. "Knowledge Complementarity and Innovation." *Small Business Economics* 6(5):363-372.

Feldman, M. P. 2001. "Assessing the ATP: Halo Effects and Added Value." In National Research Council, *The Advanced Technology Program: Assessing Outcomes.* Washington, DC: National Academy Press.

Feldman, M. P., and M. R. Kelley. 2001. *Winning an Award from the Advanced Technology Program: Pursuing R&D Strategies in the Public Interest and Benefiting from a Halo Effect.* NISTIR 6577. Washington, DC: Advanced Technology Program/National Institute of Standards and Technology/U.S. Department of Commerce.

Fenn, G. W., N. Liang, and S. Prowse. 1995. *The Economics of the Private Equity Market.* Washington, DC: Board of Governors of the Federal Reserve System.

Financial Times. 2004. "Qinetiq set to make its first US acquisition," September 8.

Fischer, E., and A. R. Reuber. 2003. "Support for Rapid-Growth firms: A Comparison of the Views of Founders, Government Policymakers, and Private Sector Resource Providers," *Journal of Small Business Management* 41(4):346-365.

Flamm, K. 1988. *Creating the Computer.* Washington, DC: The Brookings Institution.

Flender, J. O., and R. S. Morse. 1975. *The Role of New Technical Enterprise in the U.S. Economy.* Cambridge, MA: MIT Development Foundation.

Freear, J., and W. E. Wetzel Jr. 1990. "Who Bankrolls High-Tech Entrepreneurs?" *Journal of Business Venturing* 5:77-89.

Freeman, C., and L. Soete. 1997. *The Economics of Industrial Innovation.* Cambridge, MA: MIT Press.

Galbraith, J. K. 1957. *The New Industrial State.* Boston: Houghton Mifflin.

Gallagher, S. 2012. "Here Come the Inflate-a-Bots: iRobot's AIR Blow Up Bot Prototypes." ARS Technica.

Galope, R. V. 2014. "What Types of Start-ups Receive Funding from the Small Business Innovation Research (SBIR) Program?: Evidence from the Kauffman Firm Survey." *Journal of Technology Management & Innovation* 9(2):17-28.

Geroski, P. A. 1995. "What Do We Know About Entry?" *International Journal of Industrial Organization* 13(4):421-440.

Geshwiler, J., J. May, and M. Hudson. 2006. *State of Angel Groups.* Kansas City, MO: Kauffman Foundation.

Gicheva, D., and A. N. Link. 2015. "The Gender Gap in Federal and Private Support for Entrepreneurship." *Small Business Economics* (2015):1-5.

Gompers, P. A., and J. Lerner. 1977. "Risk and Reward in Private Equity Investments: The Challenge of Performance Assessment." *Journal of Private Equity* 1:5-12.

Gompers, P. A. 1995. "Optimal Investment, Monitoring, and the Staging of Venture Capital." *Journal of Finance* 50:1461-1489.

Gompers, P. A., and J. Lerner. 1996. "The Use of Covenants: An Empirical Analysis of Venture Partnership Agreements." *Journal of Law and Economics* 39:463-498.

Gompers, P. A., and J. Lerner. 1998. "Capital Formation and Investment in Venture Markets: A Report to the NBER and the Advanced Technology Program." Unpublished working paper. Harvard University.

Gompers, P. A., and J. Lerner. 1998. "What Drives Venture Capital Fund-Raising?" Unpublished working paper. Harvard University.

Gompers, P. A., and J. Lerner. 1999. "An Analysis of Compensation in the U.S. Venture Capital Partnership." *Journal of Financial Economics* 51(1):3-7.

Gompers, P. A., and J. Lerner. 1999. *The Venture Cycle.* Cambridge, MA: MIT Press.

Good, M. L. 1995. Prepared testimony before the Senate Commerce, Science, and Transportation Committee, Subcommittee on Science, Technology, and Space (photocopy, U.S. Department of Commerce).

Goodnight, J. 2003. Presentation at National Research Council Symposium. "The Small Business Innovation Research Program: Identifying Best Practice." Washington, DC, May 28.

Graham, O. L. 1992. *Losing Time: The Industrial Policy Debate.* Cambridge, MA: Harvard University Press.

Greenwald, B. C., J. E. Stiglitz, and A. Weiss. 1984. "Information Imperfections in the Capital Market and Macroeconomic Fluctuations." *American Economic Review Papers and Proceedings* 74:194-199.

Griliches, Z. 1990. *The Search for R&D Spillovers.* Cambridge, MA: Harvard University Press.

Groves, R. M., D. A. Dillman, J. L. Eltinge, and R. J. A. Little, eds. 2002. *Survey Nonresponse.* New York: Wiley.

Groves, R. M., F. J. Fowler, Jr., M. P. Couper, J. M. Lepkowski, E. Singer, and R. Tourangeau. 2004. *Survey Methodology.* Hoboken, NJ: John Wiley & Sons, Inc.

Hall, B. H. 1992. "Investment and Research and Development: Does the Source of Financing Matter?" Working Paper No. 92-194, Department of Economics/University of California at Berkeley.

Haltiwanger, J., and C. J. Krizan. 1999. "Small Businesses and Job Creation in the United States: The Role of New and Young Businesses" in *Are Small Firms Important? Their Role and Impact,* Zoltan J. Acs, ed. Dordrecht: Kluwer.

Hamberg, D. 1963. "Invention in the Industrial Research Laboratory." *Journal of Political Economy* (April):95-115.

Hao, K. Y., and A. B. Jaffe. 1993. "Effect of Liquidity on Firms' R&D Spending." *Economics of Innovation and New Technology* 2:275-282.

Hebert, R. F., and A. N. Link. 1989. "In Search of the Meaning of Entrepreneurship." *Small Business Economics* 1(1):39-49.

Held, B., T. Edison, S. L. Pfleeger, P. Anton, and J. Clancy. 2006. *Evaluation and Recommendations for Improvement of the Department of Defense Small Business Innovation Research (SBIR) Program.* Arlington, VA: RAND National Defense Research Institute.

Henrekson, M., and D. Johansson. 2009. "Competencies and Institutions Fostering High-Growth Firms." *Foundations and Trends in Entrepreneurship* 5(1):1-80.

Himmelberg, C. P., and B. C. Petersen. 1994. "R&D and Internal Finance: A Panel Study of Small firms in High-Tech Industries." *Review of Economics and Statistics* 76:38-51.

Hong, S., S. Myung. 2007. "Nanotube Electronics: A Flexible Approach to Obility. *Nature Nanotechnology* 2(4):207-208.

Hubbard, R. G. 1998. "Capital-Market Imperfections and Investment." *Journal of Economic Literature* 36:193-225.

Huntsman, B., and J. P. Hoban Jr. 1980. "Investment in New Enterprise: Some Empirical Observations on Risk, Return, and Market Structure." *Financial Management* 9(Summer):44-51.

Institute of Medicine. 1998. "The Urgent Need to Improve Health Care Quality." National Roundtable on Health Care Quality. *Journal of the American Medical Association* 280(11):1003, September 16.

Jacobs, T. 2002. "Biotech Follows Dot.com Boom and Bust." *Nature* 20(10):973.

Jaffe, A. B. 1996. "Economic Analysis of Research Spillovers: Implications for the Advanced Technology Program." Washington, DC: Advanced Technology Program/National Institute of Standards and Technology/U.S. Department of Commerce).

Jaffe, A. B. 1998. "Economic Analysis of Research Spillovers: Implications for the Advanced Technology Program." Washington, DC: Advanced Technology Program/National Institute of Standards and Technology/U.S. Department of Commerce.

Jaffe, A. B. 1998. "The Importance of 'Spillovers' in the Policy Mission of the Advanced Technology Program." *Journal of Technology Transfer* (Summer).

Jarboe, K. P., and R. D. Atkinson. 1998. "The Case for Technology in the Knowledge Economy: R&D, Economic Growth and the Role of Government." Washington, DC: Progressive Policy Institute. http://www.ppionline.org/documents/CaseforTech.pdf.

Jewkes, J., D. Sawers, and R. Stillerman. 1958. *The Sources of Invention.* New York: St. Martin's Press.

Johnson, W. 2004. Delivering combat power to the fleet, Naval Engineers Journal, Fall 2004, pp. 3-5.

Johnson, T., and L. Owens. 2003. "Survey Response Rate Reporting in the Professional Literature." Paper presented at the 58th Annual Meeting of the American Association for Public Opinion Research. Nashville, TN. May.

Kaplowitz, M.l D., T. D. Hadlock, and R. Levine. 2004. "A Comparison of Web and Mail Survey Response Rates." *Public Opinion Quarterly* 68(1):94-101.

Kauffman Foundation. 2014. About the Foundation. http://www.kauffman.org/foundation.cfm.

Kleinman, D. L. 1995. *Politics on the Endless Frontier: Postwar Research Policy in the United States.* Durham, NC: Duke University Press.

Kolosnjaj, J., H. Szwarc, and F. Moussa. 2007. Toxicity studies of carbon nanotubes, *Advances in Experimental Medicine and Biology* 620:181-204.

Kortum, S., and J. Lerner. 1998. "Does Venture Capital Spur Innovation?" NBER Working Paper No. 6846, National Bureau of Economic Research.

Krugman, P. 1990. *Rethinking International Trade.* Cambridge, MA: MIT Press.

Krugman, P. 1991. *Geography and Trade.* Cambridge, MA: MIT Press.

Lanahan, L. 2015. "Multilevel Public Funding for Small Business Innovation: A Review of US State SBIR Match Programs." *The Journal of Technology Transfer* (2015):1-30.

Lanahan, L. 2015. "The Multilevel Innovation Policy Mix: Three Essays on State SBIR Matching Programs." Available at SSRN <http://ssrn.com/abstract=2598229>.

Lanahan, L., and M. P. Feldman. 2015. "Multilevel Innovation Policy Mix: A Closer Look at State Policies that Augment the Federal SBIR Program." *Research Policy* 44(7):1387-1402.

Langlois, R. N., and P. L. Robertson. 1996. "Stop Crying over Spilt Knowledge: A Critical Look at the Theory of Spillovers and Technical Change." Paper prepared for the MERIT Conference on Innovation, Evolution, and Technology. Maastricht, Netherlands, August 25-27.

Langlois, R. N. 2001. "Knowledge, Consumption, and Endogenous Growth." *Journal of Evolutionary Economics* 11:77-93.

Lebow, I. 1995. *Information Highways and Byways: From the Telegraph to the 21st Century.* New York: Institute of Electrical and Electronic Engineering.

Lerner, J. 1994. "The Syndication of Venture Capital Investments." *Financial Management* 23-(Autumn):16-27.

Lerner, J. 1995. "Venture Capital and the Oversight of Private Firms." *Journal of Finance* 50:301-318.

Lerner, J. 1996. "The Government as Venture Capitalist: The Long-Run Effects of the SBIR Program." Working Paper No. 5753, National Bureau of Economic Research.

Lerner, J. 1998. "Angel Financing and Public Policy: An Overview." *Journal of Banking and Finance* 22(6-8):773-784.

Lerner, J. 1999. "The Government as Venture Capitalist: The Long-Run Effects of the SBIR Program." *Journal of Business* 72(3):285-297.

Lerner, J. 1999. "Public Venture Capital: Rationales and Rvaluation." In *The SBIR Program: Challenges and Opportunities.* Washington, DC: National Academy Press.

Levy, D. M., and N. Terleckyk. 1983. "Effects of Government R&D on Private R&D Investment and Productivity: A Macroeconomic Analysis." *Bell Journal of Economics* 14:551-561.

Liles, P. 1977. *Sustaining the Venture Capital Firm.* Cambridge, MA: Management Analysis Center.

Link, A. N. 1998. "Public/Private Partnerships as a Tool to Support Industrial R&D: Experiences in the United States." Paper prepared for the working group on Innovation and Technology Policy of the OECD Committee for Science and Technology Policy, Paris.

Link, A. N., and J. Rees. 1990. "Firm Size, University Based Research and the Returns to R&D." *Small Business Economics* 2(1):25-32.

Link, A. N., and J. T. Scott. 1998. "Assessing the Infrastructural Needs of a Technology-Based Service Sector: A New Approach to Technology Policy Planning." *STI Review* 22:171-207.

Link, A. N., and J. T. Scott. 1998. *Overcoming Market Failure: A Case Study of the ATP Focused Program on Technologies for the Integration of Manufacturing Applications (TIMA).* Draft final report submitted to the Advanced Technology Program. Gaithersburg, MD: National Institute of Technology. October.

Link, A. N., and J. T. Scott. 1998. *Public Accountability: Evaluating Technology-Based Institutions.* Norwell, MA: Kluwer Academic.

Link, A. N., and J. T. Scott. 2005. *Evaluating Public Research Institutions: The U.S. Advanced Technology Program's Intramural Research Initiative.* London: Routledge.

Link, A. N., and J. T. Scott. 2012. "The Exploitation of Publicly Funded Technology." *The Journal of Technology Transfer* 37(3):375-383.

Link, A. N., and J. T. Scott. 2012. "The Small Business Innovation Research Program." *Issues in Science and Technology* 28(4):89-92.

Longini, P. 2003. "Hot Buttons for NSF SBIR Research Funds." Pittsburgh Technology Council. *TechyVent.* November 27.

Malone, T. 1995. *The Microprocessor: A Biography.* Hamburg, Germany: Springer Verlag/Telos.

Mankins, J. C. 1995. *Technology Readiness Levels: A White Paper.* Washington, DC: NASA Office of Space Access and Technology. Advanced Concepts Office.

Mann, D., Q. Wang, K. Goodson, and H. Dai. 2005. "Thermal Conductance of an Individual Single-Wall Carbon Nanotube Above Room Temperature." *Nano Letters* 6(1):96-100.

Mansfield, E. 1985. "How Fast Does New Industrial Technology Leak Out?" *Journal of Industrial Economics* 34(2).

Mansfield, E. 1996. *Estimating Social and Private Returns from Innovations Based on the Advanced Technology Program: Problems and Opportunities.* Unpublished report.

Mansfield, E., J. Rapoport, A. Romeo, S. Wagner, and G. Beardsley. 1977. "Social and Private Rates of Return from Industrial Innovations." *Quarterly Journal of Economics* 91:221-240.

Martin, Justin. 2002. "David Birch." *Fortune Small Business* (December 1).

McCraw, T. 1986. "Mercantilism and the Market: Antecedents of American Industrial Policy." In C. Barfield and W. Schambra, eds. *The Politics of Industrial Policy.* Washington, DC: American Enterprise Institute for Public Policy Research.

Mervis, J. D. 1996. "A $1 Billion 'Tax' on R&D Funds." *Science* 272:942–944.

Morgenthaler, D. 2000. "Assessing Technical Risk," in *Managing Technical Risk: Understanding Private Sector Decision Making on Early Stage Technology-Based Project.* L. M. Branscomb, K. P. Morse, and M. J. Roberts, eds. Gaithersburg, MD: National Institute of Standards and Technology.

Mowery, D. 1998. "Collaborative R&D: How Effective Is It?" *Issues in Science and Technology* (Fall):37-44.

Mowery, D., and N. Rosenberg. 1989. *Technology and the Pursuit of Economic Growth.* New York: Cambridge University Press.

Mowery, D., and N. Rosenberg. 1998. *Paths of Innovation: Technological Change in 20th Century America.* New York: Cambridge University Press.

Mowery, D. C. 1999. "America's Industrial Resurgence (?): An Overview." in National Research Council. *U.S. Industry in 2000: Studies in Competitive Performance*. D.C. Mowery, ed. Washington, DC: National Academy Press, p. 1.

Murphy, L. M., and P. L. Edwards. 2003. *Bridging the Valley of Death—Transitioning from Public to Private Sector Financing*. Golden, CO: National Renewable Energy Laboratory. May.

Myers, S., R. L. Stern, and M. L. Rorke. 1983. *A Study of the Small Business Innovation Research Program*. Lake Forest, IL: Mohawk Research Corporation.

Myers, S. C., and N. Majluf. 1984. "Corporate Financing and Investment Decisions When Firms Have Information That Investors Do Not Have." *Journal of Financial Economics* 13:187-221.

The National Academies of Sciences, Engineering, and Medicine. 2015. *Innovation, Diversity, and the SBIR/STTR Programs*. Washington, DC: The National Academies Press.

National Aeronautics and Space Administration. 2002. "Small Business/SBIR: NICMOS Cryocooler—Reactivating a Hubble Instrument." *Aerospace Technology Innovation* 10(4):19-21.

National Aeronautics and Space Administration. 2005. "The NASA SBIR and STTR Programs Participation Guide." http://sbir.gsfc.nasa.gov/SBIR/zips/guide.pdf.

National Institutes of Health. 2003. "Road Map for Medical Research." http://nihroadmap.nih.gov/.

National Research Council. 1986. *The Positive Sum Strategy: Harnessing Technology for Economic Growth*. Washington, DC: National Academy Press.

National Research Council. 1987. *Semiconductor Industry and the National Laboratories: Part of a National Strategy*. Washington, DC: National Academy Press.

National Research Council. 1991. *Mathematical Sciences, Technology, and Economic Competitiveness*. Washington, DC: National Academy Press.

National Research Council. 1992. *The Government Role in Civilian Technology: Building a New Alliance*. Washington, DC: National Academy Press.

National Research Council. 1995. *Allocating Federal Funds for R&D*. Washington, DC: National Academy Press.

National Research Council. 1996. *Conflict and Cooperation in National Competition for High-Technology Industry*. Washington, DC: National Academy Press.

National Research Council. 1997. *Review of the Research Program of the Partnership for a New Generation of Vehicles: Third Report*. Washington, DC: National Academy Press.

National Research Council. 1999. *The Advanced Technology Program: Challenges and Opportunities*. Washington, DC: National Academy Press.

National Research Council. 1999. *Funding a Revolution: Government Support for Computing Research*. Washington, DC: National Academy Press.

National Research Council. 1999. *Industry-Laboratory Partnerships: A Review of the Sandia Science and Technology Park Initiative*. Washington, DC: National Academy Press.

National Research Council. 1999. *New Vistas in Transatlantic Science and Technology Cooperation*. Washington, DC: National Academy Press.

National Research Council. 1999. *The Small Business Innovation Research Program: Challenges and Opportunities*. Washington, DC: National Academy Press.

National Research Council. 2000. *The Small Business Innovation Research Program: An Assessment of the Department of Defense Fast Track Initiative*. Washington, DC: National Academy Press.

National Research Council. 2000. *U.S. Industry in 2000: Studies in Competitive Performance*. Washington, DC: National Academy Press.

National Research Council. 2001. *The Advanced Technology Program: Assessing Outcomes*. Washington, DC: National Academy Press.

National Research Council. 2001. *Attracting Science and Mathematics Ph.D.s to Secondary School Education*. Washington, DC: National Academy Press.

National Research Council. 2001. *Building a Workforce for the Information Economy*. Washington, DC: National Academy Press.

National Research Council. 2001. *Capitalizing on New Needs and New Opportunities: Government-Industry Partnerships in Biotechnology and Information Technologies.* Washington, DC: National Academy Press.

National Research Council. 2001. *A Review of the New Initiatives at the NASA Ames Research Center.* Washington, DC: National Academy Press.

National Research Council. 2001. *Trends in Federal Support of Research and Graduate Education.* Washington, DC: National Academy Press.

National Research Council. 2002. *Government-Industry Partnerships for the Development of New Technologies: Summary Report.* Washington, DC: The National Academies Press.

National Research Council. 2002. *Making the Nation Safer: The Role of Science and Technology in Countering Terrorism.* Washington, DC: The National Academies Press.

National Research Council. 2002. *Measuring and Sustaining the New Economy.* Washington, DC: National Academy Press.

National Research Council. 2002. *Partnerships for Solid-State Lighting.* Washington, DC: The National Academies Press.

National Research Council. 2004. *An Assessment of the Small Business Innovation Research Program: Project Methodology.* Washington, DC: The National Academies Press.

National Research Council. 2004. *Productivity and Cyclicality in Semiconductors: Trends, Implications, and Questions.* Washington, DC: The National Academies Press.

National Research Council. 2004. SBIR—Program Diversity and Assessment Challenges: Report of a Symposium. Washington, DC: The National Academies Press.

National Research Council. 2004. *The Small Business Innovation Research Program: Program Diversity and Assessment Challenges.* Washington, DC: The National Academies Press.

National Research Council. 2006. *Beyond Bias and Barriers: Fulfilling the Potential of Women in Academic Science and Engineering.* Washington, DC: The National Academies Press.

National Research Council. 2006. *Deconstructing the Computer.* Washington, DC: The National Academies Press.

National Research Council. 2006. *Software, Growth, and the Future of the U.S. Economy.* Washington, DC: The National Academies Press.

National Research Council. 2006. *The Telecommunications Challenge: Changing Technologies and Evolving Policies.* Washington, DC: The National Academies Press.

National Research Council. 2007. *Enhancing Productivity Growth in the Information Age: Measuring and Sustaining the New Economy.* Washington, DC: The National Academies Press.

National Research Council. 2007. *India's Changing Innovation System: Achievements, Challenges, and Opportunities for Cooperation.* Washington, DC: The National Academies Press.

National Research Council. 2007. *Innovation Policies for the 21st Century.* Washington, DC: The National Academies Press.

National Research Council. 2007. *SBIR and the Phase III Challenge of Commercialization.* Washington, DC: The National Academies Press.

National Research Council. 2008. *An Assessment of the SBIR Program at the Department of Defense.* Washington, DC: The National Academies Press.

National Research Council. 2008. *An Assessment of the SBIR Program at the Department of Energy.* Washington, DC: The National Academies Press.

National Research Council. 2008. *An Assessment of the SBIR Program at the National Science Foundation.* Washington, DC: The National Academies Press.

National Research Council. 2009. *An Assessment of the SBIR Program at the Department of Defense.* Washington, DC: The National Academies Press.

National Research Council. 2009. *An Assessment of the SBIR Program at the National Aeronautics and Space Administration.* Washington, DC: The National Academies Press.

National Research Council. 2009. *An Assessment of the SBIR Program at the National Institutes of Health.* Washington, DC: The National Academies Press.

National Research Council. 2009. *Revisiting the Department of Defense SBIR Fast Track Initiative.* Washington, DC: The National Academies Press.

National Research Council. 2009. *Venture Capital and the NIH SBIR Program.* Washington, DC: The National Academies Press.

National Research Council. 2010. *Managing University Intellectual Property in the Public Interest.* Washington, DC: The National Academies Press.

National Research Council. 2011. *Building the 21st Century: U.S.-China Cooperation on Science, Technology, and Innovation.* Washington, DC: The National Academies Press.

National Research Council. 2011. *Expanding Underrepresented Minority Participation: America's Science and Technology Talent at the Crossroads.* Washington, DC: The National Academies Press.

National Research Council. 2011. *Growing Innovation Clusters for American Prosperity.* Washington, DC: The National Academies Press.

National Research Council. 2011. *The Future of Photovoltaics Manufacturing in the United States.* Washington, DC: The National Academies Press.

National Research Council. 2012. *Building Hawaii's Innovation Economy.* Washington, DC: The National Academies Press.

National Research Council. 2012. *Building the Arkansas Innovation Economy.* Washington, DC: The National Academies Press.

National Research Council. 2012. *Building the U.S. Battery Industry for Electric-Drive Vehicles: Progress, Challenges, and Opportunities.* Washington, DC: The National Academies Press.

National Research Council. 2012. *Clustering for 21st Century Prosperity.* Washington, DC: The National Academies Press.

National Research Council. 2012. *Meeting Global Challenges: German-U.S. Innovation Policy.* Washington, DC: The National Academies Press.

National Research Council. 2012. *Rising to the Challenge: U.S. Innovation Policy for the Global Economy.* Washington, DC: The National Academies Press.

National Research Council. 2013. *Building the Illinois Innovation Economy: Summary of a Symposium.* Washington, DC: The National Academies Press.

National Research Council. 2013. *Building the Ohio Innovation Economy: Summary of a Symposium.* Washington, DC: The National Academies Press.

National Research Council. 2013. *Competing in the 21st Century: Best Practice in State and Regional Innovation Initiatives.* Washington, DC: The National Academies Press.

National Research Council. 2013. *Strengthening American Manufacturing: The Role of the Manufacturing Extension Partnership—Summary of a Symposium.* Washington, DC: The National Academies Press.

National Research Council. 2014. *SBIR at the Department of Defense.* Washington, DC: The National Academies Press.

National Research Council. 2014. *The Flexible Electronics Opportunity.* Washington, DC: The National Academies Press.

National Science Board. 2005. *Science and Engineering Indicators 2005.* Arlington, VA: National Science Foundation.

National Science Board. 2006. *Science and Engineering Indicators 2006.* Arlington, VA: National Science Foundation.

National Science Board. 2014. *Science and Engineering Indicators 2014.* Arlington, VA: National Science Foundation.

National Science Foundation. 1999. *1999 SBIR/STTR Phase I Program Solicitation and Phase II Instruction Guide.* Arlington, VA: National Science Foundation.

National Science Foundation. 2004. *Federal R&D Funding by Budget Function: Fiscal Years 2003-2005 (historical tables).* NSF 05-303. Arlington, VA: National Science Foundation.

National Science Foundation. 2006. "SBIR/STTR Phase II Grantee Conference, Book of Abstracts." Office of Industrial Innovation. May 18-20, 2006. Louisville, Kentucky.

National Science Foundation. "Committee of Visitors Reports and Annual Updates." http://www.nsf.gov/eng/general/cov/.
National Science Foundation. "Emerging Technologies." http://www.nsf.gov/eng/sbir/eo.jsp.
National Science Foundation. "Guidance for Reviewers." http://www.eng.nsf.gov/sbir/peer_review.htm.
National Science Foundation. "National Science Foundation at a Glance." http://www.nsf.gov/about.
National Science Foundation. National Science Foundation Manual 14, *NSF Conflicts of Interest and Standards of Ethical Conduct.* http://www.eng.nsf.gov/sbir/COI_Form.doc.
National Science Foundation. 2006. "SBIR/STTR Phase II Grantee Conference, Book of Abstracts." Office of Industrial Innovation. May 18-20, 2006, Louisville, Kentucky.
National Science Foundation. 2006. "News Items from the Past Year." Press Release. April 10.
National Science Foundation, Office of Industrial Innovation. 2005. Draft Strategic Plan. June 2.
National Science Foundation, Office of Legislative and Public Affairs. 2003. SBIR Success Story from News Tip. Web's "Best Meta-Search Engine." March 20.
Nelson, R. R. 1982. *Government and Technological Progress.* New York: Pergamon.
Nelson, R. R. 1986. "Institutions Supporting Technical Advances in Industry." *American Economic Review, Papers and Proceedings* 76(2):188.
Nelson, R. R., ed. 1993. *National Innovation System: A Comparative Study.* New York: Oxford University Press.
Office of Management and Budget. 2004. "What Constitutes Strong Evidence of Program Effectiveness." http://www.whitehouse.gov/omb/part/2004_program_eval.pdf.
Office of the President. 1990. *U.S. Technology Policy.* Washington, DC: Executive Office of the President.
Organization for Economic Cooperation and Development. 1982. *Innovation in Small and Medium Firms.* Paris: Organization for Economic Cooperation and Development.
Organization for Economic Cooperation and Development. 1995. *Venture Capital in OECD Countries.* Paris: Organization for Economic Cooperation and Development.
Organization for Economic Cooperation and Development. 1997. *Small Business Job Creation and Growth: Facts, Obstacles, and Best Practices.* Paris: Organization for Economic Cooperation and Development.
Organization for Economic Cooperation and Development. 1998. *Technology, Productivity and Job Creation: Toward Best Policy Practice.* Paris: Organization for Economic Cooperation and Development.
Organization for Economic Cooperation and Development. 2006. "Evaluation of SME Policies and Programs: Draft OECD Handbook." *OECD Handbook.* CFE/SME 17. Paris: Organization for Economic Cooperation and Development.
Perko, J. S., and F. Narin. 1997. "The Transfer of Public Science to Patented Technology: A Case Study in Agricultural Science." *Journal of Technology Transfer* 22(3):65-72.
Perret, G. 1989. *A Country Made by War: From the Revolution to Vietnam—The Story of America's Rise to Power.* New York: Random House.
Poland, C. A., R. Duffin, I. Kinloch, A. Maynard, W. A. H. Wallace, A. Seaton, V. Stone, and S. Brown. 2008. "Carbon Nanotubes Introduced into the Abdominal Cavity of Mice Show Asbestos-Like Pathogenicity in a Pilot Study." *Nature Nanotechnology* 3(7):423.
Porter, Michael E. 1998. "Clusters and Competition: New Agendas for Government and Institutions." In Michael E. Porter, ed. *On Competition.* Boston, MA: Harvard Business School Press.
Powell, W. W., and P. Brantley. 1992. "Competitive Cooperation in Biotechnology: Learning Through Networks?" In *Networks and Organizations: Structure, Form and Action.* N. Nohria and R. G. Eccles, eds. Boston, MA: Harvard Business School Press. Pp. 366-394.
Price Waterhouse. 1985. *Survey of Small High-tech Businesses Shows Federal SBIR Awards Spurring Job Growth, Commercial Sales.* Washington, DC: Small Business High Technology Institute.
Reid, G. C., and J. A. Smith. 2007. *Risk Appraisal and Venture Capital in High Technology New Ventures.* New York: Routledge.

Roberts, E. B. 1968. "Entrepreneurship and Technology." *Research Management* (July):249-266.

Rogelberg, S., C. Spitzmüeller, I. Little, and S. Reeve. 2006. "Understanding Response Behavior to an Online Special Survey Topics Organizational Satisfaction Survey." *Personnel Psychology* 59:903-923.

Romer, P. 1990. "Endogenous Technological Change." *Journal of Political Economy* 98:71-102.

Rosa, P., and A. Dawson. 2006. "Gender and the Commercialization of University Science: Academic Founders of Spinout Companies." *Entrepreneurship & Regional Development* 18(4):341-366. July.

Rosenberg, N. 1969. "The Direction of Technological Change: Inducement Mechanisms and Focusing Devices." *Economic Development and Cultural Change*, 18:1-24.

Rosenbloom, R., and W. Spencer. 1996. *Engines of Innovation: U.S. Industrial Research at the End of an Era.* Boston, MA: Harvard Business School Press.

Rubenstein, A. H. 1958. *Problems Financing New Research-Based Enterprises in New England.* Boston, MA: Federal Reserve Bank.

Ruegg, R., and P. Thomas. 2007. *Linkages from DoE's Vehicle Technologies R&D in Advanced Energy Storage to Hybrid Electric Vehicles, Plug-in Hybrid Electric Vehicles, and Electric Vehicles.* Washington, DC: U.S. Department of Energy/Office of Energy Efficiency and Renewable Energy.

Sahlman, W. A. 1990. "The Structure and Governance of Venture Capital Organizations." *Journal of Financial Economics* 27:473-521.

Saxenian, A. 1994. *Regional Advantage: Culture and Competition in Silicon Valley and Route 128.* Cambridge, MA: Harvard University Press.

Schacht, W.H. 2008. "The Small Business Innovation Research (SBIR) program: Reauthorization efforts," Congressional Research Service, Library of Congress.

Schell, J. Krist, and N. Berente. 2014."Avoiding the Valley of Death: A Cross-Case Analysis of SBIR Innovation Processes." *Academy of Management Proceedings* 2014(1):16828.

Scherer, F. M. 1970. *Industrial Market Structure and Economic Performance.* New York: Rand McNally College Publishing.

Schumpeter, J. 1950. *Capitalism, Socialism, and Democracy.* New York: Harper and Row.

Scotchmer, S. 2004. *Innovation and Incentives.* Cambridge MA: MIT Press.

Scott, J. T. 1998. "Financing and leveraging public/private partnerships: The hurdle-lowering auction." *STI Review* 23:67-84.

Scott, J. T. 2000. "An Assessment of the Small Business Innovation Research Program in New England: Fast Track Compared with Non-Fast Track." In National Research Council. *The Small Business Innovation Research Program: An Assessment of the Department of Defense Fast Track Initiative.* Washington, DC: National Academy Press.

Sheehan, K. 2001. "E-mail Survey Response Rates: A Review." *Journal of Computer Mediated Communication* 6(2).

Siegel, D., D. Waldman, and A. Link. 2004. "Toward a Model of the Effective Transfer of Scientific Knowledge from Academicians to Practitioners: Qualitative Evidence from the Commercialization of University Technologies." *Journal of Engineering and Technology Management* 21(1-2).

Silverman, I. M., J. M. Dawicki-McKenna, D. W. Frederick, C. Bialas, J. R. Remsberg, N. L. Yohn, N. Sekulic, A. B. Reitz, and D. M. Gross. 2015. "Evaluating the Success of the Small Business Innovation Research (SBIR) Program: Impact on Biotechnology Companies in Pennsylvania." *Technology* 2(1):5.

Silverstein, S. C., H. H. Garrison, and S. J. Heinig. 1995. "A Few Basic Economic Facts about Research in the Medical and Related Life Sciences." FASEB 9:833-840.

Society for Prevention Research. 2004. "Standards of Evidence: Criteria for Efficacy, Effectiveness and Dissemination." http://www.preventionresearch.org/softext.php.

Sohl, J. 1999. *Venture Capital* 1(2).

Sohl, J., J. Freear, and W..E. Wetzel Jr. 2002. "Angles on Angels: Financing Technology-Based Ventures—An Historical Perspective." *Venture Capital: An International Journal of Entrepreneurial Finance* 4(4).

Solow, R. S. 1957. "Technical Change and the Aggregate Production Function." *Review of Economics and Statistics* 39:312-320.

Specht, D. C. 2009."Recent SBIR Extension Debate Reveals Venture Capital Influence." *Procurement Law* 45(2009):1.

Stam, E., and K. and Wennberg. 2009. "The Roles of R&D in New Firm Growth." *Small Business Economics* 33:77-89.

Stiglitz, J. E., and A. Weiss. 1981. "Credit Rationing in Markets with Incomplete Information." *American Economic Review* 71:393-409.

Stokes, D. E. 1997. *Pasteur's Quadrant: Basic Science and Technological Innovation.* Washington, DC: The Brookings Institution.

Stowsky, J. 1996. "Politics and Policy: The Technology Reinvestment Program and the Dilemmas of Dual Use." Mimeo. University of California.

Tassey, G. 1997. *The Economics of R&D Policy.* Westport, CT: Quorum Books.

Thurber, K. J. 2011. *Big Wave Surfing*. Edina, MN: Beaver Pond Press.

Tibbetts, R. 1997. "The Role of Small Firms in Developing and Commercializing New Scientific Instrumentation: Lessons from the U.S. Small Business Innovation Research Program." In *Equipping Science for the 21st Century.* J. Irvine, B. Martin, D. Griffiths, and R. Gathier, eds. Cheltenham UK: Edward Elgar Press.

Tirman, J. 1984. *The Militarization of High Technology.* Cambridge, MA: Ballinger.

Tyson, L., T. Petrin, and H. Rogers. 1994. "Promoting Entrepreneurship in Eastern Europe." *Small Business Economics* 6:165-184.

University of New Hampshire Center for Venture Research. 2007. *The Angel Market in 2006.* http://wsbe2.unh.edu/files/Full%20Year%202006%20Analysis%20Report%20-%20March%202007.pdf.

U.S. Congress, House Committee on Science, Space, and Technology. 1992. *SBIR and Commercialization: Hearing Before the Subcommittee on Technology and Competitiveness of the House Committee on Science, Space, and Technology, on the Small Business Innovation Research [SBIR] Program.* Testimony of James A. Block, President of Creare, Inc. Pp. 356-361.

U.S. Congress. House Committee on Science, Space, and Technology. 1992. *The Small Business Research and Development Enhancement Act of 1992.* House Report (Rept. 102-554) Part I (Committee on Small Business).

U.S. Congress. House Committee on Science, Space, and Technology. 1998. *Unlocking Our Future: Toward a New National Science Policy: A Report to Congress by the House Committee on Science, Space, and Technology.* Washington, DC: Government Printing Office. http://www.access.gpo.gov/congress/house/science/cp105-b/science105b.pdf.

U.S. Congress. House Committee on Science, Space, and Technology. Subcommittee on Technology and Innovation. 2007. Hearing on "Small Business Innovation Research Authorization on the 25th Program Anniversary." Testimony by Robert Schmidt. April 26.

U.S. Congress. House Committee on Small Business. Subcommittee on Workforce, Empowerment, and Government Programs. 2005. *The Small Business Innovation Research Program: Opening Doors to New Technology.* Testimony by Joseph Hennessey. 109th Cong., 1st sess., November 8.

U.S. Congress. Senate Committee on Small Business. 1999. *Small Business Innovation Research (SBIR) Program.* Senate Report 106-330. August 4, 1999. Washington, DC: U.S. Government Printing Office.

U.S. Congress. Senate Committee on Small Business. 1981. Small Business Research Act of 1981. S.R. 194, 97th Congress.

U.S. Congress. Senate Committee on Small Business. 1999. *Small Business Innovation Research (SBIR) Program.* Senate Report 106-330. August 4. Washington, DC: U.S. Government Printing Office.

U.S. Congress. Senate Committee on Small Business. 2006. *Strengthening the Participation of Small Businesses in Federal Contracting and Innovation Research Programs.* Testimony by Michael Squillante. 109th Cong., 2nd sess., July 12.

U.S. Congressional Budget Office. 1985. *Federal Financial Support for High-Technology Industries.* Washington, DC: U.S. Congressional Budget Office.

U.S. Department of Education. 2005. "Scientifically-Based Evaluation Methods: Notice of Final Priority." *Federal Register* 70(15):3586-3589.

U.S. Food and Drug Administration. 1981. Protecting Human Subjects: Untrue Statements in Application. 21 C.F.R. §314.12.

U.S. Food and Drug Administration. "Critical Path Initiative." http://www.fda.gov/oc/initiatives/criticalpath/.

U.S. General Accounting Office. 1987. *Federal Research: Small Business Innovation Research Participants Give Program High Marks.* Washington, DC: U.S. General Accounting Office.

U.S. General Accounting Office. 1989. *Federal Research: Assessment of Small Business Innovation Research Program.* Washington, DC: U.S. General Accounting Office.

U.S. General Accounting Office. 1992. *Federal Research: Small Business Innovation Research Program Shows Success But Can Be Strengthened.* RCED–92–32. Washington, DC: U.S. General Accounting Office.

U.S. General Accounting Office. 1997. *Federal Research: DoD's Small Business Innovation Research Program.* RCED–97–122, Washington, DC: U.S. General Accounting Office.

U.S. General Accounting Office. 1998. *Federal Research: Observations on the Small Business Innovation Research Program.* RCED–98–132. Washington, DC: U.S. General Accounting Office.

U.S. General Accounting Office. 1999. *Federal Research: Evaluations of Small Business Innovation Research Can Be Strengthened.* RCED–99–114, Washington, DC: U.S. General Accounting Office.

U.S. General Accounting Office. 1999. *Federal Research: Evaluations of Small Business Innovation Research Can Be Strengthened.* T-RCED–99–198, Washington, DC: U.S. General Accounting Office.

U.S. Government Accountability Office. 2006. *Small Business Innovation Research: Agencies Need to Strengthen Efforts to Improve the Completeness, Consistency, and Accuracy of Awards Data.* GAO-07-38. Washington, DC: U.S. Government Accountability Office.

U.S. Government Accountability Office. 2006. *Small Business Innovation Research: Information on Awards Made by NIH and DoD in Fiscal years 2001-2004.* GAO-06-565. Washington, DC: U.S. Government Accountability Office.

U.S. Public Law 106-554, Appendix I–H.R. 5667—Section 108.

U.S. Small Business Administration. 1992. *Results of Three-Year Commercialization Study of the SBIR Program.* Washington, DC: U.S. Government Printing Office.

U.S. Small Business Administration. 1994. *Small Business Innovation Development Act: Tenth-Year Results.* Washington, DC: U.S. Government Printing Office.

U.S. Small Business Administration. 1998. *An Analysis of the Distribution of SBIR Awards by States, 1983-1996.* Washington, DC: Small Business Administration.

U.S. Small Business Administration. 2003. "Small Business by the Numbers." SBA Office of Advocacy. May.

U.S. Small Business Administration. 2006. "Frequently Asked Questions." http://www.sba.gov/advo/stats/sbfaq.pdf.

U.S. Small Business Administration. 2006. "Small Business by the Numbers." SBA Office of Advocacy. May.

U.S. Small Business Administration. 2012. "SBIR Policy Directive." October 18.

Venture Economics. 1988. *Exiting Venture Capital Investments.* Wellesley, MA: Venture -Economics.

Venture Economics. 1996. "Special Report: Rose-Colored Asset Class." *Venture Capital Journal* 36 (July):32-34.

VentureOne. 1997. *National Venture Capital Association 1996 Annual Report.* San Francisco: VentureOne.

Wallsten, S. J. 1996. "The Small Business Innovation Research Program: Encouraging Technological Innovation and Commercialization in Small Firms." Unpublished working paper. Stanford University.

Wallsten, S. J. 1998. "Rethinking the Small Business Innovation Research Program." In *Investing In Innovation.* L. M. Branscomb and J. Keller, eds., Cambridge, MA: MIT Press.

Warner, E. E. 2014. "Factor Analysis of Disruptive Technology Approaches and Company Demographics in Defense SBIR Phase 1 Competition." Ph.D. Dissertation. Capella University.

Washington Technology. 2007. "Top 100 Federal Prime Contractors: 2004." May 14.

Weiss, S. 2006. "The Private Equity Continuum." Presentation at the Executive Seminar on Angel Funding, University of California at Riverside, December 8-9, Palm Springs, CA.

Whalen, P. S., S. S. Holloway, and I. D. Parkman. 2015."Navigating the 'Valley of Death': an Investigation of Which Marketing Competencies Lead Toward Successful Technology Commercialization." In *Proceedings of the 2008 Academy of Marketing Science (AMS) Annual Conference.* Springer International Publishing. Page 184.

Yu, M-F., O. Lourie, M. J. Dyer, K. Moloni, T. F. Kelly, and R. S. Ruoff. 2000. "Strength and Breaking Mechanism of Multiwalled Carbon Nanotubes Under Tensile Load." *Science* 287(5453):637-640.